Modern Approaches in Solid Earth Sciences

Volume 21

Series Editors

Yildirim Dilek, Department of Geology and Environmental Earth Sciences, Miami University, Oxford, Ohio, USA

Franco Pirajno, The University of Western Australia, Perth, Western Autralia, Australia

Brian Windley, Department of Geology, The University of Leicester, Leicester, United Kingdom

T0181069

Neil Phillips

Formation of Gold Deposits

 Springer

Neil Phillips
University of Melbourne
Melbourne, Victoria, Australia

Stellenbosch University
Stellenbosch, South Africa

Responsible Series Editor: F. Pirajno

ISSN 1876-1682   ISSN 1876-1690   (electronic)
Modern Approaches in Solid Earth Sciences
ISBN 978-981-16-3083-5   ISBN 978-981-16-3081-1   (eBook)
https://doi.org/10.1007/978-981-16-3081-1

This Springer imprint is published by the registered company Springer Nature Singapore Pte Ltd.
The registered company address is: 152 Beach Road, #21-01/04 Gateway East, Singapore
189721, Singapore

Keywords: Gold · Discovery · Exploration · Yilgarn · Archean · Carlin · Pacific Rim · Petrology · Metamorphism · Magmatism · Hydrothermal · Geochemistry · Alteration · Genesis

Preface

How gold deposits form is a question at the heart of human curiosity in a similar way to the evolution of mankind or the beginning of the universe. Gold gives great joy by its colour, lack of tarnish and use in jewellery. In many civilisations, gold has been a symbol and store of wealth. Today, gold is a vital component of electronics from smart phones to space probes. A better understanding of gold deposits can lead to increased exploration success and commercial advantage for the annual US$100 billion-plus global gold industry.

This book is designed for readers with an interest in gold, science, the mineral industry, and natural curiosity. It is based upon the author's career researching and teaching about the formation of gold deposits; and is written particularly for geoscience undergraduates, postgraduates, and professional geologists from the mineral industry. The book will be relevant to company leaders, stockbrokers, curious prospectors, and retired scientists.

This is a scientific detective story beginning with observations about gold on many scales. There is a focus on what, in the author's opinion, are the more important pieces of information, and the test of this subjective selection process can be judged in the quality of the final story. The highest-level of deposit classification into *gold-only* and *gold-plus* deposit types is based upon whether there are economic base metals like Cu; this classification is practical to apply as well as having a strong theoretical basis.

The book combines theory with observations drawn globally. There are examples of gold deposits formed from magmatic processes – though not necessarily the classic examples! For many deposits it is hot water in the metamorphic environment that concentrates gold.

Field and underground mapping of multiple deposits from microscopic to regional scale has been integrated with contemporary advances in igneous and metamorphic petrology, and the science of fluid generation and migration in the Earth's upper crust. Experimental and theoretical petrology have been combined with the geochemistry of element transport and partitioning in magmatic and hydrothermal systems to better understand the ore and non-ore forming processes. Knowledge of advanced mathematics and thermodynamics is not essential to read this book though these disciplines have been important in developing many of the concepts. Radiometric dating and stable isotope geochemistry results are reported where such data are diagnostic and thus have a practical application rather than out of any obligation for completeness. The book is designed to be read in totality although early chapters might be skimmed by those familiar with the gold industry.

The author has devoted much of his research career to understanding the Witwatersrand goldfields of South Africa but has deliberately only provided a brief overview of the Wits here. Ultimately, it is hoped that the rapid decline of the Witwatersrand industry might be reversed by a less introspective approach to its geology and exploration that starts by understanding the full behaviour of gold both in the laboratory and in deposits globally.

This book arises from the author's research on gold formation since the 1970s. Its content and sequencing draw upon the Melbourne Geology of GOLD course, of which he has been a leader for 25 years. Examples come from his studies of deposits in Canada, USA, Brazil, South Africa, Namibia, Zimbabwe and various States of Australia supplemented by numerous geological visits particularly within New Zealand, Northern Ireland, Indonesia - Irian Jaya, Russia, China, Wales, Portugal, Czechia and Eire. Neil's career spanning academia, government and industry has provided the opportunity to contribute to the generating of new science and then to demonstrate commercial value through discovery and improved gold mining.

Knowing how gold deposits form is not the same as knowing where they form but knowing the former certainly helps in finding them.

Melbourne, VIC, Australia   Neil Phillips

Terms and Abbreviations

Fundamental subdivision
Gold-only: gold deposits lacking economic base metals
Gold-plus: gold deposits containing economic base metals particularly copper
Base metals: copper, lead and zinc in the context of gold deposits

Measurements
Troy ounce: always used for gold
1 Moz: one million ounces, also 31.103 metric tonnes; approximated as 30 tonnes
1 metric tonne: 32,151 troy ounces; approximated as 30,000 troy oz
Avoirdupois ounce: not used to measure gold, approximately 90% of a troy ounce
ppb: parts-per-billion, also 10^{-9}
ppm: parts-per-million, also 10^{-6}
1 gram per tonne, or g/t, is the same as 1 part per million
tpa: tonnes per year (as in gold production)
Endowment, Reserves, Resources: see Appendix A
CRIRSCO, JORC or NI43-101: reporting standards; see Appendix A
EDR: Economic Demonstrated Resources
All-time production: a cumulative figure since mining started in a region or mine
Age in years—it occurred at 1 Ma or 1 million years ago (abbreviation is Ma)
Period of time in years—it lasted 1 myr or for 1 million years (abbreviation is myr)

Characteristics
Provinciality: characteristic of goldfields to form in clusters
Enrichment: ratio of an element in ore compared to its average crustal value
Segregation: characteristic of decoupling of gold and base metals
Timing: used in the sense of when a geological event happened such as deposit formation
Partitioning: the unequal separation of a component between two phases: an example might be NaCl between ice and water; almost all the NaCl is in the water

Rocks, minerals and structures
Tholeiite: type of basaltic rock; a mafic rock that generally has elevated Fe
Dolerite: mafic intrusive rock; used to describe differentiated tholeiitic (dolerite) sills
BIF: banded iron formation rock
Hornfels: fine grained metamorphic rock formed through thermal metamorphism
Magmatic: pertaining to silicate melt, and, where specified, to sulfide melt
Minerals: pyrite - FeS_2; pyrrhotite - FeS; stibnite – Sb_2S_3; arsenopyrite – FeAsS
Fault: generic term for planar break in rock unit including a shear zone
Rheology: the study of the flow behaviour (of rocks) during deformation
Competence: generic term describing the behaviour of a heterogeneous rock sequence undergoing deformation; less competent rocks tend to flow whereas more competent rocks are rigid
Hydraulic fracture: breakage of rock under high fluid pressure
Regolith, BOCO (base of complete oxidation), BOA (base of alluvium), weathering: see Appendix B, everything from fresh rock to fresh air
Nugget: piece of gold typically 4 mm or greater

Deposit types
IOCG: iron oxide copper gold deposit
VMS: volcanogenic massive sulfide deposit (contains Cu, Zn, Pb, Ag and Au)
Wits or Rand: shortened forms for the Witwatersrand goldfields in South Africa

Chemical terms
Aqueous fluid: water based liquid or gaseous phase
Hydrothermal fluid: literally hot water, particularly in the Earth's crust
Ligand: an ion or molecule that can bond with a cation such as gold
Redox state: refers to oxidation and reduction
Isochemical: refers to a geological process such as metamorphism to indicate no loss or gain of elements; when using the term isochemical, it is usual to disregard migration of volatiles like H_2O and CO_2

Science approach
Descriptive: observed characteristics of a gold deposit
Genetic: to do with how a gold deposit forms
Definition broadening: the act of expanding a definition to accommodate new and unexpected information or observations. Applies to the modifying of definitions of deposit types and in some cases, this indicates a weakness in the classification
Unreasonably effective describes an idea that unexpectedly explains much more than was originally planned or predicted

Exploration
Brownfield exploration: close to existing mineralisation including mines
Greenfield exploration: well-removed from mines where geological data is generally lacking

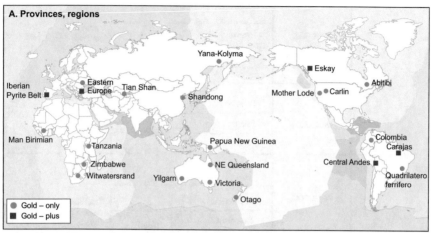

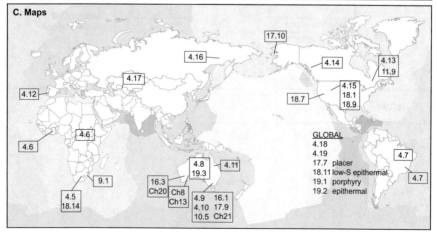

Gold provinces, deposits and map figures.

EON	ERA	PERIOD	Millions of years ago	Geologic Highlights
PHANEROZOIC	CENOZOIC	Quaternary		Andes SW Pacific
		Neogene	3	
			23	
		Paleogene		
			65	
	MESOZOIC	Cretaceous		Yana Kolyma
			140	
		Jurassic	200	
		Triassic	250	
	PALEOZOIC	Permian	300	Tian Shan Carlin Victoria
		Carboniferous	360	
		Devonian	420	
		Silurian	440	
		Ordovician	490	
		Cambrian	541	
PROTEROZOIC	NEO-PROTEROZOIC		1000	Olympic Dam
	MESO-PROTEROZOIC		1600	
	PALEO-PROTEROZOIC		2500	
ARCHEAN	NEOARCHEAN		2800	Greenstone gold Witwatersrand Basin
	MESOARCHEAN		3200	
	PALEO-ARCHEAN			
	EOARCHEAN		3600	
			4000	
HADEAN			4500	

Time scale of Earth history.

Acknowledgements

To understand the evolution of *Formation of Gold Deposits* it is important to look on a larger scale than the book itself; I thank the many extraordinary colleagues who have shared their thoughts and time so that we could be challenged together by the geology of gold.

Much of the *Formation of Gold Deposits* story has been developed over the last 25 years leading the Melbourne Geology of Gold course during which experts in all aspects of gold geology have shared lectures, practical classes, and mine visits. Martin Hughes has been a course colleague for the duration, has a wealth of knowledge on all aspects of gold geology, and is the expert on the Victorian Gold Province. Roger Powell has collaborated on the formation of gold deposits since the 1980s, been generous with his sharing of ideas about science and has added a hard petrology edge to the economic geology. Contributors attending the Geology of Gold course have included Rod Boucher, Dave Craw, Kim Ely, Katy Evans, Nikolai Goryachev, Janet Hergt, Simon Jowitt, Reid Keays, Jonathan Law, Dave Phillips, Iain Pitcairn, Andy Tomkins, Don Thomson, and co-founder Peter Arden. Much appreciated support for the gold course has been provided by Ed Eshuys, Australian Institute of Geoscientists, Australasian Institute of Mining and Metallurgy, and the Universities of Melbourne, Monash, and Stellenbosch RSA.

Several leaders in global gold geology have been continual sources of challenging scientific thought throughout my career and I learned much from their approaches. Ravi Anand, Charles Butt, Steve Cox, Dave Craw, Rich Goldfarb, David Groves, Rob Kerrich, John Ridley, Francois Robert, Ray Smith, Andy Tomkins, Julian Vearncombe and Scott Wood come to mind.

Working with experts outside classical gold geology has been especially rewarding. Not only do they teach me new aspects of geology, but these friends have ensured that I know a little about placers and sedimentology (Stephen Carey, Tony Cadle, Rod Boucher), granite (John Clemens), dolerite (Janet Hergt) and inorganic chemistry (Rod Phillips and Richard Smith).

My start in gold geology was built upon the works of major contributors to geoscience including John Ferry, Bill Fyfe, Bruce Hobbs, Richard Puddephatt, Terry Seward, Guy Travis, Ron Vernon and Vic Wall.

Julian Vearncombe has been a large part of this gold story for four decades whether it be in southern Africa, the Yandal belt, Duketon or discussing any other place where there is a gold industry. He has always provided reasoned and honest feedback without the sugar coating.

Ed Eshuys is thanked for the many opportunities he creates, for his consistent support of geology and his belief that mineral discoveries will follow from good geology. You know his corporate message has been heard when the finance team tells you that 'we are here to make a discovery'.

David Groves saw the opportunity in Yilgarn gold in the late 1970s. For me, it was a stroke of luck arriving in Western Australia in 1979, even though I found everyone interested in nickel at the time. David acquainted me with Archean metallogeny and the big picture, better approaches to writing and many economic geology skills. The field trips around Western Australia, South Africa and Namibia were educational and much fun. David realised early that good students and good supervision was a win-win combination and many of the students that we supervised made important contributions in their projects and went on to exceptional careers.

CSIRO provided an environment to solve a simple problem that had bothered me for 25 years, namely, what is the role of CO_2 in auriferous gold-only fluids. The answer seems so obvious looking back today (Chapter 14).

The Australasian Institute of Mining and Metallurgy has supported many of my activities including as editor of Applied Earth Science journal where I worked with an outstanding team including Marat Abzalov and Simon Jowitt. I also thank The AusIMM and Kristy Burt for their confidence in me as editor of their Australian Ore Deposit monograph; this was a wonderful opportunity and exposed me to hundreds of geologists and deposits. The Geological Society of Nevada, Geological Society of Australia, and Australian Institute of Geoscientists have also contributed to many of my ventures.

I do not take for granted the opportunities earlier in my career to visit gold mines around the world. I realise that the breadth of those visits would be virtually impossible today, not to mention experiences such as 'riding the bridle' in shorts and flip-flops to go underground at small gold mines in Zimbabwe. Two-years of underground mapping at the Golden Mile in Kalgoorlie was a rare opportunity to become engrossed in a single but large goldfield. There is a special excitement when the dots on your paper cross-sections join after months of underground data collection.

Rogaining colleagues globally are thanked for the opportunities to traverse on foot hundreds of mineral fields, quartz veins and structures without geological bias; these traverses became a useful complement to organised geological field trips.

Collaborators in various regions have included David Groves, Julian Vearncombe, Janet Hergt, Roger Powell (Yilgarn and Pilbara), Greg Corbett and Rich Goldfarb (Pacific Rim), Taihe Zhou and Gina Dong (Jundee and China), Jonathan Law, Terry McCarthy, Russell Myers and Judy Palmer (Witwatersrand), James Macdonald (Abitibi), Martin Hughes and Rod Boucher (Victoria, SE Australia Lachlan Orogen), Patrick Williams, Geoffrey de Jong and Peter Pollard (Cloncurry Mt Isa and NE Queensland Charters

Towers), Dorrit de Nooy (Big Bell), Donald Thomson (Carlin, Duketon), Ernst Kohler, Kim Ely and many others (Yandal), Daniel D'Oliviera, Fernando Noronha and colleagues (Portugal), Marat Abzalov and Nikolai Goryachev (northern Asia and Siberia), Bob Foster and Tim Nutt (Zimbabwe).

My special thanks to Liz Butler of Valeta Designs headquartered in Yandoit for drafting and redrafting all the figures, being patient with my modifications, and including an over-dose of humour with all her great work—thank you, Liz. Alice Coates is thanked for the exceptional artwork in Chapters 1 and 18.

The manuscript has been read by various colleagues. Julian Vearncombe provided constructive and valuable feedback on science and the arrangement of every chapter. Arthur Day came fresh from editing another book and gave me much appreciated advice on science and style throughout. Caitlin Jones read the first 15 chapters and is thanked for her helpful feedback. Martin Hughes, Jonathan Law, Peter Arden and Bjorn von der Heyden read selected chapters and are thanked.

And to my wife Jane, our four children and my wider family for their assistance, advice, and support throughout this writing period and indeed my geological career, thank you.

Contents

About the Author

Neil Phillips graduated in geoscience from the University of Melbourne (BSc) and Monash University (BSc Hons and PhD) where his thesis was on the metamorphism and geochemistry at Broken Hill, Australia. He completed the Advanced Management Program at Harvard Business School in 2004.

At the University of Western Australia, he co-founded the Archean Gold Group in 1980 with David Groves as they pioneered many of the geological principles used in gold exploration today. Two years were then spent engaged in underground mapping of the Golden Mile in the Kalgoorlie goldfield. He joined the University of Witwatersrand in Johannesburg and established research projects at all the operating Witwatersrand gold mines and continues that research focus and interest today. He was subsequently Professor of Economic Geology at James Cook University where he developed the extended field programs in the Cloncurry—Mt. Isa region.

As a consultant in Kalgoorlie for Minsaco Resources and then General Manager for Great Central Mines, he led the geological development of the Yandal Gold Province north of Kalgoorlie in a role that included teaching, research, and mentoring with teams of mine and exploration geologists. Honorary professorial appointments at the University of Melbourne, Monash University, and Stellenbosch University in South Africa have involved collaborative research, valorisation of this research through community engagement and leadership of the Melbourne Geology of Gold course since 1995 and similar courses on gold in South Africa.

He was editor of the Applied Earth Science journal for 10 years from 2010, regularly publishes in scientific journals and has written and edited several books on geoscience and cross-country navigation. He was President of the International Rogaining Federation 1989–2013 and co-author of three books on the sport of rogaining. He was editor of the AusIMM Australian Ore Deposit monograph in 2017 leading 350 authors and reviewers in a volume describing 200 mineral deposits.

Part I

Introduction

Introduction

1

Abstract

Unexpected observations and explanations provide moments of excitement for a scientist studying gold. Linkages may arise between observations and ideas—even accidentally. A key is to recognise these special moments and to not dismiss them.

Keywords

Curiosity · Commercial benefit · Unreasonable effectiveness

A day at the small Water Tank Hill gold mine 400 km northeast of Perth shaped my career. I had spent time underground at several gold mines previously, but 4th June 1981 was my first visit to a deposit in which the main rock type was banded iron formation. I was briefed to expect some beautiful gold-enriched sedimentary layers that formed on the seafloor over 2600 million years ago. Once underground, it was easy to recognise those beautiful layers, but the gold was not parallel to those layers as I had been led to expect. Instead, the gold was forming an envelope around quartz veins perpendicular to sedimentary layers. It was clear even before returning to the surface that the Water Tank Hill deposit might hold the field evidence that would revise our ideas about the formation of many gold deposits.

This is one of the small moments that rewards scientists. Waking up that morning in 1981 was normal; by the end of the day, it was evident that here was something special. It was good fortune that there was such compelling evidence at Water Tank Hill; but it was earlier personal preparation including gold geochemistry theory that helped to recognise the significance. Prior exposure to senior scientists with healthy scepticism was useful too.

A book describing the formation of gold deposits fulfils two distinct needs; one is curiosity, and the other is as a scientific contribution to the annual $100 billion gold industry. Gold stands apart from all other metals in the way it intrigues scientists and the community alike, has visual appeal and has many unrivalled lasting qualities such as its malleability, electrical conductivity, and resistance to chemical change. It is no surprise that the pinnacle of sporting achievement is called the gold medal.

Curiosity guides much of what happens in science especially with respect to science and the community. Stories of a new fossil, a revised origin for mankind, a new age for the Earth, or a different interpretation of the universe all have a broad appeal that is out of proportion to any practical value for most enquirers. Another of those universal questions is how does gold form and this might be answered in a few sentences. For completeness though, this question will need a book. Technically, gold atoms are thought to have formed from the collision of neutron stars and the fusion of lighter elements, and this makes them older than our solar system. However, we will start by accepting the existence of these gold atoms on Earth and investigate how gold deposits form in the upper few kilometres of the Earth called the crust.

Commercial enterprises benefit from knowing how gold deposits form because the knowledge aids better exploration and mining. Mineral exploration for centuries has included the use of associations; if we know how gold deposits form and what other elements and minerals can be associated with gold, then we can use those associations to our benefit. An attractive gold deposit might have a concentration of gold around 10 grams per tonne, which is 10 parts-per-million or nearly one third of an ounce per tonne, and despite the distinctive physical features of gold, any element that is this scarce is unlikely to be visible. Instead, we can use more abundant minerals like quartz in veins and pyrite (FeS_2) as indicators to help us focus any search.

Determining how a gold deposit formed is far from elementary. We are discussing a process that might have occurred 2650 million years ago, perhaps 5 km or more underground and at 350°C. The piece of land surrounding this gold deposit today has probably been moved thousands of kilometres by plate tectonics, and later undergone chemical breakdown and erosion by water and wind to remove the kilometres of rocks that used to lie above.

A hallmark of the approach used here is that gold ores and their surrounds are studied as normal rocks acknowledging that they also happen to have value. This approach is not unique, but surprisingly has not been the norm in all gold research. This book starts by taking all gold deposits (potentially 100,000 of them, possibly many more), finding some of their common characteristics, merging these with theoretical principles, then suggesting some groupings that are pragmatic and supported by theory. This approach differs from many modern studies that are based upon a single deposit which is then analysed for a wide range of trace elements and isotopes without considering the full regional geological context, rock types and geochemistry – this analytically-based approach is partly understandable given the difficulty in gaining access to multiple mines and whole gold provinces. Today the volume of data is rapidly increasing and doing so at a faster pace than our understanding of the geological context. This book attempts to synthesise some of this new and old data.

Gold deposits have been classified and subdivided into many different types providing a plethora of names and different categories that are regularly overlapping, based upon non-diagnostic criteria, and lacking adequate theoretical basis. Many of these names and groups are used sparingly in this book; the justification for overlooking many names is that gold as an element follows the same chemical principles regardless of how scientists want to classify the various deposits. Many terms are avoided here because they lack a widely understood meaning, are out of date, or lack scientific basis (see Appendix D). The highest-level classification used here of *gold-only* and *gold-plus* is grounded on sound graduate chemistry theory. It is also a classification that is practical to apply as the key question is simple, "Does this deposit contain economic base metals?".

A sub-theme of the science will be the identification of ideas that are unreasonably effective, a term borrowed from a book by Mario Livio (2010). The term reflects concepts with much wider applicability than may have originally been planned or envisioned. One example might be the combining of an economic geologist who divides deposits into gold-only and gold-plus with an inorganic chemist who is aware of the two oxidation states of gold in nature; the result is an unreasonably effective explanation of many disparate characteristics of those contrasting deposits and contrasting Au^{1+} and Au^{3+} oxidation states. Another might be the predictions from thermodynamics of a H_2O-CO_2 low salinity fluid during regional metamorphism providing an unreasonably effective match with the ore forming fluid already detected in gold-only deposits but found in virtually no other types of ore deposit.

Formation of Gold Deposits summarises the descriptive nature of gold deposits including both commonality and diversity amongst them. Then it

Fig. 1.1 Schematic cross-section of an open pit and regolith landscape. A mineralised fault zone (left), auriferous quartz vein (white in centre) and post-ore dyke (dark on right) are in fresh rock at depth and extend upward into the clay-rich regolith. Gold has been redistributed in the regolith and the settings of the 22 grains potentially give quite mis-leading information about timing of the main gold forming event. Artwork by Alice Coates reproduced with permission.

addresses their formation by processes involving magmas (i.e. molten rock) and hydrothermal fluids (literally hot waters that are in the Earth's crust). Importantly, some additional processes are discussed that modify deposits after their formation either to make the mineralisation economic, or to destroy an economic deposit (Fig. 1.1). Gold deposits with economic base metals are discussed and finally there are case histories demonstrating gold geology being applied scientifically to successful exploration and mining.

Bibliography

Livio M (2010) Is God a mathematician? Simon and Schuster, Sydney, p 308

Mining and the Nature of Gold Deposits

2

Abstract

It is a major task to make a gold bar from gold atoms which exist at a few parts per billion (0.0000001%) in the Earth's crust. This task is a two-stage process involving collaboration between Nature and mining. Nature partitions gold into small volumes of rock that become ore with several parts per million of gold – thousands of times greater than the crustal average. Mining concentrates the few parts per million of gold in that ore further by processing and refining it to produce a gold bar. These upgrades are forms of partitioning of gold between different media. An exception is the formation of specimen nuggets where nature alone achieves the 10^9 upgrade.

Keywords

Mining · Economics · Gold ore · Alteration · Waste · Ore reporting · Scale

Mining is a complex activity with many components that impact on its economic viability. The tonnage, grade and character of the material to be mined is critical and an important theme of this book. However, commodity price and many social, environmental, and governmental factors also impact mining—and these are usually dynamic. Today, gold price fluctuates daily in contrast to when it was US$20 for nearly a century and then close to US$35 for three decades from 1934. Social factors include safety and the effects of mining near towns, parks, farms, and cultural and heritage areas. Environmental factors might involve wilderness and pristine areas, national parks, water usage, air quality, or endangered animals and plants. Government factors include regulations, taxes and royalties, rehabilitation requirement and bonds. There are also the sovereign risks involving mine tenure and personal security in the case of changes of government including coups d'état. Addressing these factors needs to commence before mining and then continue for the life of a mine and beyond; each factor will constitute additional costs and may require infrastructure redesign and relocation or even reconsideration of a decision to proceed.

2.1 The Gold Mining Process

In the past, gold production may have followed a sequence of activities of finding some gold-bearing rock, digging it up then extracting the gold from the rock (Fig. 2.1). More recently, remediation may have followed. A single pass through this progression does not reflect major gold mining operations today that can be far more complex.

License-to-operate is an essential part of mining today and refers to the collaboration with community and government to ensure the mine is managed properly and its future is not in jeopardy. Mine rehabilitation, for example, has

Fig. 2.1 Early gold mines in Northern Portugal: (**a**) Três Minas gold mine with a history of mining dating back to Roman times. The size of the excavation indicates that this would have been one of the largest gold mines in Europe at that time. The geological setting is auriferous quartz veins within metasedimentary rocks. [Photo by Ms Ana Gonzales, University of Porto, reproduced with permission]; (**b**) Castromil gold mine has been mined for centuries and been operated during multiple periods. Both Três Minas and Castromil are gold-only deposits.

varied from none historically, to only at the end of mine life, to now being progressive as mining continues. Safety is a component of the license to operate, and standards have risen significantly in recent years assisted by automation, awareness and regulation.

Exploration is critical before any gold mining can begin. The initial stage of a mining operation requires exploration to identify gold through discovery, and then define or delineate its location, geometry, nature and gold grade with some quantified level of confidence. One purpose of this exploration is to reduce some of the risk once mining begins. On-going exploration remains important if the mine is to continue after the initial find has been depleted. Two end members of gold exploration are recognised today although are quite gradational. *Greenfield exploration* seeks a new goldfield away from previously known mineralisation whereas the more common *brownfield exploration* takes a known mineralised area or even a mining operation and attempts to add gold resources nearby. What constitutes *nearby* has evolved with large trucks that have the capability of transporting tonnes of ore 50 – 100 km from where it is mined to a processing plant. The benefit of brownfield additions includes the better utilisation of the capital-expensive mill and processing plant by amortising costs over an extended mine life or a diversified portfolio of operations. The benefit of a greenfield discovery is the untested potential that might become a major new find.

The mining stage involves the removal of rocks from on or near the surface via open pits and then by underground at greater depth. Access to underground may be by vertical or inclined shafts or via a decline which is essentially a roadway at an acceptably steep gradient that allows vehicles to drive down and up. Near-surface material may be unconsolidated whereas underground rock usually requires rock drills and the use of explosives. Large-scale modern mining equipment increasingly combined with automation has substantially lowered costs but does require the gold-bearing rock to be in continuous volumes that can be evaluated and mined in bulk rather than hand-picked as was done in the 19th century.

The processing of gold ore includes breaking down large rocks by various forms of milling including crushing and grinding into fine sand-sized material. Then chemicals are applied to extract gold in solution and ultimately have it converted into gold bars. Chlorination and mercury were used in 19th century gold mining, but these have been replaced by cyanide, which although toxic will oxidise to nontoxic nitrate in the surface environment after use. Sulfide roasting was used to break down sulfide minerals and expose their entrapped gold grains to chemical dissolution but is being used less for environmental reasons. Modern gold extraction is by the dissolving of gold into a sodium cyanide solution then adsorption of the gold on carbon followed by stripping of the gold from the carbon. Heap leaching is a cheap method to extract gold by spraying a cyanide solution on broken gold-bearing rock. Carbon-in-pulp and carbon-in-leach are more effective but also more expensive methods in which gold is dissolved into a cyanide solution in tanks, then adsorbed onto carbon, and finally stripped from that carbon.

Gold mining has taken many different forms over the centuries with one of the earliest being basic operations collecting gold grains from active creek and river systems by panning and sluicing of soils and unconsolidated rocks (Fig. 2.2 a-d). A logical progression was to trace this alluvial gold upstream to its source in hard rock where further extraction required

digging of small pits and shafts that could be operated by one or two prospectors. A small fraction of these shafts warranted major expansion by extending the mine to much greater depths and more people and capital would become necessary to achieve this. Starting in the 20th century, major open pit mining offered a substantial advantage for near-surface extraction by taking advantage of larger-scale machinery. This approach required less labour intensity and reduced the digging and removal costs which have made large low-grade deposits economic to mine. Several open pit gold mines are over 500 m deep including the deepest, Bingham Canyon copper-gold mine in Utah USA, at 1.2 km deep and 4 km long. For greater depths, underground mining becomes necessary and the largest underground gold mining operations involve multiple interconnected shafts that can reach 4 km below the surface.

Fig. 2.2 Various styles of mining operation in the late 20th century: (**a**) Several groups mining alluvial gold from a river near Cuiabá in Brazil; (**b**) Perseverance mine headframe and Great Boulder mill, Golden Mile, Kalgoorlie in 1982 prior to the Superpit; (**c**) open pit mine at Fosterville in Central Victoria [photo by Gary Wallis, reproduced with permission]; (**d**) the shaft system and mining complex of the Western Deep Levels mine in the Carletonville goldfield in South Africa: this goldfield has the world's five deepest mine shafts including Mponeng at 4100 m deep, Tau Tona at 4000 m and Savuka at 3777 m.

2.2   Economics of a Gold Deposit

Gold mining operations vary from being highly profitable to marginal and in some cases loss-making. Using a 100 t, or 3 Moz, gold deposit as an example, if all its contained gold could be extracted and sold at a gold price of $1500 per ounce then the total revenue would be $4.5 billion. That appears to be a very attractive proposition for a large deposit that is far from being a giant. However not all the identified gold in this hypothetical deposit can ever be mined, extracted from the ore and sold so there will be a shortfall in the $4.5 billion revenue figure. Costs need to be considered including initial establishment of the mine which may exceed $100 million. To the capital (establishment) cost needs to be added day-to-day operating costs for the life of the mine, which might be divided into some major categories like administration, mining, and processing. Many factors will determine whether the operating cost is several $100s or over $1000 per ounce of gold produced. Finally, no person or company will consider starting a gold mining operation if the expected outcome will be to out-lay $100s million to simply get their money back in ten years if all goes well. They will consider the decreased value of money in ten years, i.e. inflation, and expect some recompense (profit) to cover the risk that not all operations go exactly to plan. That same money could be invested in long term government bonds with virtually no risk to gain a small return as interest. Investing that same money into the gold operation involves some risk and for this risk a better return is expected than for the zero-risk government bond interest. As an example, consider:

- Resource = 3 Moz = 100 tonnes of gold
- Recovery = 90% (meaning 10% of the gold was not commercial to extract)
- Selling price = $1500 per ounce
- Revenue for life of the mine = $4.05 billion
- Original cost of establishing the mining operation = $400 million
- Cost to mine and process the gold = $1200 per ounce produced

- Profit = $410 million. Does this adequately recompense for the commercial risks?
- How significant is it if the gold price is now $2000 per ounce?
- There is no accounting for inflation and especially the inflation in the years ahead
- How reliable are the numbers above?

This book will keep coming back to a hypothetical deposit of 3 Moz of gold because this is a realistic target today that might reasonably be economic, it is useful as a standard for discussion and comparison, and it makes for simple approximations, i.e. it is 100 t of contained gold.

The grade of ore will make a significant difference to the economics. Some large mines operate successfully by mining ore that is less than 1 g/t Au and many operate at 1 to 10 g/t using current 2020s mining methods. A small group of mines have the luxury of ore that is over 10 g/t; for example, Fosterville gold mine in southeast Australia was producing at the rate of 20 tonnes or 0.6 Moz of gold per year in 2019 at a grade above 30 g/t Au. Several Witwatersrand gold mines in South Africa have produced 1 to 2 Moz per year in their past, or at a grade over 20 g/t, or both and were therefore highly profitable at their peaks.

A different proposition is a small gold mine opened after 2000 with a predicted grade of 1.7 g/t, planned production of 3000 oz per month for a two-year life assuming a gold price of $700 per ounce. The eventual gold grade, determined from the mass of ore processed and actual gold produced, was only two thirds of what was planned, and local costs were much higher than predicted. During the mine's life the operating costs were up to $1700 per ounce in some months, there was no month of profitable operation, and this loss-making situation was clearly not sustainable. A decision point was a review of the operation, and this led to the next month's forecast figures for administration, mining and processing being lowered by 30% in the spreadsheet, and once recalculated, this generated a forecast profit. Nothing materially was changed at the mine (except the forecast input figures), the same loss continued, and the mine closed before its cumulative production ever reached 30,000 oz. There are various lessons in this example including the small initial scale of the operation combined with the lack of the exploration required to grow that scale for a longer life. There was a separate review of the processing stage of the operation to better understand why it was necessary to be regularly transporting the carbon offsite for re-activating, how to account for a difference of nearly 10,000 oz of gold between that predicted and recovered, and why the costs for some essential consumables appeared excessive. A decade after this small mine opened and then closed the gold price was 200 percent higher. It is likely that the history of this mine may have been quite different should it have opened a decade later.

A reflection of changing economics of gold mining is provided by the Golden Mile at the heart of the Kalgoorlie goldfield in Western Australia. The Golden Mile operated for over 80 years from 1893 producing 1120 t Au until it became uneconomic as an underground mine and closed in the late 1970s; the average grade for the first 84-year period was 11.9 g/t Au. A completely new mine plan for a large open cut then followed and has led to the Golden Mile becoming a profitable operation since 1980s at less than 5 g/t Au and with a production of 0.5 – 1 Moz Au per year. All-time production is 70 Moz (2200 t Au) for the Kalgoorlie goldfield which includes the Golden Mile (2000 t) and adjacent Mt Charlotte mine (200 t). There is a Resource of 600 t, making an endowment of 2800 t for the Kalgoorlie goldfield (terms such as endowment are explained in Terms and Abbreviations, and in Appendix A).

2.3   Gold Ore and Waste

The production of gold is a process of separating two rocks – retaining the valuable ore and discarding the uneconomic waste. Once the ore is separated, mining and metallurgical methods can be used to extract the gold. Both ore and waste are rocks that provide information about how gold deposits formed, and both need to be studied. In isolating ore from waste, the former always includes uneconomic minerals referred to as gangue.

Ore and waste are not necessarily static classifications as a new more economic method to extract elements or a change of commodity price can cause the re-classification of rock material from being ore to waste, or vice versa, with some minerals becoming either ore minerals or gangue minerals. Quantitative analysis today allows a more detailed separation of ore into high gold grade, normal gold ore, and marginal material. Higher-grade ore might be processed immediately, waste being discarded, and marginal material stockpiled in case economic circumstances or recovery technologies improve. Mineralisation is a general term to include ore (i.e. economic material) but also material in which there is gold at levels well above background but currently uneconomic. Many ore deposits contain more than one element of value as either co-products (if of quasi-equal value) or as by-products if they are subordinate to a major commodity and not economic on their own. For example, silver is a by-product of almost all gold mining, and gold is a co-product of some copper mining.

The minerals that constitute both ore and waste have considerable scientific value in determining how gold deposits have formed. The only valuable mineral in many gold deposits is either gold itself (Au) or electrum which is an alloy of gold and silver. A common occurrence is that of gold grains within quartz. Yet other deposits have sub-micron sized gold grains trapped within pyrite and arsenopyrite grains, and although these sulfide minerals are worthless, they need to be isolated to liberate the gold. Gold can even be bound in pyrite at the ionic level. Many other gangue minerals may have gold grains physically trapped within them and require fine crushing to release the gold, and this will impact recovery costs. Small amounts of gold-bearing telluride minerals are found in several gold deposits, and are abundant in a few, such as Cripple Creek deposit in Colorado USA, and Kalgoorlie Australia. Maldonite (AuBi) and aurostibite (AuSb$_2$) have gold chemically bonded in these relatively rare minerals (Table 2.1).

Table 2.1 Ore minerals containing gold.

Mineral	Composition	Comments	
Gold	Au	Easy to extract the gold	Widespread
Electrum	Au-Ag	Easy to extract the gold	Widespread
Sylvanite	$AgAuTe_4$	Extraction of gold is more difficult	Quite common
Calaverite	AuTe	Extraction of gold is more difficult	Uncommon
Aurostibite	$AuSb_2$	Extraction of gold is more difficult	Uncommon
Maldonite	AuBi	Extraction of gold is more difficult	Uncommon
Krennerite	$AuTe_2$	Extraction of gold is more difficult	Uncommon
Petzite	Ag_3AuTe_2	Extraction of gold is more difficult	Uncommon
Kostovite	$AuCuTe_4$		Uncommon
Gangue minerals that may contain substantial fine-grained gold			
Pyrite	FeS_2	Extraction of gold is more difficult	Widespread
Arsenian pyrite	FeAsS	Extraction of gold is more difficult	Widespread

The gangue minerals mostly come from four main mineral groups; these are the silicates, sulfides, carbonates and oxides. Within each group are several different minerals and every one of these minerals is stable under a specific set of physical and chemical conditions. Each mineral holds some information as to how the gold deposits form and they become even more useful when recorded in combination as assemblages of several co-existing minerals. Despite the large total number of minerals recorded from gold deposits, the subset at a specific mine is much more limited. In practical terms for a mine geologist, having the knowledge of one dozen minerals is a good start for most day-to-day activity at a single operation; moving to a different mine may require learning a further set of minerals.

The former occur in the upper part of the crust, the hydraulic veins form deeper than 5 km when rocks are under lithostatic load from the rock mass above. Veins often follow faults in the rock mass or may be terminated at them.

Faults are planar structures across which there is some fracture and offset. Conceptually, these can be a brittle fracture with abrupt offset and referred to as faults, or ductile with more continuous deformation across some of the offset and referred to as shear zones. In practice, structures with brittle fault offset and some ductile margins are widespread and technically are brittle – ductile shear zones. For much of the time, all these structures are referred to here as faults. The brittle – ductile transition in the Earth's crust is usually around 10 – 15 km depth and 300 – 400°C.

2.3.1   Veins and Faults

Veins are sheet-like bodies within the rock mass and made up of one or more minerals (Fig. 2.3). Veins of quartz and gold are a major component of many gold deposits and form by precipitation of these components from aqueous solutions. There are two main types of quartz veins corresponding to different depths of formation:

- Open space filling veins
- Hydraulic veins characterised by crack seal textures.

2.3.2   Fluid-wallrock Alteration, and Alteration Haloes

It is common for the rocks immediately surrounding auriferous quartz veins to be quite different to the more distal rocks, and this proximal envelope around the ore is referred to as the alteration halo or alteration zone. The alteration halo forms when water (possibly an ore-forming solution) is out of chemical equilibrium with the adjacent country rocks and the two can react chemically. Alteration is important because it has scientific value informing about an

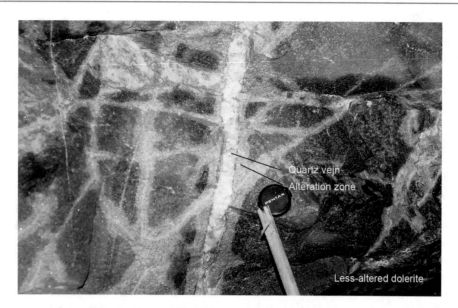

Fig. 2.3 White quartz vein (2 cm thick and vertical, left of lens cap) in dark green dolerite from Mt Charlotte mine in the Kalgoorlie goldfield of Western Australia. Numerous cream veins at various orientations are a network of fractures along which fluid has infiltrated. The addition of CO_2 from the veins during the gold mineralising event has altered the dolerite forming carbonate minerals and making it cream in colour adjacent to the veins.

ore-forming solution, and because it is usually anomalous in its gold concentration and may constitute ore.

The alteration halo can be centimetres to hundreds of metres in width; and depending upon its constituent mineral assemblages it may be quite visually distinct (Fig. 2.3). The proximal and the distal parts of the alteration halo may include any of ore, marginal ore, or waste. As alteration haloes form from the reaction of ore-bearing solutions with neighbouring country rock, they are an important guide during exploration because they can present a larger target than the interval of ore itself. The diverse array of minerals comprising an alteration halo is especially important in understanding how gold deposits form.

The country rock exchanges elements with the ore-forming solution gaining some and losing other elements to become an altered rock, i.e. the alteration halo. With a new mix of elements, the alteration zone develops different minerals and these, in turn, confer colour differences. Some of the elements added from solution to form alteration will be negligible in the country rock so their presence in alteration will be a strong guide to a potential ore forming process; these elements are sought during exploration geochemistry. Those elements that do not exchange between country rock and fluid can be used in whole rock geochemistry to aid in identifying the original country rock if only the alteration is available.

The solution is also modified by its reaction with the country rock with an important possibility that the change involves a decrease in the solubility of the gold. Sufficient precipitation of gold can then lead to formation of ore. The fluids studied by investigating the quartz veins will not necessarily be the original ore-bearing solution but will be slightly modified. The alteration process and its value in gold geology is described in more detail in Chapter 12.

2.3.3 Ore Reporting

The characterisation of ore and waste is a major role for the mine geologist on modern gold mines, and this activity follows on from centuries of practice by miners of visually separating rock into that which is valuable (ore) and that which is not (waste). The geometry and physical properties of ore and waste are important for mining, and chemical properties are also important during processing.

The communications to the public around ore has changed significantly with the development of national reporting codes under the Committee for Mineral Reserves International Reporting Standards (CRIRSCO) framework. In Australia, the Joint Ore Reserves Committee (JORC) was established in 1971, published as the JORC Code in 1989 and has played a crucial role in initiating the development of standards definitions for these codes and guidelines. Most other countries, except China, have similar systems for their mining industries. As well as addressing exploration results, these codes identify Mineral Resources when there is a reasonable chance of economic extraction of material. The component of the Mineral Resource that can be mined economically taking account of several modifying factors is reported as the Ore Reserve. Relevant factors may include mining costs, expected recovery effectiveness and commodity prices.

2.4   Ore on a Larger Scale

A common theme will be the use of many scales of study in determining how gold deposits form, and several geological and mining terms describe areas of ore and waste larger than single samples or individual mining areas (Table 2.2).

An orebody is contiguous mineralisation including ore with some common geological characteristics. There are many synonymous

Table 2.2 Terminology related to gold distribution on the district and regional scale.

Term	Explanation	Dimensions	Example
Province	Several geographically related goldfields with geological similarities	Up to several 100 km	Carlin, Abitibi, Victoria, Witwatersrand
Goldfield	Semi-continuous mineralised area that may include multiple styles	Surface footprint of 10–20 km by 1–5 km generally	Bendigo, Kalgoorlie
Orebody/ deposit	Continuous mineralisation with implication of being economic	Up to 1–2 km by 1–2 km by 1–20 m	Barton Deeps at Jundee mine
Ore shoot	Higher grade component of orebody	100s m by 10s m by few m May follow intersection of two planar orebodies, or fold axis	Oroya Shoot at Kalgoorlie
Primary mineralisation			
Gold-only	Gold deposit with no economic base metals	Grade: historically 10–30 ppm Au	Carlin, Victoria, Witwatersrand Archean greenstone
Gold-plus	Gold as a co-product or by-product in a deposit with economic base metals particularly with Cu	Grade: 0.5–3 ppm Au less commonly over 10 ppm Au	Cu-Au, VMS, IOCG, hi-S epithermal
Secondary mineralisation			
Alluvial	Formed by fluvial processes involving detrital gold grains rarely marine	Linear, kms in length, 10–100s m wide, few m thick usually within 5–10 km of primary source gold grains 0.5–5 mm diameter	
Supergene	Formed or strongly modified by regolith processes	Sheet-like and 100s m across, few m thick	Enrichment blankets in regolith profile
All-time production	Cumulative production for a goldfield since mining commenced, and may include artisanal working		
Endowment	Total of all-time production plus Reserves and reasonable Resources, and/or Economic Demonstrated Resources		

Abbreviations: *VMS:* volcanogenic massive sulfide, *IOCG:* iron oxide copper gold deposit, *hi-S:* high sulfidation epithermal
The term camp has been used to describe some goldfields

mining terms that are region-specific having been developed at one mine or group of mines. Examples include *lode* and *reef*, but the usage of these terms is inconsistent globally as a reef at one location may bear no similarity to a reef at another goldfield. For example, an auriferous quartz vein is likely to be called a reef at Bendigo in Victoria, and an auriferous carbon seam is a reef in the Witwatersrand goldfields of South Africa. Many gold orebodies are planar, up to some tens of metres thick and one hundred metres or more in length and width. Part of an orebody that is truly ore (i.e. economic) may be referred to as an ore shoot, with a connotation of an ore shoot generally being linear or nearly so. A common situation might be of two non-parallel planar orebodies that intersect to produce a linear ore shoot that may be either higher grade or increased thickness (or both) at their intersection. Deposit is a less formal but very widely used term for an orebody, or a potential orebody.

Goldfields can be tens of kilometres in length and contain many orebodies sharing geography and some geological features, but also with significant variability. Their known depth is commonly limited by the depth of economic mining.

Gold provinces are areas of 100s km^2 containing a high concentration of goldfields that share some common geological features. Examples might include Quadrilátero Ferrífero in Minas Gerais State of Brazil, Carlin Gold Province of Northern Nevada USA, Abitibi Province of Eastern Canada, Eastern Goldfields Province of the Yilgarn Craton of Western Australia, and the Victorian Gold Province of south-east Australia. The term province can also be used for other mineral commodities.

Snapshot

- Mining is an activity based on economics.
- Those economics are dynamic and influenced by society, environment, and government.
- Mine geology influences economics of a mine well beyond grade and commodity price.
- The distribution of gold from microscopic to continent scale inform on how deposits form.
- The minerals of ore, alteration and waste all inform about deposit formation.

Bibliography

Abzalov M (2016) Applied mining geology. Springer, New York, p 448. https://doi.org/10.1007/978-3-319-39264-6

Pohl WL (2020) Economic geology, principles and practice: metals, minerals, coal and hydrocarbons – an introduction to formation and sustainable exploitation of mineral deposits, 2nd edn. Schweizerbart Science Publishers, Stuttgart, p 755

Revuelta MB (2018) Mineral resources—from exploration to sustainability assessment. Springer, New York, p 653. https://doi.org/10.1007/978-3-319-58760-8

Ridley J (2013) Ore deposit geology. Cambridge University Press, New York, 398pp

Robb L (2005) Introduction to ore-forming processes. Blackwell Publishing, Oxford, p 373. https://doi.org/10.1144/1467-7873/05-073

Data to Processes, Examples, and Discovery

3

Abstract

Determining how gold deposits form is facilitated by the wealth of information available on the Earth's surface, in underground mines and open pits, through characterisation and experiment in research laboratories, through libraries, and even through mining company annual reports. Designing and collecting the more useful information is one stage, but then it is necessary to decide what to retain and how it should be arranged and prioritised. It is difficult to argue against the complexity and variability within and between deposits, but there is value in occasionally pausing the data collection and reviewing the similarities.

Keywords

Scale · Gold-only · Gold-plus · Modification of deposits · Sequence of chapters

There are significant difficulties associated with determining how gold deposits have formed because we are trying to interpret processes that occurred millions of years ago and kilometres beneath the Earth's surface at temperatures around 300 – 500°C. Experiments to replicate these situations are possible but quite limited. However, the problem is not as intractable as it at first appears if the considerable information contained in and around deposits is used to the fullest. The approach here is to apply standard petrological, geochemical, and structural methods to gold ores and surrounding alteration haloes as has been outlined by Robb (2005) and Pohl (2020); they emphasise that mineral deposits are rocks that have value, and therefore their study using petrological processes has an important role in determining their formation.

3.1 Utilising Scale of Study, Classification, and Modification of Deposits

Two approaches to scale are adopted to help understand gold deposit formation. One approach is to study from global to microscope scale and involve the integration of tectonic setting, host rock sequence, structure, stable mineral assemblages, and the timing of mineral growth. Such an approach can ensure that a conclusion based on one scale is at least compatible with observations at other scales. Another approach to scale is to study many different deposits to identify what appears essential amongst them, and what occurs sporadically and is only accidental. In many parts of the world, the ability to visit and study large numbers of deposits is much more limited today than it was at the start of the 1980s global gold boom (see Appendix D).

Classifying gold deposits with numerous historic and modern terms is minimised in the design of this book especially in early chapters. In the literature today, many subdivisions are

© The Author(s), under exclusive license to Springer Nature Singapore Pte Ltd. 2022
N. Phillips, *Formation of Gold Deposits*, Modern Approaches in Solid Earth Sciences 21,
https://doi.org/10.1007/978-981-16-3081-1_3

Fig. 3.1 Subdivision of all-time gold production into two main classes using the presence, in economic amounts, of the base metals of Cu, Zn and Pb. Gold production has been dominated by the gold-only deposits that lack economic base metals. Gold-plus deposits include those with copper and gold as co-products, and the volcanogenic massive sulfide (VMS) deposits in which gold is a by-product of the mining of base metals. This book focuses on the gold-only deposits and to a lesser extent the copper-gold deposits.

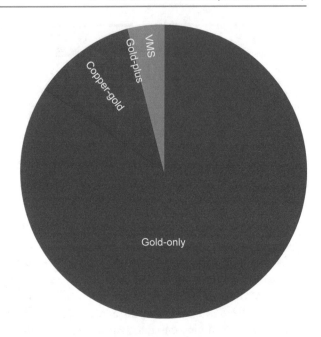

overlapping which invites debate as to where a deposit should be placed, some schemes lack diagnostic characteristics that distinguish types, some schemes are impractical to apply with debates lasting half a century or more, and many schemes lack a theoretical basis so will ultimately become fraught with modifications, caveats, and definition broadening. The two-way split into gold-only and gold-plus is introduced early and justified later because of its strong scientific basis and ultimately because it appears *unreasonably effective*, i.e. the ability of a hypothesis to predict additional outcomes (Fig. 3.1). The gold-plus grouping recognises those rich in copper-gold, and the VMS (volcanogenic massive sulfide) deposits of generally subordinate by-product gold.

Modification of gold deposits after their formation has played a major role in determining their appearance and dominant characteristics, so much so that the modifications may be mistaken as the distinguishing character of a deposit or type (Fig. 3.2; Chapters 16–19). Such characteristics may bear no relationship to the process of deposit formation and need to be isolated as might be done automatically when collection in the field avoids weathered samples.

Specialist skills may be required to unravel the modifications including metamorphic petrology, structural geology and regolith science.

3.2 Sequence of Coming Chapters

The chapters ahead can be subdivided into the gathering of some critical data, examination of alternative genetic processes, application of the genetic ideas to some global examples, and then case histories of discovery (Fig. 3.3).

Chapters 4 to 9 summarise many of the characteristics of gold deposits with emphasis on *provinciality, enrichment, segregation, timing, and ore fluid types*. These five characteristics are selected as particularly important in evaluating any models for deposit formation.

Chapters 10 to 17 explore magmatic and then hydrothermal processes in the generation of deposits and describe modifications to deposits after they have formed.

Chapters 18 and 19 describe examples of deposits and their likely formation.

Chapters 20 and 21 are case studies in which understanding the formation of gold deposits has assisted in discovery. One example is from the

Fig. 3.2 Modification clock showing a sequence of events leading up to gold deposit formation and then further events that modify the deposit. Small numbers refer to the relevant chapters.

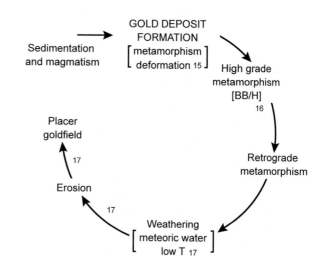

Fig. 3.3 Sequence of major components of this book from data assembly to potential genetic processes that might form deposits, their subsequent modification, and finally some global examples and some discovery case histories.

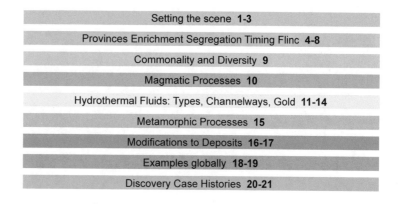

Archean Yilgarn Craton, the other from the Paleozoic metasedimentary succession of the Victorian Gold Province.

The Appendices C-E record some of the philosophy of science and softer skills. The importance of these is commonly underestimated in gold geology research and exploration success.

Knowing how a gold deposit forms is not the end of the story. The exploration geologist standing beside a drill rig in the desert or tundra might easily say "Don't tell me how deposits form, tell me where they form?". For this most important of questions, structural geometry of the various rock units is all important. Some structural features are discussed such as faults, shear zones, breccias and unconformity surfaces to understand the mechanisms of auriferous fluid flow along these channelways. Major gold deposits form where

there are multiple parallel structures or structures of many orientations and types, i.e. structural complexity, but the full story of *where* is a topic for another book.

Snapshot

- Standard petrological, geochemical and structural methods are applied to gold ores including any surrounding alteration halo.
- Five characteristics of all global gold deposits are prioritised before detailed classification: these are *provinciality, enrichment, segregation, timing and ore fluid type.*

(continued)

- Scale includes integrating global to microscopic studies from multiple deposits.
- Classification schemes for deposits should be as simple as possible, be practical to apply, and have a sound theoretical basis.
- Early classification of deposits is minimised to avoid silos separating examples.
- Modification of deposits after their formation may strongly influence their characteristics.

Bibliography

Killick AM (2003) Fault rock classification: an aid to structural interpretation in mine and exploration geology. South Afr J Geol 106:395–402. https://doi.org/10.2113/106.4.395

Phillips GN, Powell R (2015) A practical classification of gold deposits, with a theoretical basis. Ore Geol Rev 65:568–573. https://doi.org/10.1016/j.oregeorev.2014.04.006

Pohl WL (2020) Economic geology, principles and practice: metals, minerals, coal and hydrocarbons – an introduction to formation and sustainable exploitation of mineral deposits, 2nd edn. Schweizerbart Science Publishers, Stuttgart, p 755

Robb L (2005) Introduction to ore-forming processes. Blackwell Publishing, Oxford, 373 pp. https://doi.org/10.1144/1467-7873/05-073

Five Characteristics of Gold Deposits That Reveal Their Formation and Lead to Exploration Success

Abstract

There is a strong tendency for gold ore bodies to be clustered as goldfields, and goldfields to be clustered as provinces. This tendency to occur in provinces is the **first** of five characteristics that are described and given importance when determining how gold deposits form. Goldfields are clusters up to 10 – 20 km in length. Provinces can be several 100 km, contain multiple goldfields, and indicate one of the scales of study necessary to understand gold deposit formation.

Keywords

Global gold production · Global gold provinces · Gold through time

The occurrence of multiple gold deposits in geographic clusters was known well before the gold rushes of California in 1849 and Victoria in 1851 (Table 4.1). Examples are the Jales district in Northern Portugal including the Três Minas deposit, and the Morro Velho district near Belo Horizonte in Brazil. It became accepted wisdom that where some gold was found there was likely to be more. Clusters of deposits that extended for a few kilometres with some similarities of geology became known as goldfields (e.g. the Kalgoorlie goldfield in Australia, and Timmins goldfield in Canada); much larger areas extending 10s to 100s km with many goldfields became known as gold provinces (e.g. Carlin gold

province in Nevada USA; Victorian gold province in southeast Australia). This provinciality is one critical piece of scientific evidence that is important in determining how gold deposits form as it places constraints on the scale of the processes of formation.

Initially, it might sound a little trite but there are two possibilities for the distribution of gold deposits, one is that the distribution of gold metal is homogeneous so that there are either just a few large deposits, or there are many small ones. The alternative possibility is that special areas exist where there are many large, medium, and small deposits together. The overwhelming evidence globally is in favour of the latter, and these areas of concentration lead to the concept of gold provinces. So evident is this clustering that prospectors are advised that "The best places to look for gold are where others have found some." (Schwartz 1980, p. 37). This principle is even given the name *nearology* in exploration and being close to major mines or new discoveries has distinct commercial appeal.

4.1 Gold Production Records

Given that there has been up to 6000 years of gold production, records are necessarily incomplete but facilitated by some unusual characteristics of gold trading and usage. Unlike most elements, the durability and value of gold means that much of the earliest gold produced is still in use. There has

N. Phillips, *Formation of Gold Deposits*, Modern Approaches in Solid Earth Sciences 21,
https://doi.org/10.1007/978-981-16-3081-1_4

Table 4.1 History of gold discoveries and production.

	Discovery or development
	Sumerians in Mesopotamia, maybe current Iran, 3800 years ago
	Upper Nile and Nubian desert, 3500 years ago
1000	Global production 2 tpa
1100	Shandong/Jiaodong China gold mining
1690	Brazil gold rush to Minas Gerais
1835	Morro Velho Brazil
1848	California USA
1851	Victoria Australia
1861	Otago NZ
1876	Homestake USA
1886	Witwatersrand discovered in South Africa
1893	Kalgoorlie discovered in Western Australia
1897	Ashanti Ghana
1897	Klondike northwestern Canada
1909	Porcupine Timmins Canada
1935	Carletonville then Vaal Reef and Welkom discovered in Witwatersrand
1951	Evander - Last discovery of a Witwatersrand goldfield
1957	Blue Star Carlin discovered, then Carlin deposit in 1961, USA
1958	Muruntau, Uzbekistan
1967	Grasberg, Irian Jaya, Indonesia
1970	Peak production from Witwatersrand: 1000 tpa Au
1971	Start of a 50 year decline of Witwatersrand production
1980	Gold boom of 1980s especially in Australia and USA
2010	China and Australia outproduce South Africa
2010 on	Australia and/or China are leading gold producers (see Appendix A)

always been a high degree of recycling and storage of gold, which means that little is deliberately discarded or destroyed. The estimates of historical mining take account of this preservation of gold, the scale of ancient gold mining operations, the history of gold trading and transport, and the estimated amount of gold in global circulation at different times.

Different terms are used in the geological literature to convey the quantum of gold production (see also Appendix A):

- All-time production is a cumulative figure since mining of an area began
- Annual production applies to a particular year
- Reserves and Resources are estimates of gold that is yet to be mined
- Endowment is one of the better descriptions of the size of a deposit from a geological perspective and combines all-time production with Reserves and Resources. This is essentially a

measure of the deposit without the effects of depletion through mining
- Where a significant proportion of production has come from alluvial gold and the primary source is relatively certain because of present and past geography, production figures for placer gold may be added back into the endowment with a statement explaining the method of calculations used. Adding back placer gold is subjective but can provide a better indication of the original endowment of a large goldfield.

All-time production and annual production can be quite accurate, but they are constantly changing during production. Endowment is an estimate, but only changes when Reserves and Resources are adjusted. For an understanding of how gold deposits form, it is endowment that is particularly useful, whereas production, Reserves and Resources have considerable commercial significance. Figures are commonly rounded to one or two significant digits for presentation unless additional precision is necessary. Compilation has been in tonnes of Au mostly, but an approximate conversion to million ounces is provided, i.e. Homestake deposit USA (40 Moz / 1250 t Au).

4.1.1   All-time Gold Production

All-time global gold production to the start of 2020 is estimated at 194,300 t so expected to reach ~6000 Moz (200,000 t) in the second half of 2021. The most important contributing countries have been South Africa (1500 Moz / 50,000 t), and Russia and USA (each 600 Moz / 20,000 t), Australia and Canada (500 Moz / 15,000 t each).

Twenty-one countries have an all-time gold production of 60 Moz / 2000 t, i.e. each has contributed more than 1% of the global total: in this regard, the sources of gold are much more diverse politically and geographically than many other commodities (cf. U, Mo, rare earth elements). For many countries, their gold production is reflected by the size of their landmass such that the largest countries by area dominate the list of all-time gold production. The global average gold production is 0.05 Moz / 1.4 t Au per 1000 km^2 of landmass.

There are some striking exceptions to this generalisation with South Africa being the standout (Fig. 4.1; Fig. 4.2). USA, Australia, Colombia, Mexico and Egypt have slightly above production for their area. Particularly notable producers for their size are Ghana, New Zealand, Papua New Guinea, Uzbekistan, Philippines, Spain / Portugal and Zimbabwe (one might add Dominican Republic and Fiji whilst noting their small land areas). There is a risk in over-interpreting some smaller countries. If an area the size of Ghana was superimposed over parts of the Abitibi province of Canada or Yilgarn Craton of Western Australia these latter two areas might appear at least as rich in gold as Ghana. There are also many small countries that do not feature here as they have produced negligible gold. Large countries that appear under-represented in their all-time gold production are India, Brazil, Argentina, Kazakhstan, Algeria and Congo DRC, though for some the issue may be politics and sovereign risk. The above calculations do not include the area of Antarctica. As a hypothetical projection, if Antarctica was gold-bearing to the same extent as the global average, its endowment might be near 600 Moz (20,000 t) of gold.

4.1.2   Annual Gold Production Globally

Annual global production in 2019 was 110 Moz / 3300 t, and the leading producers were Australia, Russia, USA and Canada (Fig. 4.3). For countries such as Australia, USA, South Africa and Canada, transparency is expected such that their national figure can be confirmed by summing production from each State and Province; the results can then be verified by summing production from the company reports to stock exchanges covering over one hundred gold mines that observe a globally recognised CRIRSCO reporting code such as JORC or NI43-101.

Throughout history, it has always been possible for a very small gold mine to be operated economically by one or two persons but even with thousands of such mines during past gold rushes their production has been subordinate to the achievements of today's larger mines. The large national gold outputs today were never matched in the past from many small operations. For example, Victoria, which had 7000 or more mine workings that operated during the second half of the 19th century, almost achieved 3 Moz / 100 tpa in its peak year but mostly produced at half this rate. More recently, Serra Palada mine in Brazil operated through the 1980s with over 50,000 artisanal miners producing a total for that period estimated around 2 Moz / 60 t or less than 10 tpa average.

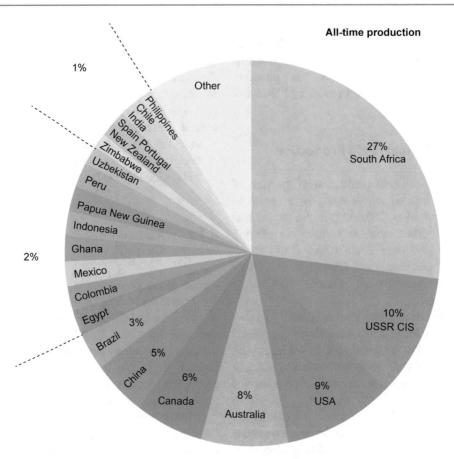

Fig. 4.1 All-time gold production by country showing the dominance of South Africa; many countries have significant all-time gold production.

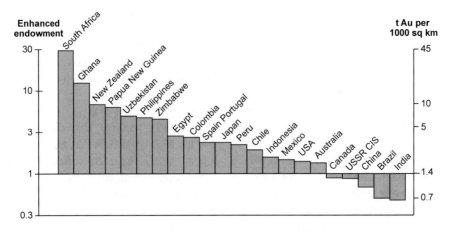

Fig. 4.2 All-time gold production by land area with a log vertical axis. A value of 1 on the left axis corresponds to 0.05 Moz (1.4 t) of gold per 1000 km^2 which is the average counting all countries. South Africa dominates this figure, followed by some gold-producing countries of small area. The six largest countries by area are close to the global average.

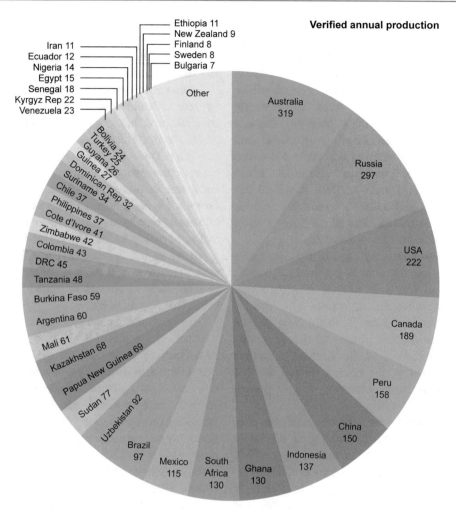

Fig. 4.3   Annual gold production in tonnes for various countries for 2018 based upon verified national figures. The figure for China is an unverified estimate only; it may be too low and is not strictly comparable with the verified figures.

4.1.3   Different Figures for Gold in China

For China, the gold production and resource figures do not allow any like-on-like comparisons with other countries. The figures are not generated according to CRIRSCO, JORC or NI43-101 equivalent reporting standards that almost all other mining countries use (see Appendix A). To provide some comparison with other countries, an estimate of 150 tpa is used for China annual production, but a high degree of uncertainty should be noted.

4.2   Largest Goldfields

Globally there are almost 350 goldfields around the world with an endowment exceeding 3-4 Moz (100 t Au), and this number is reached by treating the Witwatersrand of South Africa as five semi-discrete goldfields (Evander, Rand, Carletonville, Klerksdorp, Welkom. Note that it is also common to refer to seven by recognising West, Central and East Rand goldfields).

Major goldfields are well-represented in the Archean, somewhat less so in the Proterozoic, and are widespread in the Paleozoic and younger regions of the world (Table 4.2). None of the major goldfields are in rocks deposited in the first quarter of Earth history. The oldest sequence that hosts major goldfields is the 2900 – 2700 Ma upper Witwatersrand succession of South Africa. Somewhat younger but much more widely distributed are the 2700 – 2600 Ma Archean greenstone belts as found in Australia and Canada. The goldfields in host rocks older than 2900 Ma are rare and small; examples include Barberton, the Murchison Belt, the 3080 Ma Dominion conglomerate, and the lower Witwatersrand succession all of South Africa, and the Pilbara Craton of Australia.

This distribution with age of host sequences partly reflects the scarcity of the oldest rocks because of their poor chance of preservation contrasting with the abundance of Cenozoic rocks on the Earth's surface today. Nevertheless, the younger Archean regions are clearly anomalously rich in gold, i.e. the Archean comprises less than 10% of the Earth's surface and has produced the largest goldfields, and 40 percent of production overall.

None of the largest Archean goldfields have by-product Cu except for Boddington in Western Australia, and the important and numerous VMS deposits of the Abitibi province of Canada are not endowed with enough gold to feature in Table 4.2. Gold deposits in the Proterozoic can be large, and in many there is co-product Cu to make up for lower gold grades. Major Paleozoic goldfields include several in clastic metasedimentary sequences variably termed slate-belt, shale-greywacke or turbidite-hosted (Fig. 4.4). Many large Cenozoic goldfields show a spatial association with volcanically active areas or shallow intrusive rocks of various compositions.

4.3 Major Gold Provinces

4.3.1 Synthesising Whole Provinces

Regional-scale studies of multiple goldfields provide an important contribution to understanding how deposits form. Integrated studies that involve many gold deposits play a particularly important role in separating accidental from essential deposit features. Ideally the regional syntheses inform the context for on-going detailed field and analytical work on individual deposits.

Regional-scale studies however have several inherent difficulties which mean that they are comprehensive in some provinces and less so in others. Access difficulties due to security, lack of research personnel or funds, cultural differences, or even too many researchers all focused on small-scale studies can make a regional synthesis difficult. Timing is important also as these factors of access can be easily changed such as by national security issues. Given that the regional syntheses require a team, and usually just one team in a province, it requires special leadership to avoid the development of groupthink in these teams (see Appendix E).

The most comprehensive synthesis of gold provinces on a global scale is by colleagues Rich Goldfarb of the US Geological Survey and David Groves of University of Western Australia; they provide compilations of all-time production and resource figures, and most importantly, their figures are provided within the context of regional geological settings. At the global scale few others have attempted summarising and classifying deposits to the extent achieved by Goldfarb and Groves who have included all continents and Archean to Phanerozoic examples. They have mostly confined their study to what they refer to as orogenic gold deposits without detailed coverage of other gold-only deposits such as Witwatersrand, Carlin and some epithermal and porphyry deposits.

What follows is a summary of some of the largest gold provinces, but also some others that display useful features for understanding gold deposit formation. The endowment figure attributed to any province is necessarily subjective with the limitation of historical record keeping, and judgement of its geographic limit.

4.3.2 Witwatersrand Goldfields

The discovery by George Harrison of the Central Rand goldfield in 1886 was the start of the greatest gold exploration and mining endeavour

Table 4.2 Major goldfields and their estimated endowment.

Goldfield	Country	Age of host sequence	Endowment (Moz Au)	Endowment (t Au)	Other economic metals
600–0 Ma					
Grasberg	Indonesia	Phanerozoic	90	2800	Cu
Yanacocha	Peru	Phanerozoic	70	2200	Cu
Bingham	USA	Phanerozoic	60	1900	Cu
Porgera	Papua New Guinea	Phanerozoic	25	800	
Lihir	Papua New Guinea	Phanerozoic	40	1200	
Pueblo Viejo	Dominica	Phanerozoic	40	1200	Cu
Cripple Creek	USA	Phanerozoic	30	900	
Round Mtn	USA	Phanerozoic	25	800	
600–0 Ma (recent discoveries)					
Colosa	Colombia	Phanerozoic	20	680	
Livengood	Alaska USA	Phanerozoic	20	600	
Pebble	Alaska USA	Phanerozoic	50	1500	Cu
KSM Brucejack Eskay Ck	Western Canada	Phanerozoic	50	1500	Cu, Ag and others (VMS)
600–200 Ma					
Muruntau	Uzbekistan	Paleozoic	170	5300	
Sukhoi Log	Siberia Russia	Paleozoic	60	1900	
Natalka	Siberia Russia	Paleozoic	60	1800	
Olympiada	Russia	Paleozoic	50	1500	
Goldstrike	Nevada USA	Paleozoic	50	1500	
Cadia	NSW Australia	Paleozoic	50	1500	Cu
Kalmakyr	Uzbekistan	Paleozoic	50	1500	Cu
Donlin Creek	Alaska USA	Paleozoic	25	800	
Bendigo	Victoria Australia	Paleozoic	25	700	
Gold Quarry	Nevada USA	Paleozoic	20	650	
Grass Valley	California USA	Paleozoic	20	600	
2500–600 Ma					
Olympic Dam	South Australia	Proterozoic	95	3000	Cu
Homestake	S Dakota USA	Proterozoic	40	1250	
Telfer	Western Australia	Proterozoic	20	600	Cu
Older than 2500 Ma					
Rand	Wits, South Africa	Archean	680	21,000	
Welkom	Wits, South Africa	Archean	350	11,000	
Carletonville	Wits, South Africa	Archean	290	9000	
Klerksdorp	Wits, South Africa	Archean	230	7000	
Evander	Wits, South Africa	Archean	50	1600	

(continued)

Table 4.2 (continued)

Goldfield	Country	Age of host sequence	Endowment (Moz Au)	Endowment (t Au)	Other economic metals
Kalgoorlie	Western Australia	Archean	90	2800	
Ashanti	Ghana	Archean/ Proterozoic	70	2100	
Hollinger (Timmins)	Abitibi Canada	Archean	30	1000	
Kirkland Lake	Abitibi Canada	Archean	30	1000	
Boddington	Western Australia	Archean	25	850	minor Cu
Kolar	India	Archean	25	850	

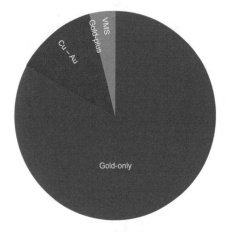

Fig. 4.4 Subdivision of the 6000 Moz (200,000 t) all-time global gold production as of mid-2021 highlighting: (**left**) a three-fold subdivision based on whether there are any economic base metal co-products particularly Cu; (**right**) some common examples within the three-fold subdivision. Some assumptions have been made in producing this figure: placer gold which is mostly historical is derived from gold-only deposits, production from the Carlin gold province in Northern Nevada USA is included in 'Slate' (see Chapter 18), and gold-only deposits account for about 80% of all-time production. Subdivisions used in the right figure are in common usage but not well defined, not always mutually exclusive, and reflect the author's estimates.

in history. One quarter of the world's all-time gold production has come from the Witwatersrand (commonly referred to as Wits), a peak of 1000 tpa production was achieved in 1970, and there has been a 50-year decline since then. The early 20th century production was sustained by further exploration success finding new goldfields from 1934 to 1951 but no new goldfields have been discovered since.

Near-surface mining from 1886 especially near Johannesburg was followed in the early 20th century by property amalgamations and a period of mid-depth mines requiring additional infrastructure (Fig. 4.5). This second phase of mining was followed by further amalgamation and increased infrastructure to mine at depths below 2 km. After 1930, the Carletonville, Welkom and Evander goldfields were discovered beneath younger sequences and were major deep mines from their outset. At the centenary of Witwatersrand mining there were over 40 operating mines producing a total of 600 t Au pa in seven discrete goldfields in a 400 km arc with its long axis of 300 km. By 2020, many

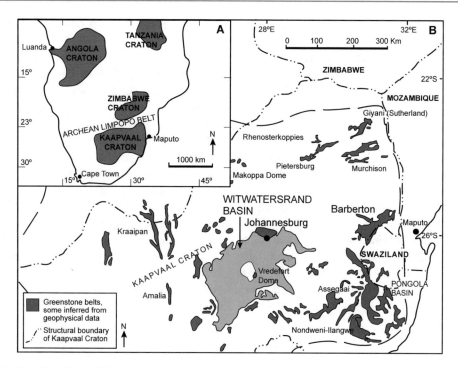

Fig. 4.5 Map of southern Africa showing Archean areas including the Kaapvaal and Zimbabwe Cratons, and the Witwatersrand basin which is host to 25% of all-time global gold production. There are numerous Archean greenstone belts within the Kaapvaal Craton but only Barberton and Murchison have significant gold.

of these mines had closed and the Witwatersrand was barely producing 100 tpa Au.

Most of the gold has been mined from unconformity surfaces in the upper Witwatersrand sequence (Central Rand Group) and all are within packages of 100s m thickness of altered metasedimentary rocks. Each goldfield is adjacent to one or more thrust faults dipping out of the basin. There are considerable differences between various metasedimentary host rocks from pyrite-bearing quartz pebble conglomerate, carbon seam, pyritic sandstone and polymictic (i.e. with abundant rock clasts and fewer quartz pebbles) conglomerate. Subordinate mineralisation is associated with arsenopyrite and pyrrhotite.

Regional syntheses of the Witwatersrand goldfields are rare because of the scale of the goldfields and the limited sharing of data between adjacent mines even within one company. Until the early 1980s the main synthesis was that of Pretorius who had focused on the Rand goldfields. The synthesis of Phillips and Law (2000) was based on all operating gold mines and goldfields, and integrated the synthesis of Pretorius, much Witwatersrand research since 1980, and modern global gold geoscience.

4.3.3 Barberton Greenstone Belt of South Africa

The Barberton greenstone belt in eastern South Africa is famed for its outstanding Archean geology but its gold has been overshadowed by the Wits. Barberton has more than a century of gold mining history, 300 old mines, and multiple high-grade long-life deposits exploited through the Sheba, Fairview, Agnes and Consort mines with an all-time production of 11 Moz / 350 t Au for the province. The Barberton greenstone belt is over 100 km in length, and itself 350 km east of Johannesburg and the Central Rand (Fig. 4.5).

The main deposits form a cluster within 10 km of one another and are in structurally complex settings. They differ in their metamorphic grade from greenschist to amphibolite facies, and specific host rocks include ultramafic rocks, shale, sandstone, and banded iron formation. Mineralisation transgresses three major stratigraphic packages in the greenstone belt all of Archean age. The Murchison belt, a neighbouring greenstone north of Barberton has widespread gold mineralisation and likely significant endowment, but, to date, not the gold production profile to match Barberton.

Apart from Barberton and Murchison, other greenstone belts around South Africa have remarkably low gold endowments with none exceeding the 2 Moz / 60 t Au of the Tati greenstone belt which is 600 km north of Johannesburg.

4.3.4   Zimbabwe Gold Province

Zimbabwe is a small country dominated by Archean granite and auriferous greenstone belts with a long history of gold production. A synthesis of Zimbabwe gold geology was led by Dr Bob Foster with a group of postgraduates of the University of Zimbabwe culminating in the hosting of Gold'82 Conference and Proceedings in Harare. A characteristic of the deposits is their large number, widespread distribution throughout so many greenstone belts, and generally modest size of goldfields. The granitic rocks are mostly barren of gold with small deposits in the granite but near the margins with greenstone belts. Much primary production of agriculture and minerals including that of gold virtually ceased in Zimbabwe during the Mugabe dictatorship from the mid-1980s. Relative to many other countries, there has been no intense application of modern exploration techniques since 1980, and when this does occur Zimbabwe production could rise significantly. The two largest goldfields of Cam & Motor and Globe & Phoenix each have an all-time production above 3 Moz / 100 t.

4.3.5   West African Craton

Gold has been known in West Africa for many centuries and mined from deposits within the Archean Man Craton and Paleoproterozoic Birimian gold belt. Southeast Ghana is particularly well endowed with the Obuasi-Ashanti deposit of 70 Moz (2100 t Au; Fig. 4.6). There are several other goldfields within a province that is 300 km long with parallel belts over a 200 km width comprising Paleoproterozoic Birimian greenstone belts and clastic metasedimentary rocks surrounded by granitic rocks. The same province contains large goldfields in conglomerate host rocks including Tarkwa, Iduapriem and Damang, of 300 t, 200 t and 100 t, respectively. Farther west there has been historic artisanal working and increasing late 20th century exploration interest in Mali, Burkina Faso, Senegal and Ivory Coast. This has led to large greenfield and brownfield discoveries including Morila, Syama, Siguiri, Sadiola, Sabodala, Loulo and Yatela. Production has increased in recent years though sovereign risk remains an issue. The Tasiast goldfield is in Mauretania and producing gold from an Archean greenstone belt.

4.3.6   Quadrilátero Ferrífero region of Brazil

Gold has been produced from many parts of Brazil including Quadrilátero Ferrífero, Carajás, Goiás and Amazonia (Fig. 4.7). The Quadrilátero Ferrífero area adjacent to the city of Belo Horizonte comprises Archean meta-sedimentary and igneous sequences of a greenstone belt surrounded by granite. It is the best example globally of multiple Archean gold deposits hosted in banded iron formation including Morro Velho (over 15 Moz / 500 t), Cuiabá (10 Moz / 300 t), Raposos and São Bento (each 3 Moz / 100 t or more). All of these comprise ore shoots in banded iron formation and the epigenetic mineralisation is inferred to be 2600 – 2700 My.

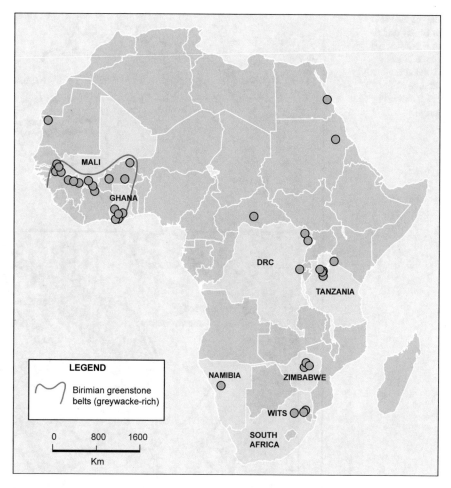

Fig. 4.6 Goldfields in Africa with endowment of 3 Moz / 100 t Au or more. West Africa is a major province in Archean to Proterozoic rocks, Zimbabwe, Lake Victoria area of northwest Tanzania and northeast DRC all have multiple major goldfields. Eritrea and Sudan have produced significant gold prior to any detailed recording. Navachab (6 Moz / 200 t Au) in Namibia stands out as a large deposit that is not in a province of many goldfields. The Witwatersrand is marked here as a single dot that reflects 100 large mines.

4.3.7 Yilgarn Craton of Western Australia

Western Australia comprises the two large Archean Pilbara and Yilgarn Cratons. So far, little gold has been found in the former, but the Yilgarn is one of the world's most gold-mineralised areas with an endowment over 350 Moz / 11,000 t of gold and the site of exceptional exploration successes since 1980 (Fig. 4.8). It is a deeply weathered terrain of limited relief, with lake and sand dune cover. It comprises linear greenstone belts and extensive intervening granitic batholiths. Hundreds of gold discoveries were made in the Yilgarn in the late 1800s, but there were decades of dwindling production after 1903. Exploration from 1980 was based on new techniques and new scientific ideas leading to important discoveries in new (greenfield) areas and many more in already-known (brownfield) gold districts. Virtually all gold deposits are either in the greenstone belts or, if within granite, are within a kilometre of greenstone. Immediate host rocks for deposits include basalt, dolerite, other

Fig. 4.7 Map of the main goldfields in Brazil, and the Quadrilátero Ferrífero area with its multiple large deposits in banded iron formation. Source of lower map: redrawn using Ribeiro-Rodrigues et al. 2007.

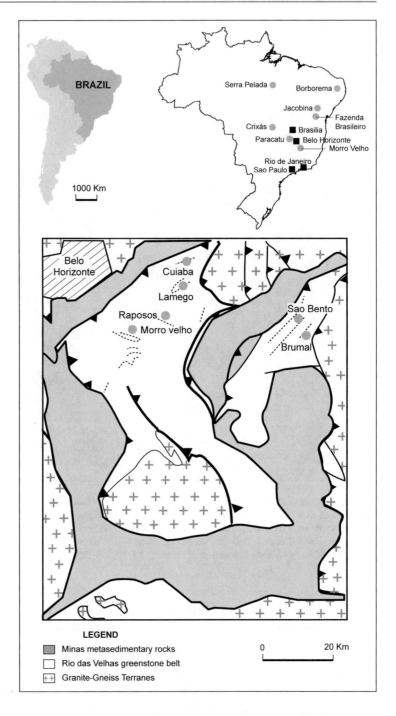

igneous rocks, banded iron formation and clastic metasedimentary rocks especially black shale. Examples from the Yilgarn Craton are used through this book including a case history of the giant Kalgoorlie goldfield in Chapter 13, and an overview of discoveries since 1980 in Chapter 20.

4.3.8 Victorian Gold Province

The discovery of gold in 1851 led to the Victorian Gold Province becoming the largest gold producer globally for a few years, and now has an all-time production of 3000 t. Early production

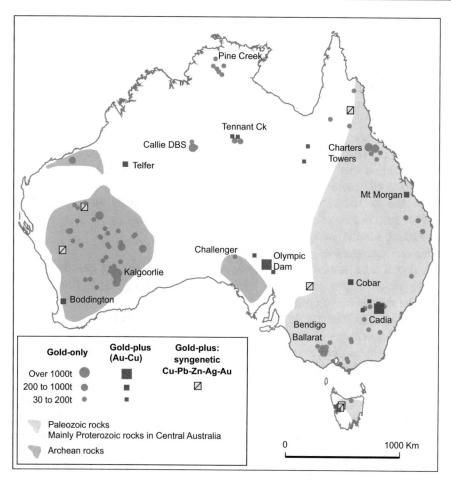

Fig. 4.8 Map of goldfields in Australia with endowments of 1 Moz / 30 t Au or more. The greatest endowment is within the Yilgarn Craton which might be considered as several distinct provinces. Other important clusters include Callie (Dead Bullock Soak) and the Tanami region, Pine Creek, Tennant Creek, NE Queensland Charters Towers, Central New South Wales, and Bendigo – Ballarat region of the Victorian Gold Province. Reprinted from Phillips 2017.The addition in the Pilbara Craton west of Telfer is the Mallina goldfield in which a Resource of 9 Moz has been announced in 2021 following the Hemi discovery.

was predominantly from placer and paleoplacer deposits, but as these became depleted, underground mining flourished especially in the Bendigo goldfield. The latter became famous for its quartz veins that defined saddle reefs, but it also contained many other types of mineralisation. Government records for Victoria suggest that there have been at least 7000 old mines and shafts across an area of 600 km by 200 km. By 1910 much gold mining had finished with small production at isolated mines until a small resurgence at the end of the 20th century.

The highlight of this resurgence has been the discovery of Fosterville which is the case history of Chapter 21.

A synthesis of the Victorian goldfields has been undertaken over many years with Martin Hughes by driving to hundreds of old workings including all the larger historical operations. This work was supplemented by much walking of the intervals between lines of mineralisation and through extensive review of the literature dating back to the late 19th century when many mines were active. Such work is never complete but the uniform field-based approach across all parts of the province provides insights that are quite difficult to replicate elsewhere. The distribution of goldfield size and number, and the nature of sulfide mineral assemblages associated with gold ores warrant some elaboration here (Fig. 4.9).

One interesting insight comes from plotting the size of deposits for each of the nine geological zones mapped in the Paleozoic rocks of Victoria (these were defined on their stratigraphy, structural and lithological character, and inferred deformational history, but independent of any contained gold mineralisation). All-time gold production is highest for the Bendigo geological zone and decreases progressively in other zones going east and west from there. A similar pattern applies to the single largest goldfield in each zone, and to the number of goldfields over 30 t Au and reef mines of at least 1 t Au. Together these maps are strong evidence of the coincidence of large goldfields where there are many small goldfields (rather than the alternative of many small deposits OR a single large one).

These maps also help to understand how provinces terminate. In the case of Victoria, the limiting extent north is relatively thin Cenozoic fluvial sediment; and to the south is thicker Mesozoic to Cenozoic sequences. To east and west, the Victorian gold province coincides with the four structural zones that have at least one goldfield of 30 t, and the termination of the province is marked by significant decrease of auriferous veins across major fault zones but remains somewhat subjective (Fig. 4.9e). However, the Early Paleozoic clastic metasedimentary host rock sequence is relatively unchanged in the east across these province boundary faults.

A different insight comes from analysing the information from the thousands of gold occurrences based upon their hypogene minerals and the geochemical characteristics of the sulfide assemblage (Fig. 4.10). Five mineral assemblages are used to subdivide these occurrences with the main minerals being pyrite, arsenopyrite, stibnite, various uneconomic base metal sulfides and more complex sulfide minerals. The mineralogical domains are over one hundred kilometres in length parallel to the regional structural trend,

and tens of kilometres in their east – west width. Most domains are spatially associated with major faults which flatten into a zone of duplexed greenstones overlying older basement rocks in the deeper crust. The variation in mineralogy in Victorian gold occurrences indicates that ore fluid compositions differed significantly between adjacent domains. This in turn informs on predictions about ore fluids including their origins and pathways. The characteristics of the mineralogical domains assist genetic interpretation especially of the scale at which deposit-forming processes operate.

4.3.9   Northeast Queensland Charters Towers Gold Province

Gold deposits in Paleozoic rocks are known along the eastern margin of Queensland, and in Proterozoic rocks in the northwest near Cloncurry - Mt Isa. The NE Queensland (Charters Towers) province refers to a concentration of larger gold deposits south and southwest of Townsville; the province is 100 km inland and parallels the coast for 250 km (Fig. 4.11). The total endowment around 30 Moz / 1000 t is based on a production history dating back to the 1860s and substantial brownfield discoveries since 1980. Significant goldfields from the late 19th century include Charters Towers (8 Moz / 240 t endowment) and Ravenswood (8 Moz / 230 t due to its brownfield additions since 1980). Important discoveries since 1980 included Kidston (4 Moz / 110 t), Mt Leyshon (3 Moz / 100 t), Pajingo (4 Moz / 120 t), Mt Wright (1 Moz / 34t), Wirralee (1 Moz / 34t), Mt Carlton (1 Moz / 40 t) and Camel Creek. These deposits are exclusively gold producers though some such as Pajingo have significant Ag (by weight, but not by value). These goldfields are situated in a variety of Paleozoic igneous and metasedimentary sequences and include auriferous quartz veins of varied textures and breccia pipes.

Quite discrete from the Charters Towers gold province is the volcanogenic massive sulfide (VMS) Mt Windsor Thalanga base metal belt which trends east-west for 200 km and is within

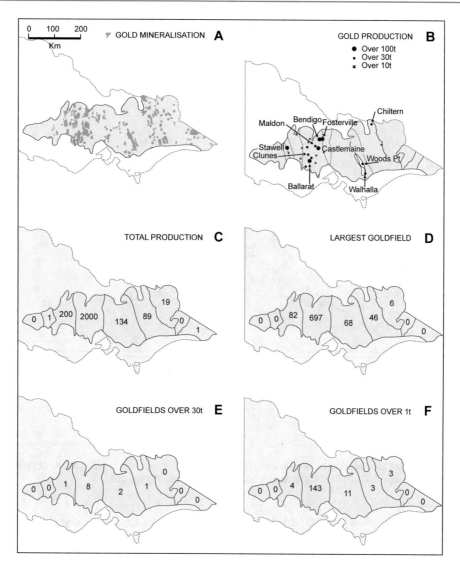

Fig. 4.9 The outcrop of Paleozoic igneous and metasedimentary rocks dominates Central Victoria and these are subdivided into nine geological zones: (**a**) location of major goldfields based on endowment of 30 t Au or more. All these occurrences lack economic base metals; (**b**) main gold producing areas as north – south belts throughout a 600 km by 200 km width of the Paleozoic succession of Central Victoria; (**c-f**) distribution of gold in each Paleozoic structural zone of Victoria showing, for each zone; (**c**) the total all-time production; (**d**) the largest goldfield by all-time production; (**e**) the number of goldfields over 30 t; and (**f**) the number of goldfields over 1 t. Production in tonnes of gold; 1 Moz is 30 t. Source includes Hughes and Phillips 2015.

an early Paleozoic sub-aqueous felsic and mafic volcanic and sedimentary sequence. These VMS deposits have copper, zinc, lead, silver and subordinate gold, i.e. they are gold-plus with a total of 0.5 Moz / 15 t Au.

The Paleozoic of Central New South Wales has an endowment exceeding 70 Moz / 2100 t and includes Cadia – Ridgeway (50 Moz / 1500 t Au with co-product copper; Fig. 4.8). Cowal (8 Moz / 240 t Au), Northparkes (4 Moz

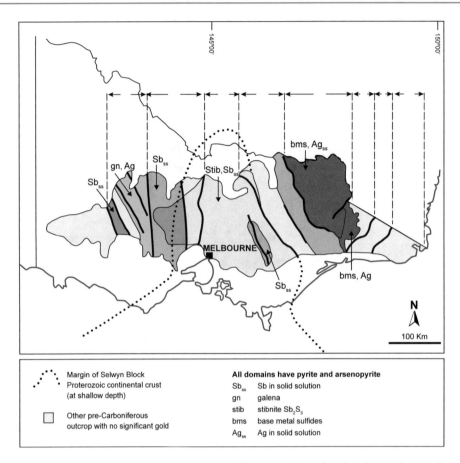

Fig. 4.10 Distribution of mineralogical domains across the Victorian gold province based upon the ore mineralogy of the numerous small deposits in each domain. Source includes Hughes and Phillips 2015.

/ 120 t Au and co-product copper) and Temora (Cu-Au, 2 Moz / 60 t Au) are within a 200 km radius. Cobar (6 Moz / 200 t Au and co-product copper) is a farther 200 km NW. The deposits of Central NSW comprise auriferous quartz veins in both igneous and metasedimentary rocks and include gold-only and gold-plus examples.

4.3.10   Portugal and Spain

Portugal and Spain have historically been important European producers of gold from auriferous quartz veins in Early to Mid Paleozoic metasedimentary and igneous rocks (Fig. 4.12). The earliest gold mines date back to Roman times and include Três Minas in Northern Portugal which comprises three open pits and the removal

of 10 Mt of ore and waste; one pit is 500 m long and 100 m wide and deep. Current gold production in SW Europe is minor though there has been the Salsigne district in southern France with an endowment over 3 Moz / 100 t.

Quite discrete from the modest gold deposits in the northern and central part of Portugal, are significant mines in which gold has been a by-product from the VMS deposits of the Iberian Pyrite belt or South Portuguese Zone. This Devonian – Carboniferous volcanic sedimentary complex runs eastward for 200 km from southern Portugal into Spain and the Rio Tinto deposit. Although grades of gold have been low, the large volumes of ore have yielded 30 Moz / 900 t Au, i.e. 1700 million tonnes of ore from 80 deposits averaging: Cu (1.3%); Zn (2.0%); Pb (0.7%); Ag (26ppm) and Au (0.5ppm).

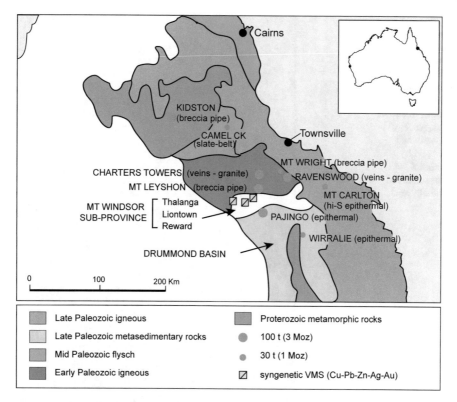

Fig. 4.11 Significant goldfields of the NE Queensland Charters Towers province inland from Townsville. The several large goldfields are separate from, but juxtaposed against, the Mt Windsor VMS belt. There has been considerable effort subdividing these deposits on a descriptive basis, but there has been no coherent synthesis explaining how the province has formed.

4.3.11   Abitibi Gold Province, Eastern Canada

Canada is dominated by Archean granitic rocks and greenstone belts that contain many goldfields. The main gold producing region is the 2700 Ma Abitibi belt in the SE of the country where there are numerous large and many small goldfields. Much of the production of 150 Moz / 4500 t Au from the Abitibi belt has come from goldfields arranged along three east—west deformation zones (variably described as breaks or faults; Fig. 4.13). The larger two, the Porcupine—Destor zone and Kirkland Lake—Larder Lake—Cadillac zone, are approximately 300 km in length and 50 km apart. Major goldfields include Timmins including Hollinger, McIntyre and Dome (60 Moz / 1800 t), Kirkland Lake (30 Moz / 1000 t), Malartic—Val d'Or (20 Moz / 600 t) and Larder Lake (15 Moz / 500 t). The Horne deposit has been a significant copper producer and yielded 10 Moz / 300 t Au.

The Abitibi greenstone belts also contain many base metal VMS deposits with a total endowment of almost 15 Moz / 400 t. It was these VMS deposits that strongly influenced the application of syngenetic seawater exhalative models to Archean gold deposits of the Abitibi belt and globally, i.e. Brazil. It was only after the documentation of structural control at the BIF-hosted Geraldton goldfield in Canada and analogies in Australia such as Water Tank Hill that an epigenetic model was adopted for both volcanic- and sedimentary-hosted gold deposits.

Beyond the Abitibi Province are further significant goldfields in the Archean cratons of Canada including Red Lake (22 Moz / 700 t), Hemlo (25 Moz / 800 t) and Yellowknife (6 Moz /

Fig. 4.12 Map of Portugal and western Spain showing the Iberian Pyrite belt (South Portuguese zone) with many VMS deposits that have by-product gold. To the north of the Pyrite belt are various mid to late Paleozoic sequences of metasedimentary and igneous rocks with numerous modest gold-only deposits related to auriferous quartz veins.

200 t); and others in younger Cordilleran sequences (see Eskay Belt, Fig. 4.14), all contributing to Canada's 400-plus Moz / 12,000 t all-time production.

4.3.12   Eskay Belt, NW Canada

The Eskay belt of north-western British Columbia became a significant gold province in the late 20th Century based on a diverse group of high-grade deposits with numerous associated commodities. The province is 400 km long in a NW- trending belt of Mesozoic host rocks. The dominant mineralisation has been interpreted as VMS-style based upon the Au—Ag—Cu—Zn—Pb enrichment and the stratigraphic package of deformed bimodal basalt – rhyolite succession with clastic metasedimentary rocks. Some further prospects in the Eskay belt are notionally por-phyry- and epithermal-style deposits. The total resource in the region of 60 Moz / 2000 t Au is supplemented by substantial Cu and Ag resources. The Eskay Creek deposit itself was discovered in 1988 and has produced 100 t Au at 45 g/t average, and 5000 t Ag also at the very high-grade above 2000 g/t Ag; it has a minor remaining resource. Larger discoveries have followed including the 1100 t Au Kerr Sulphurets Mitchell Iron Cap deposits (KSM on Fig. 4.14).

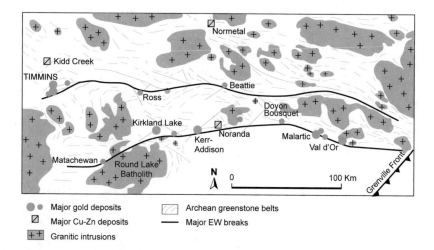

Major gold deposits
Major Cu-Zn deposits
Granitic intrusions
Archean greenstone belts
Major EW breaks

Fig. 4.13 Granite and greenstone belts of the Abitibi Gold Province in Ontario, southeast Canada, two major fault zones (locally called breaks), larger gold deposits aligned with those fault zones, and a few of the many VMS deposits. For reference, Toronto is 500 km to the south of Timmins.

Amongst VMS provinces globally, this appears to have the highest gold endowment and some of the highest Au and Ag grades.

4.3.13   Carlin Gold Province, Nevada USA

Western USA including Nevada was already well-known for gold in the 1800s with the Mother Lode and its related placers in California, Cripple Creek in Colorado, Juneau and Nome in Alaska, and Homestake deposit in South Dakota. It was only in 1961 that microfine gold was discovered near Carlin in Northern Nevada leading to the opening of the Carlin gold mine in 1965. This was the first of many large mines that have been the mainstay of USA gold production for the last half century.

The Carlin gold province comprises clastic metasedimentary rocks deformed in the Paleozoic and subsequent orogenic events. Around 30 – 40 Ma there was extension and emplacement of voluminous volcanic and intrusive rocks; and the later landscape evolution has been dominated by Basin and Range tectonics around 10 Ma and uplift of the Sierra Nevada mountains in the last five million years. All these events have played a part in the formation and modifications of the deposits of Northern Nevada that are discovered and mined today.

Goldfields are distributed on the NNW-trending Carlin and Battle Mtn—Eureka Trends that are 100 km in length and 50 km apart, and shorter E-W Getchell and Independence trends (Fig. 4.15). The largest goldfield, Goldstrike – Betze – Post, has an endowment of 50 Moz (1500 t) and is part of the Carlin Trend (120 Moz). The footprint of individual goldfields is typically 5 km by 10 – 20 km.

For many researchers, the Carlin deposits have proven difficult to characterise despite an enormous amount of research and mining access over the last half century. The deposits have some characteristics that overlap with other types (they were previously referred to as epithermal, and more recently as being continuous with distal disseminated and porphyry types). The lack of any focus on diagnostic characteristics may explain the poor return from exploring for similar deposits globally. Carlin deposits are not as well integrated into global gold metallogeny, nor gold geology compared to the successes globally for greenstone belts and slate belts. Instead, Carlin deposits are usually slotted into on-going western USA geology and terminology with its century of

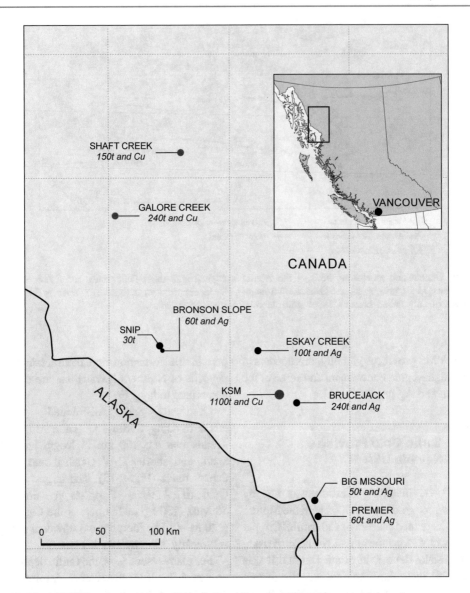

Fig. 4.14 Map of NW Canada showing the Eskay belt and its major gold and base metal deposits.

tradition. In contrast there are some other regions where the 1980s were a new start for gold with new science and skill sets applicable to fluid flow in the crust, metamorphic petrology and regolith science.

4.3.14 Siberian Gold in Russia

Eastern Siberia (east of the Yenesei River) comprises the Siberian Craton including the

high metamorphic grade granite, gneiss and amphibolite of the Aldan shield (Fig. 4.16). There are extensive deformed and metamorphosed Proterozoic to Phanerozoic sequences and volcanic belts surrounding the craton. Eastern Siberia has been a major gold mining region with all-time production exceeding 150 Moz / 5000 t and significant resources at some large deposits particularly near Lake Baikal and north of Magadan. A feature of the gold distribution is that the largest goldfields

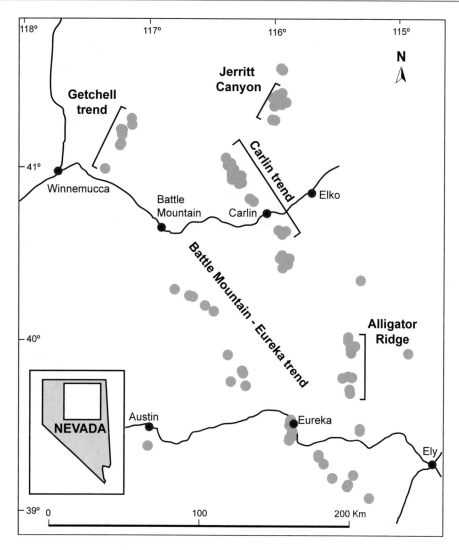

Fig. 4.15 Map of Northern Nevada showing many Carlin gold deposits (dots) along the five different trends. These are mostly within deformed Paleozoic metasedimentary rocks.

(endowments of Olympiada 50 Moz / 1500 t, Sukhoi Log 60 Moz / 1900 t, Natalka 60 Moz / 1800 t, Nezhdaninskoe 30 Moz / 1000 t and Kubaka 3 Moz / 100 t) are not within cratonic areas but generally close to the cratonic margins in Phanerozoic and younger orogenic belts. Peschanka (10 Moz / 400 t Au and co-product Cu and Ag) is in the far northeast of Siberia, as is Kupol (3 Moz / 100 t Au and co-product Ag). Also, in far eastern Siberia adjacent to the Okhotsk Sea are several deposits with the less common Au – Sn and Au – Mo association.

Natalka and Nezhdaninskoe have ores elevated in Pt and lesser Pd correlated with Au and As.

A visit in 2013 to Nezhdaninskoe, Natalka and several other goldfields of the Yana-Kolyma region with Nikolai Goryachev, professor at Magadan University, was quite an experience and a special treat. The two-week field trip started at Yakutsk, the largest city built on permafrost, traversed the infamous Highway of Bones built under the rule of Stalin following World War 2 traversing 1000 km of forest and very few towns before ending in Magadan on the Pacific

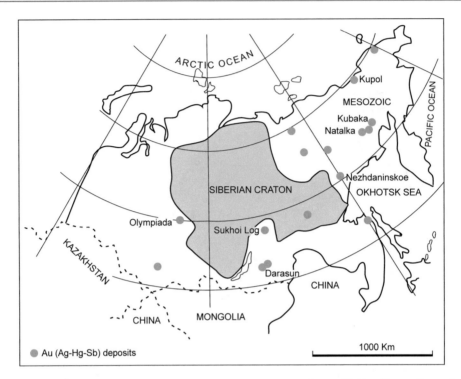

Fig. 4.16 Map of eastern Russia with some of the major goldfields covering several provinces. The large goldfields of Siberia are within Mesozoic successions in the east rather than within the Siberian Craton. Source including Goryachev and Pirajno 2014; and Goldfarb et al. 1997, 2001, 2014.

Ocean. This road provided access to several large primary and alluvial goldfields. The Yana—Kolyma region comprises Mesozoic fold belts of turbiditic metasedimentary rocks, granitic rocks, and gold in quartz veins. The Kolyma Gold Province includes hard rock deposits near steep-sided valleys with placer gold workings for 20 km or more downstream from their primary sources. This region is one of the few parts of the world with major alluvial gold mining today.

4.3.15 Gold Provinces in China

China has a long history of mining gold from many different provinces and geological settings. The Jiaodong gold province SE of Beijing is the most important producing area of the 21st century. In the far NW there are deposits along the northern margin of the Tarim Craton, in the SE there is by-product gold-plus deposits of the Yangtze

Craton, and in the centre and immediately east of Xian there are deposits of Xiao Qinling. Most gold deposits are generally not documented in accord with the main global standards for reporting Resources, Reserves, and production, and many do not have integrated regional and mine geology documentation underpinning any estimates (Appendix A).

The Jiaodong (Shandong) gold province is on a peninsula, 500 km SE of Beijing. It may have an endowment around 3000 t and comprises about a quarter of China's annual gold production from over 100 deposits. The geological setting of these deposits is mostly along NNE-trending normal faults adjacent to Mesozoic intrusions with areas of Archean rocks on the Jiaodong Peninsular.

Although China comprises extensive Archean rocks in three major cratons (North China, South China, and the Tarim Craton in the west), the country lacks the large greenstone-hosted gold deposits so common in other countries. However, the reworked margins of these cratons are especially important for gold districts. The mid Paleozoic (Variscan) deformation was particularly important for gold deposit formation including sequences in the far west that are a continuation of the Tian Shan province of Kyrgyzstan and Uzbekistan. For a major gold producing country the largest mines are quite small by global standards and all appear to be below 1Moz Au pa.

4.3.16   Tian Shan Belt of Uzbekistan, Kyrgyzstan and Western China: Altaid Mountain Belt

The Tian Shan gold belt from western China into Kyrgyzstan and Uzbekistan is one of the largest gold provinces outside the Witwatersrand of South Africa (Fig. 4.17). It is dominantly a metasedimentary sequence deformed in the late Paleozoic, Variscan Orogeny and intruded by granitic rocks, and with adjacent arc-related volcanic rocks.

The largest goldfield is Muruntau with an endowment of 170 Moz / 5300 t, no economic base metals but has some commercial production of platinum group elements; the deposit is structurally controlled, related to quartz veins, and hosted in clastic metasedimentary rocks. The giant Kalmakyr deposit is a Cu-Au stockwork deposit in a magmatic arc setting in eastern Uzbekistan (50 Moz / 1500 t Au, and co-product Cu) in monzonite and diorite. Near Kalmakyr are epithermal-Ag and porphyry copper gold within the Late Carboniferous arc.

In contrast to the Tian Shan gold belt, the semi-parallel Alpine – Himalayan mountain chain from western Europe to Tibet has very few large goldfields and low gold endowment overall.

4.3.17   Circum Pacific Goldfields: South America to Alaska to New Zealand

The circum Pacific region includes many goldfields unevenly distributed around the margin of the Pacific plate (Fig. 4.18). Overall, the Pacific plate is being consumed along its east, north and western margins but there are both convergent and strike slip intervals. Metasedimentary rocks and igneous rocks are host rocks for gold, and some provinces have abundant economic base metals with gold whereas others lack economic base metals.

The 5000 km long Andes mountain chain contains many gold deposits with three major provinces. The southern one of these three provinces extends 400 km northward from El Indio deposit which itself is 500 km north of Santiago. Unlike many of the provinces described so far, this one is linear and 150 km inland from the coast.

In Peru there are several goldfields within 100 km of the large Cu – Au Yanacocha deposit which is cited as a high-sulfidation type in Cenozoic volcanic rocks with an endowment around 70 Moz / 2200 t Au. In 2005, Yanacocha production peaked at just over 3 Moz / 100 t Au in that year, a feat achieved by very few gold mines in history.

In the Northern Andes to the southwest of Bogota is the 25 Moz / 680 t La Colosa deposit of gold with by-product Ag. Several deposits of 3 Moz / 100 t or more are on a 400 km long trend north of La Colosa including Titiribi (10 Moz / 300 t), Buritica (6 Moz / 200 t) and Segovia (6 Moz / 200 t). In all of these, gold is the dominant commodity but there is some by-product Ag and elevated Cu. This region yielded extensive placer gold historically.

The Central Americas are less gold endowed excepting for the large Pueblo Viejo deposit in Dominican Republic (but well off the circum Pacific margin). Mexico has some large deposits stretching along its west and trending towards the Mesquite deposit of far southern California. In Western USA and Canada, there is the 200 km long Mother Lode deposits and their extensive placers in the foothills to the west of the Sierra Nevada Range, the Carlin gold province farther east in Nevada, and large copper – gold deposits in Utah that stretch 1000 km into Canada, the Eskay province, and large deposits in the accreted terranes of Alaska.

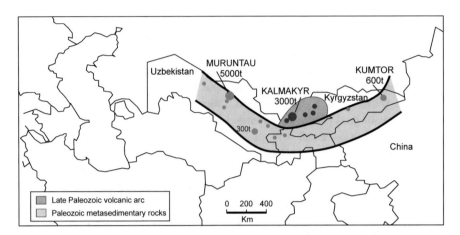

Fig. 4.17 The Tian Shan mountain belt of Central Asia contains some of the largest goldfields outside South Africa including Muruntau of 170 Moz / 5300 t. A belt of several provinces and large gold deposits is juxtaposed with polymetallic deposits in a volcanic arc that include copper and silver. Source including Zhou et al. 2002; Goryachev and Pirajno 2014; Goldfarb et al. 2014.

Fig. 4.18 Map of the larger concentrations of gold deposits around the Pacific Rim. These are not individual deposits but clusters of multiple goldfields.

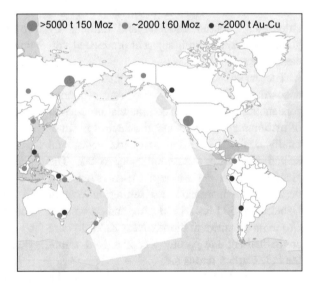

The relatively young age of events around the Pacific Ocean and the detailed understanding of its tectonic history has allowed careful dating of the age of gold deposit formation in numerous deposits. Several studied deposits spread over 200 km in southern Alaska all appear to have formed within a one-million-year period (56 – 55 My), a result that is attributed to the change of Pacific plate motion from being collisional into North America to becoming strike-slip and moving northward against North America (Goldfarb et al. 1991).

In Central Alaska, Pogo and Fort Knox are important producers with the Livengood Money Knob deposit nearby but not producing yet; all three have an endowment around 10 Moz / 300 t Au. Pogo and Fort Knox are quartz vein deposits in Cretaceous granitic hosts within metasedimentary schist belts. Donlin Creek (40 Moz / 1200 t) comprises auriferous quartz veins in plutonic rocks, and Pebble (Cu-Au-Mo, possibly 100 Moz / 3000 t Au) in SW Alaska has remained undeveloped until 2021 due to environment, community, and gold grade issues. Nome (10 Moz / 300 t Au) on the Seward peninsula near Siberia is an unusual setting of auriferous quartz veins in low hills, alluvial gold in small creek systems, and then further gold behind the beach line, along the modern beach and offshore where it has been dredged.

In the NW Pacific region, there are deposits in the young volcanic rocks of the Kamchatka peninsula in Siberia, and then copper—gold deposits farther south along the Pacific margin in southern Japan. The Philippines has major endowment from Baguio district (over 40 Moz / 1200 t Au), Far South East (over 15 Moz / 500 t Au, and Cu), and other deposits of copper and gold.

Indonesia is a tectonically active and complex part of the SW Pacific with numerous deposits of co-product Cu and Au (Batu Hijau of 10 Moz / 400 t Au, Elang 8 Moz / 250 t Au) and others with dominantly Au but significant by-product Ag (Kelian 6 Moz / 200 t, Gosowong 3 Moz / 100 t, Martabe 8 Moz / 250 t). In terms of its level of concentration, not economics, the silver can be an order of magnitude more abundant than the gold.

Irian Jaya and Papua New Guinea are tectonically active areas on the Pacific plate margin with Grasberg (90 Moz / 2800 t Au, substantial Cu), Ok Tedi (~15 Moz / 500 t Au) and Bougainville (~30 Moz / 1000 t Au) being major Cu and Au mines. Deposits producing gold without co-product Cu include Porgera, Misima and Lihir (25 Moz / 800 t, 6 Moz / 200 t, and 40 Moz / 1200 t endowments, respectively). Fiji is a group of volcanic islands with the Emperor gold mine famous for its telluride minerals and 10 Moz / 350 t of gold. New Zealand straddles the Pacific plate margin with gold deposits in the north around the Coromandel peninsula, auriferous veins in meta-sedimentary host rocks in the south, including the large Macraes goldfield of over 10 Moz / 300 t Au.

4.4 Common Features Amongst Provinces

The plate tectonic setting of the Phanerozoic gold provinces described here highlights an association with orogenic belts especially around the Pacific margin. The interpretation of the tectonic settings for older provinces has some uncertainty but a similar correlation with orogenic belts is favoured for Archean and Proterozoic provinces (Fig. 4.19). Some of the uncertainty for older provinces is not with the tectonics but understanding the gold deposition. For example, the inferred tectonic setting at the time the Carlin gold province formed depends upon the inferred age for gold introduction which is widely regarded as 30 – 40 Ma but with some conflicting evidence; establishment of an older age for introduction of the gold would lead to a different tectonic setting for Carlin gold introduction, and hence a different global exploration strategy which might be more successful. Similarly, a different tectonic setting would be indicated for formation of Witwatersrand gold depending upon whether it is inferred to be placer or hydrothermal in origin.

Important factors in generating these large gold provinces appear to be the involvement of arcs, subduction, voluminous mafic rock material and the accretion of terranes. Suggested favourable settings include the closure of large oceans (Altaids and Tian Shan), large accretionary arc complexes (Alaska) and continental arcs (Papua New Guinea and Andes and Cordillera of North America). An additional product of these settings is abundant post-tectonic granitic rocks from partial melting of crustal material. An important piece of information in the understanding of the formation of gold deposits is that large gold provinces are lacking from non-orogenic areas and some orogenic belts. The Himalaya – Alpine chain is lacking in major gold whereas the parallel Altaid Tian Shan belt to its north has many large goldfields (see Fig. 4.17).

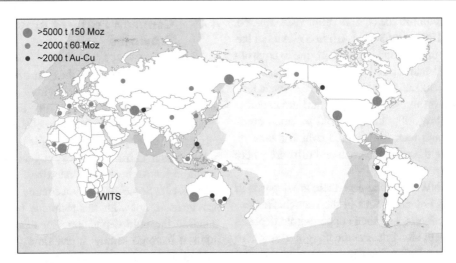

Fig. 4.19 Map of the largest clusters of goldfields globally with a concentration around the Pacific Ocean, the Central Asia Tian Shan belt, and some of the Archean cratons.

The gold production and endowment are unevenly distributed through time which is reflecting periods of global tectonic events. Apart from minor production from the Pilbara and Barberton regions there has been no significant production from rocks older than 2.7 Ga, very little from the 2.5 to 2.1 Ga period or from rocks of 1.8 to 0.6 Ga (Fig. 4.20b). This compilation did not include all parts of the world so there is some significant under-representation, e.g. Nevada USA.

Most of the world's gold production has come from *gold-only* deposits in which base metals are uneconomic. The remaining deposits are *gold-plus* and have two distinct end members. Copper—gold deposits may be large and contain economic copper, gold and commonly further elements. The volcanogenic massive sulfide deposits are characterised by many elements including Cu—Zn—Pb—Ag—Au, are generally modest in size, and occur in distinct clusters such as Noranda district in Canada or the Rio Tinto belt in Portugal and Spain. Chapters 5–18 mostly relate to gold-only deposits though several parts therein also pertain to the formation of gold-plus deposits. Chapter 19 is focused on gold-plus deposits but utilises many of the scientific principles already outlined. The volcanogenic

massive sulfide deposits and their formation are well described in the base metal literature and hardly discussed here.

This chapter is a summary from published sources, the chief of which are listed in the Bibliography section.

Snapshot

- Provinces are areas on the Earth's surface significantly over-endowed with respect to gold.
- Goldfields within a province share common geological characteristics including age.
- Provinces reflect an abundance of large goldfields, a multitude of smaller occurrences and substantial all-time gold production.
- Province dimensions of several 100 km are common, but not 1000s km.
- In younger terranes the provinces are linear for several 100 km.
- Regardless of their age, provinces are associated with active plate margins rather than anorogenic regions.

Fig. 4.20 Gold production versus time: (**a**) Witwatersrand showing age of host rocks; (**b**) most other deposits designated 'orogenic' by Groves et al. 2003. These are of Archean, Proterozoic and Phanerozoic age. There is important Phanerozoic production not included in this study. Source of data: Groves et al. 2003.

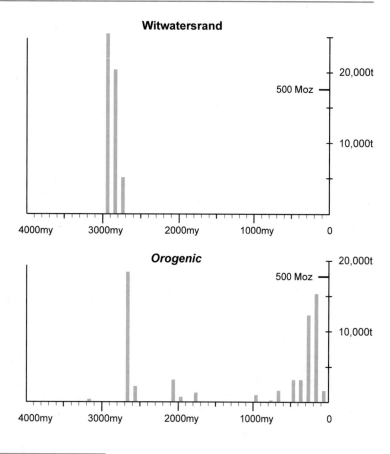

Bibliography

Birch WD (2003) Geology of Victoria. Geological Society of Australia Special Publication 23. Geological Society of Australia (Victorian Division)

Cline JS (2018) Nevada's Carlin-type gold deposits: what we've learned during the past 10 to 15 Years. Reviews in Economic Geology 20:7–37

Cline JS, Hofstra AH, Muntean JL, Tosdal RM, Hickey KA (2005) Carlin-type gold deposits in Nevada: critical geologic characteristics and viable models. SEG 100th anniversary volume, Society of Economic Geologists 451–484

Corbett GJ, Leach TM (1998) Southwest Pacific Rim gold-copper systems: structure, alteration, and mineralization. Society of Economic Geologists Special. Publication 6. https://doi.org/10.5382/SP.06

Craw D, McKenzie D (2016) Macraes orogenic gold deposit (New Zealand) Origin and development of a world class mine. Springer Switzerland 127pp

Dziggel A (2019) Kisters AFM (2019) Tectonometamorphic controls on Archaean gold mineralization in the Barberton greenstone belt, South Africa. In: Hoffmann JE, Kranendonk MJ, Bennett VC (eds) Earth's oldest rocks. Elsevier, Amsterdam, Netherlands, pp 655–671

Foster RP (1985) Major controls of Archean gold mineralization in Zimbabwe. Transactions of the Geological Society of South Africa 88:109–133

Gloyn-Jones J, Kisters AFM (2019) Ore-shoot formation in the Main Reef Complex of the Fairview Mine—multiphase gold mineralization during regional folding, Barberton greenstone belt, South Africa. Mineralium Deposita 54:1157–1178

Goldfarb R, André-Mayer A, Jowitt S, Mudd GM (2017) West Africa: The world's premier Paleoproterozoic gold province. Economic Geology 112:123–143. https://doi.org/10.2113/econgeo.112.1.123

Goldfarb RJ, Groves DI, Gardoll S (2001) Orogenic gold and geologic time: A global synthesis. Ore Geology Reviews 18:1–75

Goldfarb RJ, Phillips GN, Nokleberg WJ (1997) Tectonic setting of synorogenic gold deposits of the Pacific Rim. Ore Geology Reviews 13:185–218

Goldfarb RJ, Qiu KF, Deng J, Chen YJ, Yang LQ (2020) Orogenic gold deposits of China. Society of Economic Geologists Special Publication 22:263–324

Goldfarb RJ, Slee LW, Miller LD, Newberry RJ (1991) Rapid dewatering of the crust deduced from ages of mesothermal gold deposits. Nature 354:296–298

Goldfarb RJ, Taylor RD, Collins GS, Goryachev NA, Orlandini OF (2014) Phanerozoic continental growth and gold metallogeny of Asia. Gondwana Research 25: 48–102

Goryachev, NA (2019) Gold deposits in the Earth's history. Geology of Ore Deposits 61:6 495–511. Pleiades Publishing Ltd

Goryachev NA, Pirajno F (2014) Gold deposits and gold metallogeny of Far East Russia. Ore Geology Reviews 59:123–151

Groves DI, Goldfarb RJ, Robert F, Hart CJR (2003) Gold deposits in metamorphic belts: overview of current understanding, outstanding problems, future research, and exploration significance. Economic Geology 98:1–29

Groves DI, Phillips GN (1987) The genesis and tectonic control on Archaean gold deposits of the Western Australian Shield - A metamorphic replacement model. Ore Geology Reviews 2:287-322

Groves DI, Phillips GN, Ho SE, Henderson CA, Clark ME, Woad GM (1984) Controls on distribution of Archaean hydrothermal gold deposits in Western Australia. In: Foster RP (ed) Gold 82. Balkema, Rotterdam, pp 689–712

Hartmann LA, Delgado IM (2001) Cratons and orogenic belts of the Brazilian shield and their contained gold deposits. Mineralium Deposita 36:207–217

Ho SE, Groves DI, Bennett JM (2005) Gold deposits of the Archaean Yilgarn Block, Western Australia: nature, genesis and exploration guides. University Western Australia, Geology Department (Key Centre) and University Extension, Publication 20:407pp

Hughes MJ, Phillips GN (2015) Mineralogical domains within gold provinces. Applied Earth Science (Transactions B) 124:191–204

Leahy K, Barnicoat AC, Foster RP, Lawrence SR, Napier RW (2005) Geodynamic processes that control the global distribution of giant gold deposits. Mineral Deposits and Earth Evolution. McDonald I, Boyce A J, Butler I B, Herrington R J and Polya D A (eds) Geol Soc London Special Publications 248:119-132. doi.org/10.1144/GSL.SP.2005.248.01.06

Levitan G (2008) Gold deposits of the CIS, Xlibris Corp USA

Lobato LM, Ribeiro-Rodrigues LC, Vieira FWR (2001) Brazil's premier gold province. Part II: Geology and genesis of gold deposits in the Archaean Rio das Velhas greenstone belt, Quadrilátero Ferrífero. Mineralium Deposita 36:249–277

Markwitz V, Hein KAA, Miller J (2016) Compilation of West African mineral deposits: spatial distribution and mineral endowment. Precambrian Research 274:61–81

Phillips GN (2017) Gold in Australia, in *Australian Ore Deposits* (ed: G N Phillips). The Australasian Institute of Mining and Metallurgy, Melbourne, pp 75–82

Phillips GN, Hughes MJ, Arne DC, Bierlein FP, Carey SP, Jackson T, Willman CE (2003) In: Birch W (ed) Gold, in Geology of Victoria. Geological Society Australia, (Victorian Division), Melbourne, pp 377–432

Phillips GN, Law JDM (2000) Witwatersrand goldfields: geology, genesis and exploration. SEG Reviews 13: 439–500

Pirajno F (2013) The geology and tectonic settings of China's mineral deposits, Springer Netherlands 682pp. https://doi.org/10.1007/978-94-007-4444-8

Porter M (2020) Regular use was made of the comprehensive database of Porter Consulting, http://www.portergeo.com.au/database/mineinfo.asp?mineid=mn053

Pretorius DA (1981) Gold and uranium in quartz-pebble conglomerates. Economic Geology. 75th anniversary volume:117–138

Relvas JMRS, Barriga FJAS, Pinto A, Ferreira A, Pacheco N, Noiva P, Barriga R, Baptista R, Carvelho D, Oliveira V, Munhá J, Hutchison RW (2002) The Neves-Corvo deposit, Iberian Pyrite Belt, Portugal: impacts and future, 25 years after the discovery. Society of Economic Geologists Special Publication 9:155–176

Ribeiro-Rodrigues LC, Gouveia de Oliveira C, Friedrich G (2007) The Archean BIF-hosted Cuiabá Gold deposit, Quadrilátero Ferrífero, Minas Gerais, Brazil. Ore Geology Reviews 32:543–570

Schwartz S (1980) The complete book of gold. Horwitz Publications, Melbourne, 160pp

Seltmann R, Soloviev S, Shatov V, Pirajno F, Naumov E, Cherkasov S (2010) Metallogeny of Siberia: tectonic, geological and metallogenic settings of selected significant deposits. Australian Journal of Earth Sciences 57: 655–706. https://doi.org/10.1080/08120099.2010.505277

Stow D (2010) Vanished ocean: how Tethys reshaped the world. Oxford University Press, 300pp

Yakubchuk A, Cloe A, Seltmann R, Shatov V (2002) Tectonic setting, characteristics, and regional exploration criteria for gold mineralization in the Altaid Orogenic Collage: the Tien Shan Province as a key example. Society of Economic Geologists Special Publication 9:177–201

Zhou T, Goldfarb RJ, Phillips GN (2002) Tectonics and distribution of gold in China – an overview. Mineralium Deposita 37:249–282

Enrichment of Gold Above Background

5

Abstract

Enrichment of an element above its average abundance in the Earth's crust is required to form any ore deposit. For gold, that enrichment is around 10,000 times which is more than virtually any other element, and the high enrichment implies special partitioning in Nature. Enrichment follows provinciality as the **second** of five characteristics that needs to be accommodated when deciding how gold deposits may form.

Keywords

Enrichment · Background · Partitioning · Source volume

5.1 Enrichment is Required to Form Ore Deposits

Any ore deposit represents a substantial increase of at least one element above its crustal abundance. This increase can be expressed as the concentration of an element in ore as a ratio to its average concentration in the Earth's crust; and here we can use estimates for continental crust. The term *enrichment factor* is used for this ratio of the two concentrations. The *background concentration* level of an element in a rock refers to unenriched, primary rocks with no overlay of an ore-forming process. Background concentrations vary amongst different rock types depending upon their formation through igneous, metamorphic, and sedimentary processes; hence background concentrations differ from the average value for the crust. Here, as through much of geochemistry, gold concentration is given as parts-per-million which is the same as grams-per-tonne used in industry; tonnes are used rather than million ounces to substantially aid calculations and approximations.

For gold, the numerator in the above ratio (concentration in ore) varies considerably from mine to mine and with the type and era of mining. Long-term grade figures for mines operating before the advent of modern large-scale operations have guided the selection here of a figure of 20 g/t Au (Table 5.1). Adopting a lower or higher figure for the grade of gold ore will change the enrichment factor proportionately but does not alter the qualitative conclusions that follow. Many mines in the 21st century report a grade of 1 – 5 g/t Au but this usually reflects ore diluted by considerable waste depending upon the mining method. For example, dilution can occur both in a large-scale open pit, or following a narrow high-grade vein which is too narrow to extract without taking some of its barren wallrocks.

Average crustal values for many elements are widely available and, being estimates, vary

Table 5.1 Enrichment factors for various elements required to form an ore deposit.

Element	Crustal average	Ore grade	Enrichment factor
Al (%)	8	30	4
Fe (%)	5	60	12
Zn (ppm)	70	35,000	500
Cu (ppm)	50	20,000	400
Pb (ppm)	15	30,000	2000
Ag (ppm)	0.08	120	1500
Au (ppm)	0.002	20	10,000

slightly between different sources. Ore grade values used here are also subjective as explained for gold.

The implications of these enrichment factors for ore formation are quite profound (Table 5.1). It is relatively easy to conceptualise some processes in Nature that could achieve the enrichment of Al or Fe to generate what constitutes mineable ore. It might be expected that there could be multiple and perhaps quite unrelated processes in Nature that can achieve the required ore grades for either element. It should be noted that to mine these two bulk commodities economically today there are several important factors unrelated to their grade (e.g. deposit size, location, transport options, availability of water and power, and ore purity).

The enrichment requirements to form base metal ores are quite different as high enrichment factors in the 100s are implied. The number of ways these high enrichments might occur in Nature could be more limited than for Al and Fe.

For gold, the situation is extreme as all gold deposits imply enrichment factors in the order of 1000 – 10000 or more compared to the average crustal concentration. Extreme enrichment such as implied here suggests that the number of fundamentally different ways to form major gold deposits could be quite limited; it might follow that many gold deposits have distinctive and shared features that relate to one or two extreme enrichment processes (i.e. highly effective processes). Taking this idea further, the thousands of large and small gold deposits are unlikely to be reflecting hundreds of fundamentally different modes of formation; and the dozens of gold provinces are unlikely to each represent a fundamentally different mode of formation. It is important to understand what makes gold deposits

different from one another, but the focus here is to understand the one or two fundamental processes common to many deposits and provinces, particularly the gold-only type which form most of the economic gold deposits. The physical size of individual deposits and their distribution as gold provinces with shared characteristics indicate that *the scale of the enrichment process is likely to be kilometres rather than metres*.

5.2   Nature of Gold in Average Crustal Rocks

The scarcity of gold in normal rocks can be illustrated by a calculation based upon the 0.002 parts per million (by mass) average concentration of gold in the crust (Table 5.1). This concentration is the same as two parts per billion, and it equates to two milligrams of gold worth about 10 cents in each one tonne of rock. This amount of gold, if combined into a single lump, would be approximately 0.5 mm diameter compared to a pinhead which is one to two millimetres in diameter.

Trying to learn much about those two milligrams of gold in each tonne of average rock would be futile if the investigation employed conventional microscopic examination of the gold. However, by taking repeat samples of one to two kilograms from a large mass of rock and then analysing these rock samples we find the concentration of gold in the samples is regularly around two parts per billion, maybe ranging between the detection limit and four ppb. From this we can conclude that the gold is not present as a single 0.5 mm diameter piece in the tonne of average rock but is evenly distributed, probably on the nanometre-scale of a few atoms or less.

There is still a substantial amount of gold in the Earth's crust even at these low levels. A calculation of the amount of gold in a cubic kilometre of rock at two ppb Au can provide a useful approximation:

- One cubic kilometre is 10^9 cubic metres
- The average density of crustal rocks is ~3 g/cc which is 3 tonnes per m^3
- The mass of rock in one km^3 is $3 * 10^9$ tonnes
- Gold concentration is 2 ppb which is $2 * 10^{-9}$
- The mass of gold in one km^3 is $3 * 10^9 * 2 * 10^{-9}$ tonnes, or
- 6 tonnes of gold in one km^3 of average crustal rock.

Taking these approximations further, in 20 km^3 of rock there is 120 t Au or 3.6 Moz which is the amount of gold in a medium to large deposit, or close to our hypothetical 3 Moz deposit. This rock mass could be represented by:

- A 1 km thick slab of 4 km by 5 km, or
- A 5 km thick slab of 5 km by 2 km.

Attempting to commercially mine one of these 20 km^3 rock masses to extract the 3 Moz of gold would be futile for two reasons (Fig. 5.1). First, it would be quite uneconomic to disaggregate so much rock for that amount of gold; and second, the gold does not occur as pieces but at the nanometric-scale so it could not be mechanically collected. This example has an analogy in Nature where time and erosion might break the rock mass down at no cost to us and overcome the problem of disaggregation, but still there remains the insurmountable issue of concentrating and collecting nanometric gold by mechanical means. This all suggests that to understand how gold deposits form, we cannot rely on Nature using the mechanical means of erosion and transport alone to extract this background gold (*mechanical* and *background* are key words here). In later chapters we will return to this idea of Nature being able to extract the background nanometric gold, but there the discussion will focus on chemical methods that avoid mechanically breaking down the rock mass or mechanically transporting any gold particles.

Separately, there are also the relatively easy to recover placer (alluvial) deposits that have been a major source of gold throughout history. These deposits form from the disaggregation of rocks by Nature's forces followed by mechanical concentration of gold grains but importantly this requires that the gold is already as 0.5 – 5 mm pieces in the rock prior to breakdown of the rock. Effectively this means disaggregating gold mineralisation or gold ores; the placer enrichment process will not work if the source rocks have only background concentrations of gold and particles at the

Fig. 5.1 Cape Town, South Africa showing the scale of the issues relating to background gold: (**a**) Students from the Geology of Gold class from University of Stellenbosch looking down on Cape Town from Table Mountain; (**b**) outline of an approximate 4 km by 5 km area centred on that city. To the best of our knowledge the area around the city of Cape Town has no significant gold deposits and should approximate average background gold levels of 2 ppb. If this is the case, a 1 km thick slab centred on the city would contain 3 Moz or 100 t of gold. However, with that 100 t of gold at the nanometric scale, it is completely uneconomic for mining and quite unsuitable as a source for placer gold.

nanometric scale. To form gold placers, the gold first needs to have been naturally enriched thousands of times via non-mechanical processes.

5.3   The Variation in the Background Gold Concentrations of Different Rocks

The calculations completed above used the average crustal level for gold and take no account of the inevitable variation in background gold concentrations in different rock types. It is quite possible that rocks with higher background gold levels than the average crust might play a role in the formation of gold deposits. The range of background gold concentrations in various rocks has been facilitated by the relatively recent improvements in the ability to measure extremely low levels of gold in rocks, and compilations of results, led by Iain Pitcairn. Today, parts-per-billion level analysis is easily accessible, and a few parts-per-trillion is achievable.

There are limited differences between the background gold contents of major rock types in their primary state, i.e. rocks that are not

weathered, not mineralised, and not altered (Table 5.2). Mafic rocks may have slightly more gold than felsic rocks, and rocks from island arcs appear to have slightly more gold than mid-ocean ridge basalt (MORB). Ocean island basalt is comparable to MORB for gold-related elements including gold except for much higher sulfur in the latter. In the North Atlantic Ocean, a median for MORB of 0.3 ppb Au contrasts with up to 7 ppb Au in plume-related basalt from near Iceland. Two studies of primary granitic rocks of Archean age show negligible enrichment of gold at the igneous stage and are investigated further in Chapter 10 which includes a discussion of the spatial relationship of granitic rocks to gold deposits. There are significantly higher levels of gold in some auriferous conglomerate, but these are quite restricted in their distribution and volume globally.

The key finding here is that, despite some differences at the igneous stage, it is most improbable that the extreme enrichment factor (well over 1000 and closer to 10000) required to form a gold deposit can be explained by gold-rich source rocks alone—except perhaps in the important but local reworking of an auriferous conglomerate to form other auriferous conglomerates.

Table 5.2 Background gold concentrations in various rocks using compilation of Pitcairn (2011).

Background gold in ppb	
Some older values	
Upper continental crust	1.8
Bulk continental crust	3
Lower continental crust	3.4
Primitive chondritic mantle	0.8
Earth's core	Assumed to be similar to iron meteorites
Pitcairn values (2011)	
Normal crustal rocks	0.5–10
Extreme values	0.05–20
Sedimentary rocks	Generally low
Conglomerate with clastic gold	Elevated
With diagenetic sulfide	Elevated
With carbonaceous component	Elevated
Igneous rocks	Continental and oceanic rocks similar
Igneous rocks	Island arc greater than mid-ocean ridge
Igneous rocks	Mafic greater than felsic
Sulfide melts	Strongly enriched
Granitic rocks	Minimal correlation with fractionation
Metamorphic rocks	Decrease with metamorphic grade based on one study in Otago, NZ

5.4   The Power of Partitioning

The extreme enrichment required to form a gold deposit and the virtual absence of highly gold-rich source rocks suggest a need for some form of partitioning of gold between two or more phases so that one of these phases becomes gold-rich at the expense of the other.

Partitioning is well understood and widely employed in chemistry, mineral processing, igneous petrology, and in modern manufacturing in support of the electronics industry. It provides a mechanism to greatly increase the concentration of one (trace) element relative to the major elements in a complex multi-element system such as a rock. For example, a commercial analytical application of partitioning is illustrated by the fire assay process—one of the most widely used methods for analysing gold in rocks. In the absence of partitioning there are significant errors measuring minute amounts of gold in a rock, particularly where the gold concentrations approach the analytical detection limits of the measurement technique. The fire assay process involves melting the rock sample, and the small amount of gold migrates from this volumetrically dominant silicate melt into a small volume of liquid lead. This process relies upon the strong partitioning of gold into a metallic melt (relative to the silicate host melt) to substantially concentrate the gold before it is measured by analysis. This also relies on having a small volume of the phase into which the gold partitions compared to the silicate melt, i.e. essentially the concept of the *R factor* as used in nickel sulfide formation and igneous petrology (i.e. for a trace element partitioning between a silicate melt and a lead melt, R is the ratio of the overall mass of silicate melt compared to liquid lead). By partitioning the gold into a much smaller volume, the concentration being measured can be greatly enhanced. This allows the ultimate detection limit for gold in the original rock to be much lower and the analytical accuracy enhanced – provided the lead used has consistently low initial gold content.

Partitioning is employed in industrial processes too. For example, the tendency for trace element impurities to partition into a melt can be utilised in the process of zone refining of silicon ingots where trace element impurities enter a moving melt zone so that they are swept along in the melt. Repeating the process many times results in ultra-pure semiconductor grade silicon for use in semiconductors and solar panels.

An example from petrology would be the observation that virtually all the Mn in some medium grade metamorphic rocks is sited in garnet (Fig. 5.2). For a rock with 1% Mn and abundant garnet (i.e. low R factor meaning the mass of garnet makes up a relatively high proportion of the mass of the rock), the concentration of Mn in that abundant garnet will be quite low; however, if the rock has very few grains of garnet (high R factor), the concentration of Mn in that scarce garnet will be much higher (despite there being no change to the 1 % Mn in the rock).

There are several potential forms of partitioning of gold in nature between two phases (Table 5.3). One of the most effective of these is that between coexisting silicate rock (typically mafic to ultramafic) and a sulfide melt. The partition coefficient (D) here is very high meaning virtually all the gold will be found in the sulfide melt:

D_{Au} = gold in sulfide melt in ppb / gold in silicate melt in ppb = 10,000 or more.

Similarly, D for other elements is high for Ni (250), Cu (1000) and Pd (35000).

The principle here is to have strong partitioning towards one phase and have a small proportion of that phase so that all the targeted element is partitioned into a small volume. In the following chapters, much importance is placed on the phenomenon of gold partitioning as a way, or ways, to achieve extreme enrichment through highly effective partitioning processes in Nature. The behaviour of associated elements will inform as to the exact partitioning that may be occurring, for example, gold might mimic platinum group elements, nickel and copper in a sulfide melt but behave quite differently from these elements in some hydrothermal fluids or silicate melts.

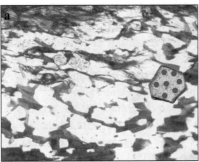

Fig. 5.2 Two biotite schists each with seven units of Mn (dots). The Mn partitions strongly into any garnet present: (**a**) minor modal garnet means the concentration of Mn in garnet is high, i.e. a high R factor; (**b**) abundant modal garnet has low concentration of Mn in garnet, i.e. low R factor.

Table 5.3 Relevance of various phases to partitioning of gold in the crust.

Phase	Relevance to partitioning
Sulfide magma	Very effective, but occurs in quite specific geological locations
Hydrothermal fluid	Very effective and widespread
Solid rock and mineral	Important to consider
Silicate magma	Important to consider
Bugs and organics	Minor at the relevant conditions for gold that are above 300 °C
Non-aqueous fluid	Unlikely to be involved in gold ore formation

Thus, partitioning can involve any of:

- mineral assemblage – silicate magma
- mineral assemblage – sulfide melt – silicate magma
- mineral assemblage – low salinity aqueous fluid; or
- mineral assemblage – saline aqueous fluid.

Each of these will affect elements differently and generate different suites of elements associated with gold (Table 5.3). Any involvement of either organics or non-aqueous fluids in partitioning involving gold would generate distinctive element patterns.

The elements associated with gold deposits provide important information with respect to process:

- element-concentrating processes that involve silicate magmas predominantly do so because of the charge and valence-linked ionic size of various elements and thus their ability to substitute in minerals.

- element-concentrating processes in aqueous fluids predominantly utilise the bonding properties of metals and ligands, and electronegativity is important.

The element associations reflecting magma versus aqueous fluid should be different from one another and different from any non-aqueous fluids.

5.5 Implications for the Scale of the Gold Deposit Formational Processes

Indications from the above data for gold are that the formation of a deposit that contains 3 Moz or 100 t of gold involves several km^3 of source rock, migration distances of at least a few kilometres, and some focusing of the gold. It is possible to use further components of major gold systems to provide confirmation of this scale.

Table 5.4 Distance to source of ore components, Kalgoorlie goldfield.

	Mass (Mt)	Source concentration (%)	Extraction efficiency (%)	Depositional efficiency (%)	Volume implied (km³)	Sides of a 5 km thick slab (km)	Distance (km)
CO_2	340	1	60	10	190	6 × 6	5–10
K	20	0.1	10	40	165	6 × 6	5–10
S	5	0.02	5	60	280	7 × 7	5–10
Au	0.0024	0.0000002	80	80	625	11 × 11	5–15

Assumed rock density of 3 g/cm³
Mapping controls and analyses are in Phillips et al. (1987)
The mass of gold at Kalgoorlie (0.0024 Mt) has been doubled since the calculations of 1987
Efficiency figures are based on our understanding of the extraction and depositional processes
Distance (from source to deposit) is approximate and assumes that the source is 5 km thick

Kalgoorlie is one of the world's largest goldfields and has already produced 2200 tonnes of gold. It is described in more detail in Chapter 13. Current mining is from a large open pit centred on the Golden Mile and the nature of the mining makes it difficult to carry out geological activities in the open pit including taking samples in any geological context. Fortunately, the underground mining from 1893 to the mid-1980s provided extensive 3-dimensional access, and 100s km of drives were mapped for rock type, alteration mineral assemblages, structure geology and mineralisation.

Other elements have been concentrated along with the gold at Kalgoorlie, and mapping of the various alteration zones around the Kalgoorlie orebodies helps to quantify the amounts of C (as in CO_2), K and S addition. These other elements provide independent estimates of the scale of the gold mineralisation and related alteration event in the Kalgoorlie goldfield including viable source areas and transport distances.

Although these calculations rely on extrapolation of underground mapping and assumptions of extraction and depositional efficiencies, they are suggesting that a 5-km thick slab of greenstone material might provide the necessary gold and related components from within a 10 – 20 km distance of Kalgoorlie (Table 5.4). The exact figures have little significance; however, what is important is that source distances of 100s km are not required even for this largest of goldfields, but conversely source distances of 1 km or less are likely to be inadequate. This bracket of distance has far-reaching implications for gold forming processes. *It means that to understand the*

formation of a gold deposit it is necessary to look well beyond the immediate host rocks.

Snapshot

- Gold occurs at extremely low abundances in all normal rocks.
- Ore deposits of any element represent enrichments of that element above its average crustal concentration.
- For gold, the necessary enrichment factor is extreme and approaching 10,000 so there is a challenge as to how enough gold can be accumulated to form a deposit.
- There is approximately 6 t of gold at background levels in a cubic kilometre of average rock.
- This background gold is evenly dispersed at the nanometric-scale.
- Higher gold in source rocks slightly eases the challenge of explaining the enrichment.
- Extreme gold enrichment is likely to be quite specific to a small number of processes.
- The gold enrichment process to form a deposit needs to operate over several kilometres, but not 100s km.
- Eroding and concentrating normal rocks with background levels of gold entirely by mechanical means cannot generate gold deposits.
- Placer (alluvial) gold deposits represent secondary reworking of gold that was

(continued)

initially concentrated as mineralisation (or pre-existing deposits) prior to its mechanical breakdown and erosion.

- It is necessary to look well beyond the immediate host rocks to understand the formation of a gold deposit given that the scale of the enrichment process is likely to be kilometres rather than metres.

Bibliography

Campbell IH, Naldrett AJ (1979) The influence of silicate: sulfide ratios on the geochemistry of magmatic sulfides. Econ Geol 74:1503–1506

Geochemical Earth Reference Model (GERM): Chemical characterization of the Earth, its major reservoirs and the fluxes between them. (n.d.). https://earthref.org/GERM

McDonough WF, Sun S-S (1995) The composition of the Earth. Chem Geol 120:223–253

Phillips GN, Groves DI, Brown IJ (1987) Source requirements for the Golden Mile, Kalgoorlie: significance to the metamorphic replacement model for Archean gold deposits. Can J Earth Sci 24:1643–1651

Pitcairn IK (2011) Background concentrations of gold in different rock types. Appl Earth Sci 120:31–38. https://doi.org/10.1179/1743275811Y.0000000021

Pitcairn IK, Warwick PE, Milton JA, Teagle DAH (2006a) A method for ultra-low level analysis of gold in rocks. Anal Chem 78:1280–1285

Pitcairn IK, Teagle DAH, Craw D, Olivo GR, Kerrich R, Brewer TS (2006b) Sources of metals in orogenic gold deposits: insights from the Otago and Alpine Schists, New Zealand. Econ Geol 101:1525–1546

Rudnick RL, Gao S (2014) Treatise on geochemistry, 2nd edn. Elsevier, Amsterdam. https://doi.org/10.1016/B978-0-08-095975-7.00301-6

Webber AP, Roberts S, Taylor RN, Pitcairn IK (2013) Golden plumes: substantial gold enrichment of oceanic crust during ridge-plume interaction. Geology 41:87–90. https://doi.org/10.1130/G33301.1

Wedepohl KH (1995) The composition of the continental crust. Geochim Cosmochim Acta 59:1217–1232

Segregation of Gold from Base Metals

6

Abstract

Segregation, or the setting apart or uncoupling, of gold from base metals has occurred during the formation of most but not all deposits of gold. Segregation is a form of partitioning and the **third** characteristic, following provinciality and enrichment, that needs to be accommodated when understanding how gold deposits may form.

Keywords

Segregation from base metals · Gold-only / gold-plus classification

Base metals such as copper, zinc and lead are much more abundant in the Earth's crust than precious metals such as gold. A geological process that enriches gold by up to four orders of magnitude above its background might be expected to substantially enrich base metals too especially as they have some chemical similarities. The observation, that in many gold deposits there is little parallel base metal enrichment with the gold, places important constraints on possible gold enrichment processes. Of additional interest, a subordinate number of gold deposits **are** enriched in base metals. These observations set up the search for processes that are highly effective in partitioning gold relative to base metals like Cu, Zn and Pb.

6.1 Gold and Base Metal Concentrations in Various Deposits Producing Gold

The enrichment of gold can be compared to the enrichment of base metals using ratios of element concentration in ores to the respective element crustal abundance. Ores from the major Kalgoorlie goldfield, a small mine in the Pilbara and a representative value from the Carlin Gold Province in USA are highly enriched in gold but have copper and zinc values close to crustal average (Fig. 6.1). In contrast, Archean, Paleozoic and modern VMS systems of Australasia are enriched in each of gold, copper and zinc. Deposits of the Canadian shield that produce gold show this same dichotomy with some being unenriched or slightly enriched in base metals (green), whereas the VMS deposits are enriched in all of gold, copper and zinc (Fig. 6.2).

For the gold deposits without matching enrichment of base metals (green) there is an important segregation of gold from base metals of two to three orders of magnitude. This segregation from gold illustrated by Cu and Zn is similar for Pb. Silver, however, has an intermediate position and is enriched but not to the extent of Au (Fig. 6.3). Examples of deposit types showing this segregation (and resorting to some of the names used in the literature) include many

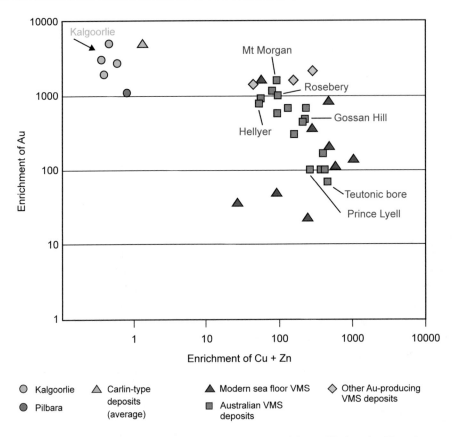

Fig. 6.1 Gold versus copper and zinc enrichments above crustal averages in ores from various gold-only deposits of different ages contrasted with gold-plus ores from modern seafloor and some Australian examples of volcanogenic massive sulfide (VMS) deposits. Note the concentration enrichments are plotted logarithmically. Modified from Phillips and Powell 2010.

greenstone-hosted gold deposits, Witwatersrand gold, Carlin gold, Paleozoic and younger slate belt gold and low-sulfidation epithermal deposits. Hence this segregation is a feature of most of the gold deposits through geological time and the characteristic transgresses many of the terms used in society for different deposit types.

Other deposits containing gold do not show this segregation including various copper—gold deposits, iron oxide copper gold (IOCG) deposits, polymetallic vein deposits, high-sulfidation epithermal gold deposits and VMS deposits. The IOCG deposits differ slightly from the VMS deposits (Fig. 6.4), but together they are both quite distinctive compared to deposits without base metals.

6.2   The Gold-only/Gold-plus Classification

An effective subdivision of most gold deposits can be made into either *gold-only* or *gold-plus*, and this classification is critical to the understanding of their formation (see also Appendix D). In the figures above, the gold-only are in green, and the gold-plus in variable mixes of red, magenta, and orange. The *gold-only* or *gold-plus* classification highlights one fundamental assumption underpinning the approach in this book and its chapters. That assumption is that the only/plus distinction made by this classification reflects the fundamental process(es) inherent in the formation of gold deposits.

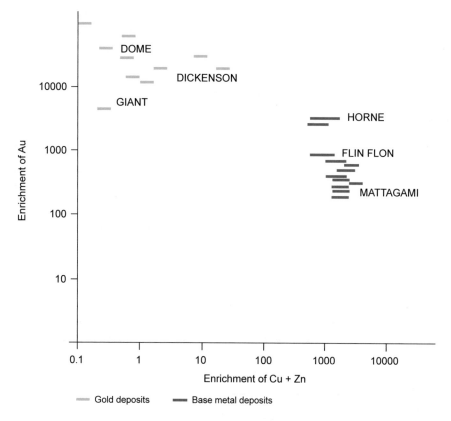

Fig. 6.2 Gold versus copper and zinc enrichments above crustal averages in deposits within Archean greenstone belts of the Abitibi Province of eastern Canada. Note the concentration enrichments are plotted logarithmically. Source includes data from Kerrich and Fryer 1981.

The classification is practical in that it can be made without undue assumptions or lengthy research—it requires answering a simple question "does this deposit have economic base metals?". The beauty of this criterion is that it can be addressed by asking a mine manager or chief financial officer or by reference to a company annual report. The second requirement of a good classification system is that it has a sound theoretical basis; this is addressed in later chapters and turns out to be especially powerful in its ability to be forward and backward modelled.

6.3   Potential Explanations for Segregation

Mechanical processes alone can cause both enrichment and segregation, but they do not appear to approach the extreme enrichment and segregation necessary to form gold-only deposits. One mechanical process is the physical removal of passive components to enrich all those components not removed. An example of this is the effective removal of clays and fine silt over time as wind blows across dry desert soils. This leads to a surface concentration of heavier pebbles. The concentration of pebbles can be substantial, as shown for the ironstone and quartz pebbles in Fig. 6.5. If 90 percent of the mass of a thin surface layer is removed as windblown clay and silt the concentrations of ironstone and quartz pebbles will rise tenfold but the ratio of ironstone to quartz remains constant, i.e. there is no segregation of quartz versus ironstone because they are both stationary. However, there is segregation of pebbles from clay; the ratio of pebbles to clay changes dramatically as the pebbles remain and

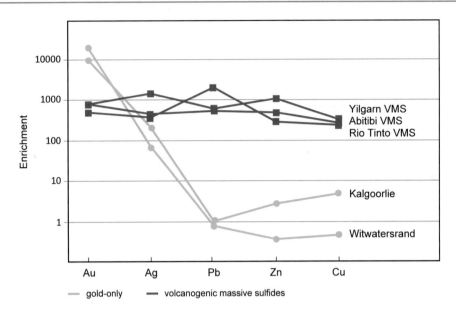

Fig. 6.3 Enrichment above crustal averages of Au, Ag, Pb, Zn and Cu for five contrasting deposits containing gold: Kalgoorlie (gold-only, Archean greenstone), Witwatersrand (gold-only, Archean), Yilgarn VMS based on Golden Grove, Abitibi VMS of eastern Canada, and Rio Tinto VMS of Portugal and Spain. Note the concentration enrichments are plotted logarithmically.

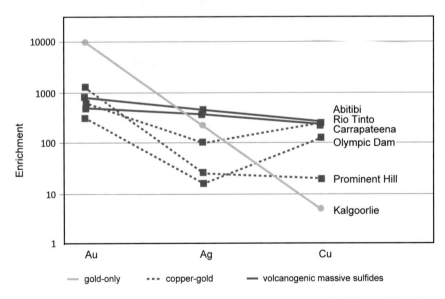

Fig. 6.4 Enrichment above crustal averages of Au, Ag and Cu for six contrasting deposits containing gold. Kalgoorlie (gold-only, Archean greenstone), Olympic Dam, Carrapateena and gold ores of Prominent Hill all in South Australia (gold-plus, IOCG), Abitibi VMS and Rio Tinto VMS.

the clay becomes airborne. If an element such as Fe is only in the ironstone pebbles and Al is only in the clay minerals, this desert process could segregate Fe from Al but typically it does not progress beyond about one order of magnitude.

Fig. 6.5 An unmineralised surface on the edge of the Simpson Desert near Birdsville, Central Australia, on which there are small quartz (white) and ironstone (red brown) rocks on the left, and much finer light-cream coloured silt and clay. Movement of the rocks only occurs during the rare rain events, but not by wind. In contrast, the wind is an ongoing factor that continually removes the fine material and, in so doing, concentrates any residual coarser denser material such as pebbles on the surface.

Further chapters will focus on any substantial gold – base metal segregation that might arise from unusual source material, selective dissolution of gold relative to base metals, or selective precipitation of gold at the deposition stage, or all of these processes combined. Both physical and chemical means are considered as potential partitioning mechanisms. Although structures such as folds and faults have an important effect on fluid flow pathways in the subsurface and may exhibit structural control on the final location of a gold deposit, they do not provide any explanation for separation of elements such as the segregation of gold from base metals. Options of melt (magmatic) involvement in the enrichment and segregation process will be considered, along with the role of hydrothermal fluids in the long-distance transport of gold.

Snapshot

- Segregation refers to the much greater enrichment of gold compared to base metals.
- Segregation of gold from copper can be of two to four orders of magnitude.

- The gold-only/gold-plus classification is of fundamental importance in understanding the formation of gold deposits.
- This classification also meets the criterion of being practical to apply because it relies on readily accessible data such as in standard company reports.

Bibliography

Kerrich R, Fryer BJ (1981) The separation of rare elements from abundant base metals in Archean lode gold deposits: implications of low water/rock source regions. Econ Geol 76:160–166

Kerrich R, Hodder RW (1981) Archaean lode gold and base metal deposits: evidence for metal separation into independent hydrothermal systems. Can Inst Mining Metall Bull Special volume 74:144–160

Phillips GN, Powell R (1993) Link between gold provinces. Econ Geol 88:1084–1098

Phillips GN, Powell R (2010) Formation of gold deposits: a metamorphic devolatilization model. J Metamorph Geol 28:689–718

Rosa DRN, Romberger SB (2003) Fluid evolution in the Jales Au district, Northern Portugal. Int Geol Rev 45:646–658

Timing of Deposit Formation

7

Abstract

To understand how gold deposits form it is useful to know the relationship between the time of deposit formation and that of sedimentation, volcanism, metamorphism, igneous intrusion, and deformation. Timing is the **fourth** characteristic that needs to be accommodated when understanding how gold deposits may form.

Keywords

Epigenetic · Syngenetic – exhalative · Quartz veins · Relative timing · Stitching pluton

Resolving whether gold deposits form at the same time as their enclosing rocks (syngenetic) or form later (epigenetic) is fundamental to understanding their genesis including what factors influence their geometry, grade, and the suite of associated valuable and deleterious elements. The issue is also fundamental to exploration for further deposits.

Two methods commonly used to determine timing are the absolute ages of two or more events, and the determination of relative ages without quantitative radiometric dating. In ideal conditions, radiometric ages can be remarkably precise as well as giving the comfort of a numeric result. The limitations of radiometric dating may lie both inside and particularly outside the laboratory and include uncertain field relationships, and several important events being protracted and essentially synchronous with uncertainty linking the radiometric age of a sample to one specific geological event. Differentiating the effects of ore-related alteration from near-surface weathering is one example of the importance of field relationships being resolved before radiometric dating. Relative timing involves the use of overprinting relationships requiring the determination of which of two geological events overprints the other; and although the time gap between the two events may not be quantified, the order of events may be quite apparent.

7.1 Syngenetic and Epigenetic Gold Deposits

A fundamental question relates to whether a deposit with gold formed at the same time as its surrounding package of sedimentary and igneous rocks (syngenetic) or the deposit formed later (epigenetic). Only if the former is the case will the environment of sedimentation relate directly to the environment in which the deposit formed.

Syngenetic gold deposits are not the main theme of this book, but it is convenient to discuss the volcanogenic massive sulfide deposits including their influence on the genetic ideas advocated for many gold-only deposits.

For epigenetic deposits, the focus is usually on the relative timing of deposit formation compared to deformation, metamorphism, and magmatic activity. Although there are well-established

N. Phillips, *Formation of Gold Deposits*, Modern Approaches in Solid Earth Sciences 21,
https://doi.org/10.1007/978-981-16-3081-1_7

criteria for relating each of these processes temporally, it is common in Nature for a major period of metamorphism to be essentially coeval, and associated with, deformation and magmatic activity. There are many examples of gold deposits that have remained virtually undeformed since the deposit formed, and there are others that may have been deformed through faulting and folding. Some epigenetic deposits form after the peak of metamorphism, some at a temperature indistinguishable from peak metamorphism, and some before the peak.

7.2   Gold Deposition Relative to the Time of Magmatism, Deformation and Metamorphism

Auriferous quartz veins are a common expression of gold mineralisation, and their emplacement constrains the timing of deposit formation. We recognise here hydraulic veins and open space filling veins, with the focus on hydraulic veins formed through multiple crack and seal events under lithostatic conditions of more than 3 – 5 km depth. The other type are the open space filling veins which form under hydrostatic conditions in the upper crust. They are also important for gold.

7.2.1   Auriferous Quartz Veins and Brittle—Ductile Fault Zones in Clastic Metasedimentary Sequences

Auriferous quartz veins are a widespread expression of gold mineralisation in clastic metasedimentary rocks particularly shale – greywacke turbidite sequences (Fig. 7.1). The crack seal growth and brittle – ductile textures of these veins demonstrate that they were emplaced into fully lithified rocks rather than into unconsolidated or diagenetic sediments. Additionally, the textures indicate that the veins were not emplaced into ductile, partially molten metasedimentary successions.

The general explanation for the veins is that the fluid pressures were high and exceeded the combination of the weight of the overlying rock, any compressive stress and the tensile strength of the rocks. The rheology or flow behaviour of various rocks during deformation describes whether they deform in a brittle or ductile way, and if they hydraulically fracture (Phillips WJ, 1976). The differences in tensile strength of rock types means that it is possible for one rock type of low tensile strength to fail and to have abundant veins whereas an adjacent rock of higher tensile strength remains intact without any fluid ingress or veins. External factors such as strain rate and temperature also dictate deformation behaviour and whether a certain rock might fail by hydraulic fracture.

A typical interpretation of such quartz veins is that they are epigenetic in origin, indicate the presence of a hydrothermal fluid, and formed in the lithostatic environment after the beginning of deformation. Some auriferous veins are folded and strongly contorted by on-going or subsequent deformation whereas others are undeformed with their geometry essentially unchanged since they formed.

7.2.2   Auriferous Quartz Veins and Brittle—Ductile Fault Zones in Igneous Rocks

Many gold deposits comprise auriferous quartz veins and shear zones in igneous rocks. Similar physical controls apply in igneous rocks as those just described for veins in metasedimentary host rocks; namely that veins can form in response to dramatically elevated pore fluid pressure particularly in rocks of low tensile strength. The conditions for hydraulic fracture are not met if the host rock is partially molten, so it can be inferred that the auriferous veins were emplaced after solidification of their hosting igneous rock (Fig. 7.2). Technically this means the veins and gold are epigenetic (formed after their host rock) even if the time gap between igneous crystallisation and gold mineralisation may be short or indeterminate.

Fig. 7.1 Several quartz veins in the Paleozoic shale and greywacke sequence at Bendigo goldfield, Victoria. On the left centre are multiple laminated quartz veins interleaved with shale and parallel to sedimentary bedding. In the upper left is a 1 cm thick vein offset by a fault. In the right centre is a more massive quartz vein. Field of view approximately 1m.

Fig. 7.2 Quartz veins in an igneous rock. The vein formed by hydraulic fracture of a lithified host rock (in this case a dolerite) and hence well after this mafic magma was solidified.

Given the constraint that the quartz veins were emplaced once the host rock was solid, the igneous rock hosting those veins was not a potential source of the auriferous fluid that formed the veins. One of several possibilities is that later igneous phases were the source of gold and fluid. Taking this further, if the auriferous fluid originated outside its immediate igneous host rock, then its source becomes unconstrained and could be removed by 100s m or kilometres; at this distance, there can be no presumption that the source of the auriferous fluid needs to be an igneous rock at all.

7.2.3   Influence of Volcanogenic Massive Sulfide Studies on Gold-only Interpretations

In the early 1980s, knowledge of the VMS systems was influential in the field of economic geology such that bedded auriferous units were inferred to be syngenetic, and auriferous quartz veins then explained as the product of remobilisation of this syngenetic gold.

The VMS deposits are poly-metallic in that they have Cu, Zn, Pb, Ag and Au as co-products and by-products, occur in mixed volcanic and sedimentary sequences of many ages and are associated with intrusive rocks. There is an extensive literature on the VMS deposits indicating that they form at or near the seafloor approximately at the time of ongoing sedimentation. Part of the evidence for their syngenetic origin comes from the analogy with naturally occurring black smokers and white smokers that vent into the modern ocean depths. These vents release metal-rich brines that mix with seawater and precipitate metal sulfides that contribute to a layer of metal-rich sediment on the ocean floor. At the same time as the smokers are releasing metals, the same process generates replacement sub-seafloor mineralisation which is technically not syngenetic (with respect to its immediate host) though will be considered essentially syngenetic to contrast these deposits with those formed much later and truly epigenetic.

Following considerable exploration success in the 1970s for VMS-type deposits in Canada, similar thinking was applied to gold deposits in greenstone belts of Canada and most other countries with Archean rocks. This thinking assumed that the gold deposits were syngenetic and formed on or near the seafloor like the Cu – Zn – Ag – Au VMS deposits. From this it followed that exploration should be guided by stratigraphy and rock units; furthermore, by this reasoning structural features like folds and faults were something imposed much later and causing re-arrangement of the geometry of deposits. What was thought to be compelling evidence for this seafloor syngenetic model was fine-scale oxide and quartz-rich layers making up banded iron formations (BIF) that merged into the auriferous sulfide—quartz layering that characterised the gold ore in these rock types. At this time, the explanation for the oxide, carbonate and sulfide facies of BIF was that they reflected increasing ocean water depth.

Part of the reason for the wide advocacy of the syngenetic model was its demonstrable success aiding base metal VMS exploration in Canada. Another factor was that the science necessary to think beyond the syngenetic model was only just emerging in geoscience in the late 1970s. Led by Bill Fyfe, the importance of metamorphic fluids

and their mode of migration through the crust were being explained by combining structural geology and metamorphic petrology principles. The introduction of hydraulic fracture science from the petroleum industry was a useful part of the fluid migration story. The second new field introduced into geoscience was that of gold chemistry and how gold and base metals were complexed in different solution types.

7.2.4 Auriferous Quartz Veins in Banded Iron Formation (BIF)

The findings at Water Tank Hill mine in 1981 facilitated the shift from a VMS-based genetic model for Archean greenstone gold deposits to an epigenetic model (see Chapter 1). The shift in genetic thinking flowed through to many deposits in metasedimentary host sequences and brought about an enhanced role for structural geology in exploration.

Water Tank Hill was only an exploratory mine in 1981 with a shaft less than 100 m deep. It was in the Mt Magnet district of the northwest Yilgarn Craton 400 km from Perth. It is 5 km southeast of the much larger Hill 50 mine in a metasedimentary succession including BIF. The intention of sinking the small shaft at Water Tank Hill was that it might lead to another major deposit like Hill 50 that had reached 900 m depth during mining of 40 t Au since 1891.

As an aside, but a relevant one, I had moved to Western Australia in the middle of 1979 to take up a postdoctoral position with a vague understanding that my brief was related to Archean gold. My arrival coincided with the field season for those working in the north and I found myself with nothing to do, or should I more correctly say, no specific tasks that I had to complete. I found a book called *Chemistry of gold* by Richard Puddephatt, and it was soon clear from reading that the formation of deposits from seawater exiting onto the seafloor was going to involve base metals—despite what was being said in the economic geology literature.

Once underground at Water Tank Hill, it was easy to recognise those beautiful centimetre-scale layers of alternating iron oxide and silica, but on a scale of metres the gold was not parallel to those layers. Instead, the gold was forming an envelope around quartz veins that were perpendicular to those sedimentary layers (Fig. 7.3). It was clear even before returning to the surface that the Water Tank Hill deposit might hold the field evidence that would revise our ideas about the formation of these gold deposits. The boundary between magnetite and pyrrhotite in each Fe-rich layer was not a gradation but a sharp interface of millimetric width (Fig. 7.4).

The combination of the gold chemistry and the mineral textures that were found underground was enough to be convinced that this BIF-hosted gold deposit was epigenetic. A follow-up underground visit with David Groves involved the collection of further samples, and it was then a matter of deciding together how to take this potentially new science forward.

The older literature threw further light on the epigenetic story as it became evident that others had noted structural control on BIF-hosted gold especially before the 1960s VMS-led era. The adjacent Hill 50 gold mine was not accessible but published cross-sections and level plans showed the epigenetic interpretation might apply there. Bullfinch, the other large BIF hosted gold deposit in the Yilgarn Craton at the time, was also closed but there were revealing cross-sections and plans suggesting structural control and a possible epigenetic origin. It still appeared that Water Tank Hill was small and might be dismissed as anomalous, so it was necessary to go overseas to some of the classic BIF-hosted gold deposits that featured in the literature.

Back in the laboratory the case for an epigenetic origin was strengthened by a third line of evidence from fluid inclusions from quartz veins collected at Water Tank Hill and other gold deposits in BIF; these fluid inclusions did not match seawater but were dominated by H_2O and CO_2.

The next question was what to do with the information. The partnership with David Groves proved critical given his immediate interest, expertise on Archean metallogeny, and knowledge on how to deliver an impactful story. Water Tank Hill became a core example in the revision of the syngenetic seawater exhalative model for Archean gold deposits. In its place, an epigenetic model (i.e. gold introduced after the rocks were deposited and deformed) appeared more viable.

One of the most memorable periods of uptake of the new epigenetic idea involved a weekend in the Barberton greenstone belt in 1982 with a group of a dozen young and keen mine and exploration geologists. David Groves and I visited three of the gold mines (Sheba, Consort and Agnes) finding epigenetic features at each, and continued making these observations working through extensive core from those deposits. It was helpful to be able to recognise and show the replacement fronts in core and underground; but equally influential in the uptake of the epigenetic ideas was being able to explain how gold-bearing fluids moved through rocks to transport and precipitate gold.

Immediately following the Barberton visit was a field excursion to many of the small gold mines in Zimbabwe led by postgraduates from the University of Zimbabwe. Several deposits were in BIF host rocks, and all had epigenetic ore textures. Importantly, the excursion included the Vubachikwe gold mine which was, at that time, internationally known for its syngenetic gold interpretation.

The International Gold-82 Conference in Harare in mid-1982 followed the field excursion and occurred as Rhodesia was transitioning to its new name of Zimbabwe. The Water Tank Hill evidence and interpretation were on display and several participants had already been exposed to the same field evidence at Barberton and the Zimbabwe goldfields (Fig. 7.5). Naturally, there was spirited questioning especially from international VMS adherents, but it was also clear that the epigenetic story had merit. Three senior geoscientists with extensive Canadian gold experience, Bill Fyfe, Rob Kerrich and Pat McGeehan, were influential at the time with their support for the new epigenetic ideas.

Uptake of epigenetic gold ideas was rapid in Western Australia especially following the Gold 82 Conference. Elsewhere, the uptake advanced at different paces that reflected individuals and their own geological experience. It was slower

Fig. 7.3 Sketch of BIF and
an auriferous quartz vein
from Water Tank Hill mine
in Western Australia.
Regular sedimentary
layering of magnetite and
quartz is preserved after
replacement of the
magnetite by pyrrhotite.
The replacement and the
quartz vein arise from
addition of Si, S and Au
from auriferous fluids along
a fracture. This replacement
is a common pattern on
many scales.

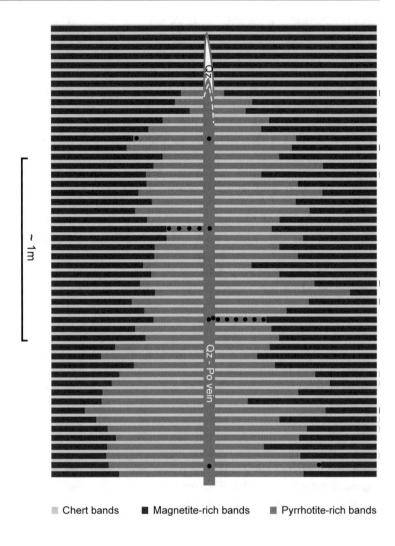

▨ Chert bands   ■ Magnetite-rich bands   ▨ Pyrrhotite-rich bands

Fig. 7.4 Polished slab of
BIF from Water Tank Hill.
The oblique boundary
between barren magnetite-
rich BIF and auriferous
pyrrhotite-rich BIF shown
by the arrow is very sharp.
Width of view is 15 cm.

Fig. 7.5 Banded iron formation from Broomstock mine in Zimbabwe. BIF bedding is dipping steeply to the left and perpendicular to a laminated quartz vein which contains pyrite and gold. Adjacent to the vein, pyrite has replaced hematite and magnetite in the BIF for limited distances along the more Fe-rich layers, whereas the silica-rich layers are unreplaced.

to be adopted, not surprisingly, in places where the VMS model had led to considerable exploration success for base metals. For example, during the same period, James Macdonald was finding similar replacement features at the large Geraldton gold deposit in BIF in Canada but in doing so was challenging a strong VMS community.

Some of the largest gold deposits hosted by BIF are situated in the Quadrilátero Ferrífero region of Brazil near Belo Horizonte including Sao Bento (Fig. 7.6), Morro Velho, Raposos and Cuiabá. Some excellent mapping showed replacement fronts and structurally controlled ore pods on the scale of 100s m at Cuiabá (Fig. 7.7).

7.2.5   Incorporation of Mineralisation in Later Intrusions and Extrusions

Dykes and larger intrusive bodies can be used for determining the relative age of mineralisation. A radiometric age of the dyke combined with knowledge of whether the dyke is gold mineralised, or barren, may be enough to provide a limit on the absolute age of gold mineralisation. Sources of error in this process may be the choice of dating method and its capabilities but might also be the field relationships and meaning of those samples used for analysis.

The situation is more complicated if there is any weathering that potentially redistributes the gold especially along the dyke margins. At the Fosterville goldfield in Victoria open pit exposures have been useful for the collection of samples and for noting field relationships, but these are also the situations in which weathering affects dyke margins and redistributed gold. In this case the weathered dyke is weakly mineralised and has been dated as Late Devonian but interpretation of the meaning of that age for the main gold introduction is problematic.

Another example comes from Homestake gold mine in South Dakota USA where a dyke is dated as Cenozoic, and it intrudes mineralised Precambrian banded iron formation. In the oxidised interval of an open pit that dyke is weakly gold mineralised. Superficially this suggests that gold at Homestake was introduced after the Cenozoic dyke crystallised and this conclusion was published as such. Once it was recognised that the mineralisation in the dyke was potentially the result of redistribution during weathering, the age of gold mineralisation was revised by over a billion years.

In principle, where dykes and sills intrude existing goldfields there is a chance that gold mineralisation may become incorporated into the igneous rock in the same way that hornfels xenoliths become incorporated; furthermore, the incorporated mineralisation may appear along trends of mineralisation in the country rocks.

Fig. 7.6 Auriferous quartz vein at Sao Bento gold mine Quadrilátero Ferrífero region in Brazil. The host rock is BIF with horizontal bedding. This is an example where the high strain nature of the shear zone (with small veins immediately above the pen) was not recognised initially and instead it was mapped as a stratigraphic unit immediately conferring (incorrectly) a syngenetic mode on gold formation.

Fig. 7.7 Underground level map at Cuiabá mine showing the ore zones in red distributed discontinuously along the BIF unit in yellow. Modified from Ribiero-Rodrigues 2007 based on original unpublished mapping by Diogenes Vial in the early 1980s.

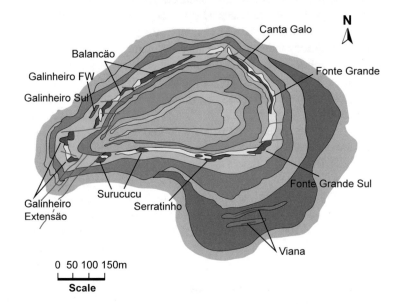

This poses a slightly difficult situation in which the late mineralisation needs to be correctly identified as trivial and late, meaning it is interpreted as accidental, and not representative of the whole gold province.

7.2.6   Stitching Plutons

Relative timing of gold introduction is provided by stitching plutons which can then be radiometrically dated for an absolute constraint. A stitching pluton is one that is emplaced across a major structure such as a fault or shear zone and is not dismembered or offset by that structure. The relative timing based on the field relations is that the pluton emplacement is after the structure. If the major structure is mineralised, there are two possibilities for gold distribution and genesis. Either gold mineralisation pervades the structure and continues through the pluton, or the mineralisation is absent from the pluton because the mineralisation has been truncated but it is present on both sides of the intrusion. In the former, gold mineralisation has been introduced after the pluton has consolidated, in the latter the gold pre-dates the pluton. Once the relative ages of the gold and pluton are determined, a radiometric age on the pluton places one limit on the age of gold introduction.

Two examples of stitching plutons from the Victorian Gold Province are adjacent to the Stawell goldfield (5 Moz, 160 t Au) and the Maldon goldfield (2 Moz, 70 t). In both cases the mineralisation is contact metamorphosed and truncated by the pluton rather than continuing straight through the granite (Fig. 7.8a and b). It is not unusual to find some anomalous gold concentrations in the granite either near the contact or along the trend of the major structure. Both plutons have been dated using zircon grains from the granitic rocks to yield 396 Ma +/- 5 (Stawell) and 361 Ma +/- 7 (Harcourt granite near Maldon) and these reflect a minimum age of gold mineralisation at Stawell and Maldon goldfields, respectively.

Textural relationships in metasedimentary hornfels adjacent to the granite can be used to confirm larger scale structural relationships. If gold mineralisation followed granite emplacement, mineralisation is likely to be out of equilibrium with distal hornfels, i.e. retrograde. However, in the examples of Maldon and Stawell, gold mineralisation preceded the hornfels event and is in equilibrium with and part of the hornfels mineral assemblages.

A slightly different example is the Goldstrike pluton midway along the 50 km NNW – SSE Carlin Trend in Nevada USA. This 5 km^2 pluton is a massive diorite of Jurassic age dominated by hornblende – plagioclase – biotite and minor quartz, that structurally appears to be a stitching pluton post-dating development of the NNW – SSE deformation. On a district and regional scale, the large mines trend towards and abut the northern and southern margins of the pluton without extending through it (Fig. 7.8 c). The Betze Post Goldstrike goldfield of 50 Moz (1500 t Au) is at the north margin of the pluton and the Genesis goldfield is immediately to the SSE. The distribution and nature of the gold mineralisation on and around the pluton is investigated further in Chapter 18.

7.2.7   Consideration of Scale in the Determination of Timing of Gold

Scale is critical for timing the relative formation of gold deposits, and scale dictates the appropriate methods of study. It is not the same to determine the age of a gold grain and the age of a gold deposit: a microscope-based study of 1 mm gold grains alone is not appropriate to constrain the age of a deposit that is five orders of magnitude larger. The attraction of a microscope-based study is understandable as the work can be completed without necessarily going underground to evaluate the context of samples, and in the extreme the approach might rely upon samples borrowed from a museum. The shortcoming is illustrated by a hypothetical study that has concluded correctly that a gold grain has formed after a late cleavage; then a late age for gold mineralisation is incorrectly inferred. More correctly, that observation of a grain post-dating the cleavage is only indicating that the last stage of growth of the gold grain was late.

Ultimately, if the timing of deposit formation is incorrect then the interpretation of its tectonic setting during formation of the deposit is likely to be incorrect also.

Snapshot

- Hydraulic veins and open space filling veins can constrain the time of gold introduction.
- Auriferous hydraulic quartz veins in metasedimentary and igneous rocks indicate gold introduction was epigenetic, at least on a local scale.
- The syngenetic seafloor model for VMS deposits does not adequately explain gold-only or copper-gold deposits.
- Stitching plutons provide field relationships than can link a radiometric age of the intrusion to place a bracket on the age of gold formation.
- Correct timing is required to correctly interpret the tectonic setting during deposit formation.

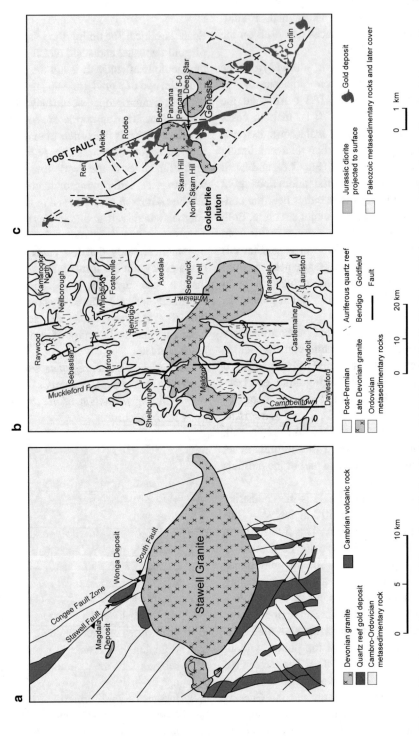

Fig. 7.8 Stitching plutons adjacent to: (a) Stawell goldfield in Victoria; (b) Maldon goldfield in Victoria: (c) Betze Post and Genesis deposits and mining operations in Nevada.

Bibliography

Arne DC, Bierlein FP, McNaughton N, Wilson CJL, Morand VJ (1998) Timing of gold mineralisation in western and central Victoria, Australia: new constraints from SHRIMP II analysis of zircon grains from felsic intrusive rocks. Ore Geol Rev 13: 251–273

Fripp REP (1976) Stratabound gold deposits in Archean banded iron-formation, Rhodesia. Econ Geol 71:58–75

Fyfe WS, Price NJ, Thompson AB (1978) Fluids in the Earth's Crust. Elsevier, Amsterdam, p 383

Lewis BR (1965) Gold deposit of Hill 50 mine. In: Mc Andrew J (ed) Geology of Australian Ore Deposits. Melbourne, Australasian Institute of Mining and Metallurgy, pp 98–100

Phillips WJ (1976) Hydraulic fracturing and mineralization. J Geol Soc London 128:337–359

Phillips GN, Groves DI, Martyn JE (1984) An epigenetic origin for the Archean banded iron-formation-hosted gold deposits. Econ Geol 79:162–171

Puddephatt RJ (1978) The chemistry of gold. Elsevier, Amsterdam, p 274

Ramsay JG, Huber M (1987) The techniques of modern structural geology, Strain analysis, vol 1. Academic Press, London

Ribeiro-Rodrigues LC, Gouveia de Oliveira C, Friedrich G (2007) The Archean BIF-hosted Cuiabá Gold deposit, Quadrilátero Ferrífero, Minas Gerais, Brazil. Ore Geol Rev 32:543–570

Ridley JR (1993) The relationship between mean rock stress and fluid flow in the crust: with reference to vein- and lode style deposits. Ore Geol Rev 8:23–37

Teal L, Jackson M (1997) Geological overview of the Carlin Trend gold deposits. Soc Econ Geol Newslett 31:1,13–1,25

Tikoff B, Blenkinsop T, Kruckenberg SC, Morgan S, Newman J, Wojtal S (2013) A perspective on the emergence of modern structural geology: celebrating the feedbacks between historical-based and process-based approaches. Geol Soc Am Spec Paper 500:5–119. https://doi.org/10.1130/2013.2500(03)

Vearncombe JR (1998) Shear zones, fault networks, and Archean gold. Geology 26:855–858

Vial DS, Groves DI, Cook NJ, Lobarto LM (2007) Special issue on gold deposits of Quadrilátero Ferrífero, Minas Gerais, Brazil. Ore Geol Rev 32:469–470

Ore Fluid Types as Recorded in Fluid Inclusions

8

Abstract

Minute inclusions in quartz have trapped ore fluids while the gold deposits formed. Study of these fluid inclusions is a direct way to determine the ore fluid composition. The nature of the ore fluid as determined from the inclusions is the **fifth** characteristic that needs to be accommodated when deciding how gold deposits may form.

Keywords

Fluid inclusions · Low salinity fluid · Saline fluid

The common quartz veins found in many gold deposits were formed by quartz crystals growing from hydrothermal ore fluids. Apart from transporting dissolved silica, these fluids are potentially the transporting mechanism for dissolved gold. As the quartz crystals grow to form veins, they trap minute amounts of that fluid as inclusions. Study of these inclusions is one of the few direct ways to determine the composition of a gold-transporting fluid.

8.1 Fluid Inclusions in Ore Deposits

As quartz crystals grow from a hydrothermal fluid, they usually have numerous irregularities in their growth pattern that trap small amounts of the surrounding fluid to become fluid inclusions. These fluid inclusions are typically smaller than 0.01 mm in diameter (a human hair which is 0.05 mm or more in diameter is as small as our unaided eye can resolve). The great value of the fluid inclusions is that they are potentially recording the actual fluid that formed many auriferous quartz veins. Careful field collection followed by microscope work helps to separate primary fluid inclusions formed during the main mineralising event from secondary fluid inclusions that may form at many later times from different and unrelated fluids.

Many analytical methods are applied to fluid inclusions in quartz veins to determine the conditions during which ore deposits formed. Non-destructive ways to heat and freeze the inclusions contained in thin slices of the vein quartz involve the use of microscopy to determine when various phases disappear, and from such measurements it is possible to make estimates of the concentration of the main salts in the hydrothermal fluid. This then allows an estimate of the temperature and pressure of formation of the ores. There are additional destructive ways of microscopically breaking open individual fluid inclusions to analyse the fluid for its main elements and isotopic composition. More recently, there have been studies on the inclusions trapped in other minerals like pyrite, and crystallised magmas with melt inclusions comprising glass or devitrified glass.

Fluid inclusions have been studied as a means of understanding ore fluids since the 19th Century

and the study advanced rapidly after 1950s led by
Ed Roedder of the US Geological Survey and
French scientists. Porphyry copper deposits
through western North America and the Andes
received considerable attention, and these
deposits had variable by-product gold and abundant fluid inclusions with common daughter
minerals. The latter are minerals formed as
precipitates from dissolved components as the
trapped fluid is cooled. Some daughter minerals
have the crystal form of salts such as halite (NaCl)
and sylvite (KCl) giving a clear indication of the
main post-depositional solutes in the hydrothermal fluid. In some cases, the primary fluid
inclusions from copper – gold porphyry deposits
and polymetallic veins were highly saline,
i.e. much more saline than seawater. A study in
the 1970s of particularly well-preserved fluid
inclusions from the Bougainville copper – gold
deposit in the SW Pacific confirmed the global
distribution of these highly saline ore fluids
beyond North America. The Bougainville fluid
inclusions are in the 0.01 to 0.04 mm size range
and have at least ten daughter mineral phases that
could be resolved optically (Eastoe 1978). The
fluid is highly saline with up to 75% combined
NaCl and KCl and 25 % H_2O. Multiple colourless
and opaque daughter minerals include various
salts, ore minerals and likely hematite (Fig. 8.1
top). The ore forming event was inferred to
initially have been above 700°C followed by an
extended period of mineralisation at declining
temperatures involving ground water influx.

In this context of saline ore fluids at porphyry
copper – gold deposits around the Pacific margin,
investigations began in the late 1970s of the ore
fluids at gold-only deposits using fluid inclusions
from Archean greenstone gold deposits in Canada
and Western Australia. The inclusions were rather
small (0.005 – 0.01 mm in diameter) and, in
contrast with those found in copper – gold
deposits, the gold-only inclusions lacked daughter minerals but had a CO_2 bubble. A detailed
study at the large Hollinger – McIntyre mine in
Canada concluded that the ore fluid was of low
salinity and dominantly H_2O-CO_2 with minor
methane and N_2; these fluid inclusions also lack
daughter minerals (Fig. 8.1 bottom).

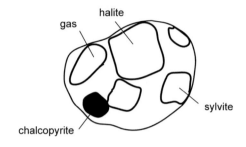

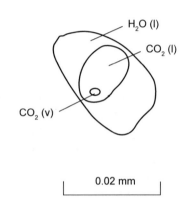

Fig. 8.1 Fluid inclusions of 0.02 mm diameter in
mineralised quartz veins from: (**top**) Bougainville PNG
copper – gold deposit showing halite (NaCl), sylvite
(KCl) and chalcopyrite ($CuFeS_2$); (**bottom**) Kanowna
gold-only deposit. The copper – gold deposit has several
large daughter minerals and very high salinity, whereas the
gold-only Kanowna example lacks daughter minerals, has
low salinity and is a three-phase inclusion of water, CO_2
bubble and gas. (l) is liquid phase; (v) is vapour phase.

8.2   The Kanowna Fluid Inclusion Study of Archean Auriferous Fluids

One of the first detailed studies comparing the
fluid inclusions from multiple Archean gold
deposits was in 1981 by Su Ho in the Kanowna –
Kalgoorlie district of Western Australia. She sampled quartz veins from gold deposits in various
host rocks and found fluid inclusions like those
being reported from gold deposits in greenstone
belts in Canada, but quite unlike those recorded
from copper – gold deposits and polymetallic
veins particularly from western USA and
Bougainville.

Kanowna is a goldfield 20 km north of Kalgoorlie in the Yilgarn Craton of Western Australia that was discovered in 1893 and produced 0.5 Moz (15 t) of gold up to 1980. Minor production through the 1980s was followed by the then modest discovery of the Kanowna Belle deposit in 1989 that grew substantially through brownfield exploration. The whole Kanowna goldfield now has an endowment around 200 t Au.

The mapping and fluid inclusion study at Kanowna included samples from quartz veins in both the Red Hill dacitic intrusion and a polymictic ultramafic conglomerate (Fig. 8.2; Fig. 8.3). The study was facilitated by relatively large inclusions at Red Hill of 0.01 – 0.04 mm, compared to most other Archean greenstone gold deposits. These fluid inclusions had distinctive bubbles but no daughter minerals and indicated an ore fluid of low salinity and dominantly $H_2O\text{-}CO_2$, and for Red Hill, only minor N_2 and H_2S were detected in a more specialised study. The inferred ore fluids are indistinguishable between remarkably different host rocks from the dacitic intrusion to the ultramafic conglomerate metasedimentary rock.

The Kanowna study was extended within the Yilgarn Craton to auriferous quartz veins in basalt, a dolerite sill, interflow carbonaceous metasedimentary rocks and BIF. The low salinity, $H_2O\text{-}CO_2$ character persisted through all host rocks for gold deposits (Fig. 8.4; Table 8.1). Fluid inclusions from further Archean gold-only deposits of Western Australia were frozen to -100°C and then heated; most examples then melted at the CO_2 melting point of -56.6°C, but those with different melting temperatures indicated an extra component. The gaseous phase in the inclusions was confirmed as pure CO_2 by the melting experiments, except for three studies in which minor amounts of methane were inferred from the depressed melting temperatures. The presence of this methane was in deposits that contained carbonaceous shales and was attributed to local reaction of the ore fluids with these shales.

The observation that the low salinity, $H_2O\text{-}CO_2$ character in gold-only deposits is global and independent of host rock type and age indicates that its origin is distal to deposits and their host rocks. This means that the local host rock does not determine the main fluid characteristic, and the scale of auriferous fluid generation and migration is much more than metres and more likely to be kilometres. The inferred conditions of 1-2.5 kb and 300-350°C for Kanowna and other Yilgarn gold deposits are also inferred for similar gold-only deposits globally and have implications for depth of formation and for high regional geothermal gradients during gold deposit forming events.

8.3 Contrasting Types of Fluid Inclusions in Deposits Containing Gold

Many subsequent fluid inclusion studies have confirmed the early findings from Kanowna and Hollinger – McIntyre in Canada, namely that fluids from Archean greenstone gold deposits are dominantly $H_2O\text{-}CO_2$ with low salinity and were trapped as inclusions around 300 – 350°C. The ratio of H_2O to CO_2 varies around 3:1 and appears independent of host rock. Minor H_2S and nitrogen are measured in some more specialised studies; and for methane there is a strong correlation between its presence and proximity to carbonaceous shale.

The fluid inclusions from quartz veins from gold deposits in Paleozoic turbidite sequences are like those from the Archean deposits, and given that carbonaceous shale is much more common, the widespread methane is expected. The pattern of similar fluid inclusions has been extended more recently to gold deposits in China and Russia as well as Carlin gold deposits and some epithermal deposits – all these are gold-only deposits, lack daughter minerals, but have low salinity fluid inclusions with CO_2.

The fluid inclusions from copper – gold deposits are quite different and are highly saline

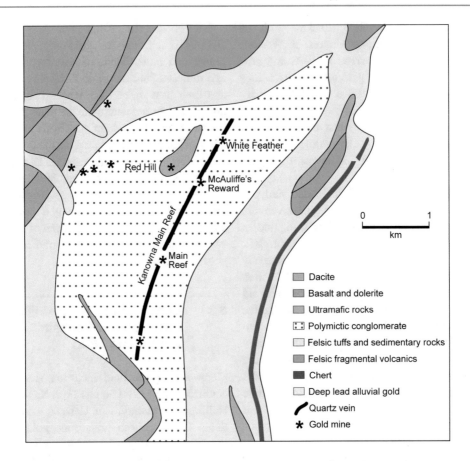

Fig. 8.2 Map of the Kanowna goldfield from the early 1980s showing the setting for some of the earliest fluid inclusion studies in the Archean Yilgarn Craton in Western Australia by Su Ho. The goldfield had produced 15 t Au at this time mainly from the Kanowna Main Reef quartz veins in polymictic conglomerate, and minor auriferous veins in the Red Hill dacite. The much larger Kanowna Belle deposit was discovered a decade later and is 2 km west of Kanowna Main Reef.

Fig. 8.3 Polymict conglomerate with felsic and ultramafic clasts. This is the main host rock for the 15 t (0.5 Moz) historic production from the Kanowna goldfield.

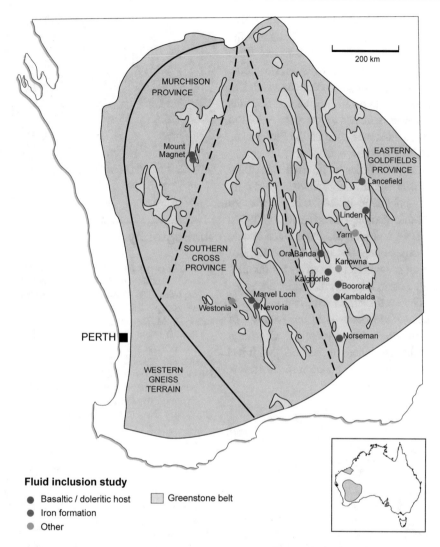

Fig. 8.4   Location of goldfields sampled by Su Ho in the early 1980s to demonstrate that the H_2O-CO_2 character of the fluid inclusions in auriferous quartz veins was regional in extent and did not vary with differing host rocks.

Table 8.1   Fluid inclusions from auriferous quartz veins (Ho et al. 1990).

Gold deposit	Host rock	Carbonaceous rocks	Mean (°C)	Median (°C)
Morning Star, Mt Magnet	Metabasalt	No	−56.6	−56.6
Lake View, Kalgoorlie	Tholeiitic dolerite sill	No	−56.6	−56.6
Mt Charlotte, Kalgoorlie	Tholeiitic dolerite sill	No	−56.6	−56.6
Main Reef, Kanowna	Polymict conglomerate	No	−56.5	−56.6
Red Hill, Kanowna	Dacitic intrusion	No	−56.8	−56.6
Hunt, Kambalda	Tholeiitic basalt	No	−56.5	−56.8
Lancefield #2	Granitic rocks	No	−56.7	−56.6
Lancefield #1	Metasedimentary rocks	Yes	−63.6	−65.0
Water Tank Hill, Mt Magnet	Banded iron formation	Yes	−60.5	−61.0
Paringa, Kalgoorlie	Tholeiitic basalt	Yes	−59.3	−59.5

Mean and median refer to the melting point of the bubble phase in the fluid inclusions
Melting point for CO_2 is −56.6 °C

with common daughter minerals and higher temperature such as reported from Bougainville. This type of saline fluid inclusion has been reported from Cu – Au deposits in ironstone, iron—oxide—copper—gold deposits (IOCG), Cu – Au porphyry deposits, epithermal deposits with economic base metals and copper – gold in quartz veins in siltstones such as the Proterozoic Telfer deposit in Western Australia.

In overview, ore fluids as inferred from fluid inclusions differ significantly between gold-only and copper – gold deposits. Much of the western world economic geology thinking in the 1960s was led by USA and included experience related to porphyry copper deposits; many foundations of fluid inclusion science came from such examples including the notion that ore fluids were necessarily saline. Then a decade of exploration success for base metal VMS deposits in the Archean of Canada re-enforced this perception. Certainly, many ore fluids are saline, but since 1980, it has emerged that gold-only deposits form from low salinity fluids.

Snapshot

- The nature of ore fluids can be inferred from fluid inclusions.
- Two types of ore fluids are suggested for deposits containing gold.
- Gold-only deposits have inclusions of low salinity, H_2O-CO_2 with inferred trapping temperatures of 300–400 °C.
- Copper—gold deposits have fluid inclusions with high salinity and inferred trapping temperatures that may exceed 500 °C.
- The low salinity, H_2O-CO_2 character is not a function of immediate host rocks for mineralisation but was derived from farther afield.
- High salinity fluid inclusions are also independent of immediate host rocks

and have been recorded in clastic and chemical metasedimentary rocks and various igneous rocks.
- Methane is a product of the reaction between ore fluids and local carbonaceous host rocks rather than being a primary component of the gold-only ore fluid.

Bibliography

Bull A, Rogers J, Ross A, Tripp G (2017) Kanowna Belle gold deposit. In: Phillips GN (ed) Australian ore deposits. The Australasian Institute of Mining and Metallurgy, Melbourne, pp 229–234

Craw D, McKenzie D (2017) Macraes orogenic gold deposit (New Zealand) Origin and development of a world class gold mine. In: Briefs in world mineral deposits. Springer, New York, p 129

Eastoe CJ (1978) A fluid inclusion study of the Panguna porphyry copper deposit, Bougainville, Papua New Guinea. Econ Geol 73:721–748

Goryachev NA (2019) Gold deposits in the Earth's history. In: Geology of ore deposits, vol 61. Pleiades Publishing Ltd, New York, pp 495–511

Ho SE (1984) Alternative host rocks for Archaean gold deposits: nature and genesis of hydrothermal gold deposits, Kanowna, Western Australia. Gold-mining, Metallurgy and Geology Conference Kalgoorlie, Australasian Institute of Mining and Metallurgy, Melbourne; pp. 405–415

Ho SE, Groves DI, Phillips GN (1990) Fluid inclusions in quartz veins associated with Archaean gold mineralization: clues to ore fluids and ore depositional conditions and significance to exploration. Stable isotopes and fluid processes in mineralization. Geology Department and Extension, University of Western Australia, Publn 23; pp. 35–50

Ridley J (2013) Ore deposit geology. Cambridge University Press, New York, 398 pp

Ridley JR, Diamond LW (2000) Fluid chemistry of orogenic lode gold deposits and implications for genetic models. SEG Rev 13:141–162

Smith TJ, Cloke PL, Kesler SE (1984) Geochemistry of fluid inclusions from the McIntyre – Hollinger gold deposit, Timmins, Ontario, Canada. Econ Geol 79: 1265–1285

Commonality and Diversity: Both Need Explanation

9

Abstract

Five characteristics have been discussed in previous chapters that are common amongst many deposits, but other geological features show diversity within and between deposits. Commonality and diversity are both informative about the way deposits form and need to be addressed in any viable genetic model.

Keywords

Common characteristics · Diversity amongst gold deposits

For some provinces, common features shared by hundreds of gold deposits can be summed up in a single sentence. For the Zimbabwe Craton, as one example, the summary statement might be "Thousands of gold mines and old workings lack economic base metals, are structurally-controlled and spatially associated with quartz veins and sulfide minerals." A remarkably similar statement might apply to gold deposits of the Yilgarn Craton in Western Australia, the Victorian Gold Province, the Abitibi Province of Ontario and Quebec in Canada, and Barberton goldfields of South Africa. These generalisations arise from the regional studies in different provinces referred to in Chapter 4.

Despite there being shared geological features within and between whole provinces there are dramatic differences even between orebodies in a single deposit and mine. The approach taken in this book is to focus on common characteristics (Chapters 4–8) but to remain aware that the differences also need explanation. The topics to date of provinces, enrichment, segregation from base metals, timing and nature of ore fluids have emphasised commonality among many but not all gold deposits. A basis for this focus is the extreme enrichment and segregation suggesting that it is quite unlikely that there could be many fundamentally different geological processes that can form gold deposits. Further geological features of deposits display an overwhelming diversity that has led to the saying "*gold is where you find it*".

9.1 Common Characteristics

Five characteristics have already been identified as common to many deposits (Table 9.1); and that these have been elevated in importance is obviously subjective. The justification for their elevation is demonstrated by the ability to forward and inverse model the genetic processes arising from these data.

This summary addresses most gold deposits by number and total production but some transitional examples are noted. Volcanogenic massive sulfide (VMS) deposits are forming today on the ocean floor (syngenetic) but some of these have significant ore interpreted to have formed in the sub-seafloor environment (technically epigenetic but part of the same overall event – still early and pre-dating any regional deformation). Although

Table 9.1 Five characteristics of deposits with gold.

	Gold-only	Gold-plus	
		Copper-gold	VMS
Provinces	Yes	Yes	Yes
Enrichment	Yes	Yes	Yes
Segregation from base metals	Yes	Moderate	No
Timing	Epigenetic	Epigenetic	~Syngenetic
Ore fluids	Low salinity	Saline	Saline
All-time gold production %: author estimate	80	15	5

the VMS deposits form clusters or provinces, these are on a scale of 10s km in length which is a smaller scale than the gold-only provinces (see Fig. 4.11; Fig. 4.12). Some gold deposits with co-product silver have elevate base metal concentrations though the base metals are still uneconomic.

9.2 Diversity and Commonality Demonstrated by Other Geological Features

This is a summary of the diversity of geological features found within gold deposits that has emerged from numerous regional studies. Rich Goldfarb and David Groves were early leaders of regional studies in Alaska and Western Australia respectively, and more recently have integrated many regional studies at a global scale. They have focused on a sub-section of the gold-only deposits, i.e. orogenic gold deposits and not including Carlin and Witwatersrand gold. Their global perspective is particularly useful as it allows extensions from individual provinces to more generalised summaries.

9.2.1 Metamorphic Grade

The metamorphic grade in and around gold deposits is quite variable. In Archean greenstone belts and Phanerozoic turbidite sequences, there are many examples of deposits in the greenschist facies, a considerable number in the amphibolite facies, and a small number in the granulite facies.

This distribution might partially be explained by the greater areas of greenschist facies rocks in some of the regions. On a global scale and throughout Earth history, the low temperature and higher-pressure metamorphic facies appear to be under-represented with respect to gold production such that major gold deposits are rare in the zeolite, prehnite – pumpellyite, glaucophane – lawsonite, and eclogite facies.

9.2.2 Structures Hosting Mineralisation

The nature and orientation of auriferous structures within gold-only deposits varies greatly from shallow to steep dipping, and includes quartz veins, faults, breccias, brittle – ductile shear zones and metamorphosed shear zones. The diversity relates to the nature of the structures and their orientations and is different from the commonality that emphasises how major gold deposits are associated with structural complexity. This complexity may be expressed as repeated anticlines mapped as saddle reefs, quartz stockworks, or shear zones at several different orientations. Gold mines with long operating lives are usually based upon multiple structures rather than a single large quartz vein.

Many gold-only deposits are described as being late in the structural history meaning that mineralised structures either formed before or contemporaneous with the mineralisation event and the orebodies have not undergone major deformation since their formation. This, however, does not preclude later reactivation along mineralised structures, mesoscopic scale folding of quartz veins or on-going metamorphism.

9.2.3   Host Rocks

The host rocks for gold deposits are highly variable and, in many terrains, virtually all rock types may be gold mineralised to some extent. Examples of hosts to some gold mineralisation include chemical and clastic sedimentary rocks and intrusive and extrusive igneous rocks. Notwithstanding this wide range, some provinces show a strong bias to one or more preferred hosts. In Archean greenstone belts, mafic rocks (especially those that are tholeiitic in composition), shales (especially where carbonaceous), and BIFs host much gold. In Phanerozoic slate belts, carbonaceous rocks such as black shales are important, with subordinate gold in igneous rocks particularly mafic rocks, e.g. magnetite-bearing diorite in the Victorian gold province. In some Phanerozoic gold provinces, the largest deposits are in igneous rocks, often collectively grouped as porphyries to include Cu-Au and Au-only porphyry deposits.

The Sheba Fairview goldfield in the Archean Barberton Greenstone Belt of South Africa (Section 4.3.3) is an exceptional example of high gold grade orebodies in different rock types (Fig. 9.1; Pintos Cerda et al., 2020). The largest orebodies are in ultramafic rocks and turbiditic shale – sandstone sequences, but there are further deposits in conglomerate, sandstone, and quartzite host rocks. The disposition of mineralisation is dictated by both fluid access (structures particularly near the Sheba Fault) and fluid-wallrock interaction (i.e. alteration processes influenced by the rock and fluid geochemistry).

This diversity of host rocks becomes more systematic when viewed in whole rock geochemical terms and amounts of gold produced. The more iron-rich igneous rocks, regardless of whether they are intrusive or extrusive, are over-represented as gold hosts in Archean greenstone belts as are the iron-rich BIFs and carbonaceous shale. In Phanerozoic slate belts, carbonaceous rocks are important, and Fe-rich rocks less so.

9.2.4   Detailed Timing of Gold Introduction Relative to Metamorphism

The late timing of gold relative to major deformation does not immediately relate to timing of gold relative to metamorphism. Many epigenetic gold deposits are approximately synchronous with regional metamorphism in that their mineral assemblages reflect the peak metamorphic conditions. In detail, some formed before the metamorphic peak, some essentially at the peak, and some slightly after the peak during the early stages of retrogression. Those formed during metamorphism but before the temperature peak are discussed in Chapter 16. Three examples from the Eastern Goldfields Province of the Yilgarn Craton illustrate the variability here.

At the Bronzewing gold deposit in the NE Yilgarn Craton the peak metamorphic temperature was 440°C established using assemblages with biotite and ankerite, but this assemblage has overprinted a gold-related alteration assemblage that formed 60 – 120°C lower (Fig. 9.2). The Kalgoorlie goldfield is described in Chapter 13 and is an example where peak greenschist metamorphism and mineralising temperatures are very similar. A contrasting example of Hunt mine in the Kambalda St Ives goldfield includes amphibolite facies metamorphic rocks adjacent to ore zones confirmed by hornblende – plagioclase bearing assemblages. At Hunt, the alteration halo around auriferous quartz veins has assemblages of biotite – chlorite – ankerite – pyrite indicating greenschist facies conditions during mineralisation and hence somewhat lower temperature than the peak of the regional metamorphism.

9.2.5   Weathering

Perhaps appearing a little out of place here, but too often overlooked, the degree of weathering of gold deposits makes a significant difference to what is seen and recorded. It is axiomatic that weathering must be recognised and accounted

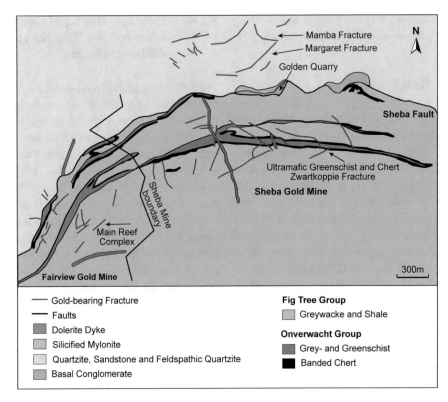

Fig. 9.1 Surface map of the Sheba Fairview goldfield in the Barberton Greenstone Belt of South Africa. Structurally controlled mineralisation, which is related to quartz veins within a kilometre of the Sheba Fault, is hosted by ultramafic rocks of the Onverwacht Group, shale and greywacke of the Fig Tree Group, and sandstone and conglomerate of the Moodies Group (prepared by Caitlin Jones using Wagener and Wiegand, 1986, Dziggel and Kisters, 2019).

Fig. 9.2 Temperature – time diagram illustrating conditions under which the Bronzewing gold deposit formed were of significantly lower temperature and earlier than the metamorphic peak.

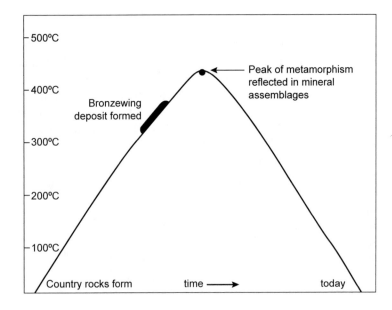

for prior to the determination of the primary geology (Chapter 17). This is commonly achieved by simply sampling fresh material in the field; however, if the weathering is pervasive, it may not be recognised and then all subsequent research and conclusions regarding the primary geology may be compromised.

9.3   Interpretation of Commonality and Diversity

The characteristics that are common to gold deposits are not compatible with an origin of the fluid or gold within metres or even 100 m of deposits. The extent of enrichment, for example, suggests that a scale of several km was more likely. Much of the diversity, however, can be directly explained by local features in and near the site of the deposit meaning on a scale of metres.

The challenge ahead in postulating ways in which gold deposits may have formed is to explain two different fluid types (gold-only, gold-plus Cu-Au), the epigenetic timing, extreme enrichment, segregation from base metals for most but not all deposits, and the tendency to form distinct provinces of 100s km dimensions that contain large, medium, and small sized deposits.

The epigenetic timing, structural settings and mineral assemblages all indicate the importance of the sub-surface environment of elevated temperature and pressure. This is the metamorphic environment and may simply be of solid rock or may include partial melts. One important issue is whether silicate magmas play a critical role in the formation of any gold deposits, and if so, what magmas form which deposit types (see Chapter 10).

Ideally, a genetic model for the formation of gold deposits would be successfully inverse modelled and forward modelled. For inverse modelling, this would mean using the observed natural features and assemblages to calculate gold-only ore-forming conditions such as temperature and fluid composition using thermodynamics. For forward modelling, it would mean using phase equilibria to predict how rocks and fluids evolve, and then compare those predictions with natural observations and refine the model further. The same modelling approach could be applied to gold-plus Cu-Au deposits.

The data that have been assembled in the first nine chapters would be familiar to many gold geologists, but these chapters are far from comprehensive and much existing data have not been discussed yet. Some will be mentioned as it is required, but much is interesting rather than imperative for the purpose so will remain undiscussed and accessible through the literature.

It then becomes an interesting question why gold geologists might have different opinions as to how gold deposits form if each can share so much public-domain information. The answer here is complex but includes how the information is prioritised which itself is influenced by personal backgrounds. Taking the extreme enrichment as an example, most scientists would acknowledge this feature as patently obvious but not all would take the next step and use it to infer that there will be a very small number of fundamentally different ways to form gold deposits. If there are very few ways in which gold deposits form, then it may not make sense to begin by subdividing and classifying them into a dozen traditional deposit types that cannot all have fundamentally different modes of formation. The simple and practical subdivision used here is of gold-only, Cu-Au gold-plus and VMS, but only after that classification has been demonstrated to be practical to apply and could be supported on a sound theoretical basis. Taking the enrichment further, the differences between gold deposits (i.e. the diversity) cannot be reflecting fundamentally different formation processes. Finally, scientists draw different conclusions based upon the less tangible background that is the make-up of each researcher including scientific strengths and weaknesses, access to mine visits and research projects worldwide, strengths and weaknesses of colleagues, groupthink within research groups, and analytical equipment capabilities and access (see Appendix E).

Snapshot

- There is commonality amongst gold deposits with respect to *provinces, enrichment, segregation from base metals, timing, and ore fluids*. These characteristics arise from gold dissolution processes in source rocks that are up to kilometres away from the rocks subsequently hosting economic deposits.
- Additional features show a diversity amongst gold deposits and arise from proximal influences such as local host rocks and structures.
- A sub-surface environment is indicated for the formation of gold deposits involving elevated temperature and pressure with or without the involvement of silicate or sulfide melts.
- Much has been written about the differences between the world's gold deposits, but understanding their formation requires some focus on the common features.

Bibliography

Dziggel A, Kisters AFM (2019) Tectonometamorphic controls on Archaean gold mineralization in the Barberton greenstone belt, South Africa. In: Hoffmann JE, Kranendonk MJ, Bennett VC (eds) Earth's oldest rocks. Elsevier, Amsterdam, pp 655–671

Elmer FL, Powell RW, White RW, Phillips GN (2007) Timing of gold mineralization relative to the peak of metamorphism at Bronzewing, Western Australia. Econ Geol 102:379–392

Goldfarb RJ, Groves DI (2015) Orogenic gold: common or evolving metal sources through time. Lithos 233: 2–26

Phillips GN, Powell R (2015) A practical classification of gold deposits, with a theoretical basis. Ore Geol Rev 65:568–573. https://doi.org/10.1016/j.oregeorev.2014.04.006

Pintos Cerda LM, Jones C, Kisters A (2020) Multi-stage alteration, rheological switches and high-grade gold mineralization at Sheba Mine, Barberton Greenstone Belt, South Africa. Ore Geol Rev 127. https://doi.org/10.1016/j.oregeorev.2020.103852

Vearncombe JR (1998) Shear zones, fault networks, and Archean gold. Geology 26:855–858

Wagener JHF, Wiegand J (1986) The Sheba gold mine, Barberton greenstone belt. In: Anhaeusser CR, Maske S (eds) Mineral deposits of Southern Africa I. Geological Society of South Africa Special Publications, Johannesburg, pp 155–161

Part III

Crustal Processes That Form and Subsequently Modify Gold-Only Deposits

Magmatic Processes that Lead to Gold-Only Deposits

10

Abstract

One of the widely suggested methods to form gold-only deposits involves magma, hence the term *magmatic gold deposits*. Many examples of such deposits have been proposed in Archean, Phanerozoic and Cenozoic terranes, in or adjacent to igneous rocks. The magmatic origin is likely to remain controversial until magmatic processes can simultaneously explain the scale of the provinciality of gold deposits, enrichment, segregation from base metals, timing, and nature of the ore fluids. Spatial association alone is not definitive, and so a more in-depth investigation of magmatic processes is initiated here that incorporates the five characteristics of many gold deposits discussed so far. Just as finding a gold deposit in sedimentary rocks does not mean that the gold originated in the sediment, so, finding a deposit in igneous rocks does not imply an inevitable role for igneous processes.

Keywords

Silicate magmas · Element partitioning · Sulfide melts · Definition broadening

Magma is molten rock material that originates beneath the Earth's surface and may be emplaced and cooled to become an igneous rock. The magma comprises the molten rock, dissolved volatiles (gases) and minerals crystallised from the melt or picked up from elsewhere in the crust. It may also include enclaves of incorporated rocks and aggregates of mineral grains. For some, the term *magma* is reserved for silicate melts, but this discussion will include sulfide melts because they represent valuable endmembers in gold-deposit formation. In much of the literature, mineral deposits formed from sulfide melts would be termed *magmatic deposits*.

Many gold deposits are now found within igneous rocks that were once magmas, *but this does not necessarily mean that magmatic processes played any role in the formation of these deposits*. The deposits hosted in igneous rocks, but without magmatic processes having been involved with the gold, are both numerous and important, and they are used as examples in several chapters of this book. However, this chapter is focused on understanding examples in which the magmatic processes were fundamental in deposit formation; these are the magmatic gold deposits in their literal sense.

For this discussion, the key component of a *magmatic* deposit is that the magma played an essential role in gold-only deposit formation. This presumably would include a form of extreme chemical partitioning of gold. There are five potential phases that might be involved in effective partitioning of gold — silicate melt, sulfide melt, crystallised minerals, enclaves of incorporated solid rock, and dissolved gases. As they are unique to the magmatic environment, it is the silicate and sulfide melts that are investigated further. The observations of

Chapters 4–8 suggest that any magmatic gold deposit requires some extreme partitioning to accomplish the enrichment, selective partitioning to achieve the segregation from base metals (for gold-only deposits) and some quite specific timing, and ore-fluid types.

Silicate magmas can be subdivided using their SiO_2 contents yielding the ultramafic (less than 45%), mafic (45 to 53%), intermediate (53% to 63%) and felsic groups (over 63%). The boundaries between these four groups are gradational and vary slightly in the literature.

10.1   Silicate Magmas and Their Potential to Partition Metals, Including Gold

Partial melting of silicate mineral assemblages becomes more common with the increase in temperature occurring at greater depths in the Earth's crust. It leads to a wide range of felsic igneous rocks, and especially granitic rocks, i.e. granites *sensu lato*. The process of crustal melting is made possible by having some form of H_2O available (as free fluid or in hydrous minerals) because its presence lowers the temperatures of melt formation into the accessible range of crustal conditions. The melting process may be followed by magma segregation, ascent, intrusion, or extrusion, and then crystallisation to form igneous rocks such as granite. The crystallisation of the magma to mainly anhydrous minerals will ultimately generate an aqueous fluid-enriched residue comprised of H_2O and other volatile components. This is one type of hydrothermal fluid, and it has a magmatic origin (Chapter 11).

Unlike the situation for a system involving sulfide melt, there are few experimental data constraining the behaviour of gold in silicate melts, especially granitic melts which are the most abundant type of crustal melt. To understand the possible behaviour of gold in felsic melts it is valuable to understand the critical roles of some important volatile components in melt chemistry, particularly H_2O. Much of this is drawn from discussions with John Clemens and Gary Stevens who are leaders in the field of igneous

experimental work at their laboratory in Stellenbosch, in South Africa (Clemens 2012).

Water is an important component of crustal melting processes, and several percent of H_2O is dissolved in many magma types. For crustal pressures and for the spectrum from mafic to granitic magmas, H_2O is much more soluble in the magma than H_2S, CO_2 or Cl.

Like H_2O, the CO_2 content of melts generally increases with pressure. As an example, at 1kb, CO_2 may dissolve at a level of several ppm, compared to ~5000 ppm CO_2 in magma at 10 kb (Mysen et al. 1976; Tamic et al. 2001). Alkaline magmas and those undersaturated with respect to silica are likely to be the types that have higher CO_2 contents, and mafic magmas generally dissolve more H_2O and CO_2 than felsic magmas. Thus, crustal (granitic) magmas have very low CO_2 contents.

Sulfur can be incorporated into melts potentially as oxidised (sulfate) or reduced (sulfide) forms. For reduced granitic magmas that include S-type and some I-type granites (S- and I- when referring to granitic rocks stand for sedimentary and igneous source rocks), reduced sulfur dominates, and the amount of sulfur is likely to be a few ppm to a few tens of ppm sulfide that will crystallise as early pyrrhotite. Mafic and intermediate magmas may have higher sulfur, particularly alkaline magmas. More oxidised magmas with oxidised sulfur species (e.g. sulfur bonded to oxygen as sulfate) may allow sulfur to build up in the silicate liquid as crystallisation progresses.

Gold contents of magmas are less well known than H_2O or CO_2 and are not usually the topic of igneous petrology experiments. Gold in melt is strongly associated with sulfide either as sulfide melt in sulfur saturated conditions (Section 10.5), or with sulfide minerals such as pyrrhotite. Au-S species are important in the silicate melts, and gold concentrations of 10s of ppm and more may be possible in some moderately oxidising magmatic – hydrothermal systems (Zajacz et al. 2013). However, the fundamental role of partitioning gold into the silicate magma to give the extreme gold enrichment and segregation from base metals is nowhere established, and magmatic explanations normally revert to an essential role for a hydrothermal component and may render magmatic partitioning as non-essential. Notwithstanding, igneous processes that lead to higher background gold should favour formation of deposits in some environments.

10.2   Gold Concentration in Common Magmas and Rocks

The progressive crystallisation of a crustal magma (typically granitic) leads to increasing amounts of solid (rock, minerals) and decreasing silicate melt. There are two ways to concentrate elements in silicate magmas, and both rely upon the valence state and ionic size of a particular element. *Compatible* behaviour occurs when elements concentrate with early-formed minerals because they can substitute due to their good match of charge (usually +2) and ionic size. *Incompatible* behaviour arises from a poor match of ionic size and charge, and concentration occurs during the late stages of magmatic evolution. As examples, Ni and Cr have similar sizes (ionic radii) and charge to Fe^{2+} and Mg^{2+}, and so are concentrated in the early-forming minerals in ultramafic rocks, whereas the incompatible element, K, is large and remains in the silicate melt component of magmas to become elevated in late-stage pegmatites. Elements that behave like K in terms of their incompatibility with early-crystallising minerals include Li, Rb, Ta, U, Th, Sn, W and Nb.

10.2.1   Ultramafic—Mafic—Intermediate—Felsic Rocks and Their Gold

Geochemical data for many rock types are available for all naturally occurring elements, including gold, but there is significant uncertainty for several elements at low natural concentrations. The division of common igneous rocks into ultramafic, mafic, intermediate, and felsic groupings illustrates trends among some compatible and

some incompatible elements when plotting the concentrations of an element in these igneous rock categories.

These log plots of element concentrations in major igneous rock types show three contrasting patterns. The first two are very well known to igneous petrologists and recognisable as compatible and incompatible element behaviours, respectively.

The compatible behaviour of Ni and Cr (Fig. 10.1 top) is conventionally explained by their partitioning into early-crystallised minerals because their ionic charges and sizes allow them to substitute in olivine (Ni) and clinopyroxene (Cr). Consequently, when early-crystallised olivine and clinopyroxene progressively separate from a mafic melt this leads to Ni and Cr depletion as the melt becomes more felsic. Predictably, Ni and Cr deposits are associated with ultramafic and mafic rocks and not felsic ones. Copper approximates the pattern of Ni and Cr (Cu in mafic to ultramafic rocks of 50-100 ppm; Cu in granite around 10 ppm).

The incompatible behaviour of Li, Th and U (Fig. 10.1 centre) is conventionally explained by their poor size and charge matches, such that they do not readily substitute in early-crystallised minerals but partition into the late melts, including pegmatites. Predictably, Li, Th and U deposits are associated with felsic rocks and concentrated in pegmatites; the latter are an important source of lithium.

The behaviour of Au follows neither the compatible nor incompatible patterns (Fig. 10.1 bottom), but instead is remarkably constant for the different igneous rock types on these log plots. Arsenic, Sb, Se and Hg and S follow the same trends as Au, and none of these elements shows compatible or incompatible behaviour within this average rock suite. It is probably significant that, among the many elements that are used by igneous petrologists to determine petrogenesis and tectonic settings, Au, As, Sb, Se, Hg or S are conspicuous by their absence. Their lack of either compatible or incompatible behaviour agrees with the observation that Au, As, Sb, Se and Hg are not generally mined from the earliest-formed

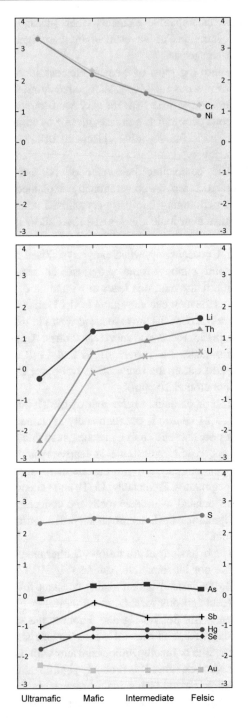

silicate magmas (ultramafic rocks) nor the latest felsic magmatic products, i.e. granite and pegmatite.

The rough groupings of rock type used here conceal some important differences, however. In tholeiitic basalt of ocean ridges, which are reduced and relatively sulfur-saturated, Au behaves as a compatible element during magmatic processes and decreases slightly with differentiation because it partitions into early formed pyrrhotite. In contrast, the calc-alkaline rocks, such as those common in island arcs, are more oxidised and sulfur-undersaturated, and Au shows incompatible behaviour increasing with differentiation as the magma evolves to more felsic compositions. However, the differences in gold contents are less than an order of magnitude, and too small to explain the four-orders-of-magnitude enrichment necessary to form gold deposits (see also Table 5.1). The different gold concentrations could make some rock types more favourable than others as *source* areas for some non-magmatic enrichment processes that lead to gold provinces.

The base metals and silver concentrations can also be shown for these four igneous rock groups, and they illustrate some variation, especially the incompatible behaviour of Pb (Fig. 10.2). The variations between the different igneous rocks are miniscule compared with the enrichment required to form either VMS or gold-only deposits. In addition, the gold-only pattern highlights the segregation of gold from base metals.

10.2.2   Lamprophyres

Elevated concentrations of gold were reported in numerous Archean lamprophyre samples from Western Australia, and, from these data, it was

Fig. 10.1 Average concentrations [in $\log_{10}$ppm] for different elements in common igneous rocks: (**top**) compatible behaviour demonstrated by Cr and Ni with enrichment in ultramafic rocks; (**centre**) incompatible behaviour demonstrated by Li, Th and U with enrichment in felsic rocks; (**bottom**) little variation with igneous rock type for gold and five of the chalcophile (affinity for sulfur)

Fig. 10.1 (continued) elements. Source of data includes the GERM database: https://earthref.org/germrd/, Mason (1966), Taylor and McLennan (1985) and Wedepohl (1995).

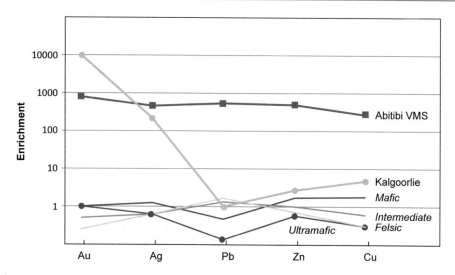

Fig. 10.2 Average enrichments in Au, Ag, Pb, Zn and Cu in common igneous rocks, and in ores from the Abitibi VMS and Kalgoorlie gold-only deposits. For each element, its enrichment is relative to the crustal average. Sources include GERM database.

concluded at the time, that the lamprophyres may have been the source for the gold in many Archean deposits. A careful study that recorded alteration and sought to determine gold contents in lamprophyres prior to any alteration showed no significant enrichment of gold in these rocks at the igneous stage. The study concluded that there was no case for suggesting that lamprophyres are important for the formation of gold deposits.

10.2.3   Overview of the Potential to Partition Gold into Magmas

There is no indication from the above for the existence of the strong partitioning of gold into silicate melts for it to reach sufficiently high concentrations to form economic deposits. There is no evidence, and indeed no theoretical reason, to expect that partitioning into a silicate melt would produce any significant segregation of gold from base metals, as would be required to form gold-only deposits. The hydrothermal fluids evolved as crustal silicate melts crystallise, will be H_2O-dominant with low CO_2. This does not match the composition of the hydrothermal fluids associated with gold-only ore deposits, as

measured in fluid inclusions (as discussed in Chapter 8).

10.3   Igneous Host Rocks for Gold Deposits

A favourite question during *Geology of GOLD* courses has been asked now for a quarter of a century, and it is posed when we reach the topic of magmatic gold deposits. Everyone is asked to individually nominate which igneous rocks are important for forming gold deposits. A significant group nominates ultramafic rocks, especially if they work in Archean terranes, and a similar number of participants favour mafic rocks. Intermediate rocks are the choice among those familiar with younger convergent margin settings, and granitic rocks commonly receive the strongest support overall. This is a fascinating outcome because when they came to the course, many participants were certain that magmatic processes formed gold deposits but there was no consensus as to which igneous rocks were important. The discussion is followed by a reminder of the extreme enrichment for gold, which tells us that an answer of "all of them" is improbable. To

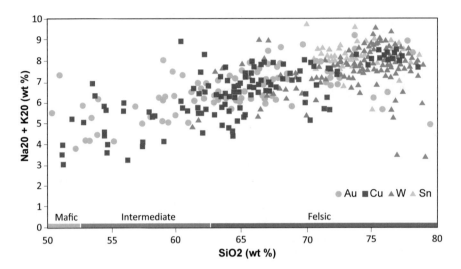

Fig. 10.3. Plot of silica versus total alkalis for various post-Archean granitic rocks (and some more intermediate to mafic types) that contain deposits of Au, Cu, W or Sn. Gold deposits occur in a wide range of igneous compositions, and it is quite unlikely that all these magma types can lead to extreme gold enrichment through partitioning into magmas. Ultramafic rocks (SiO_2 <45%) would continue off the diagram to the left (after Baker et al. 2005).

some extent, the results of this survey are not so surprising considering the data presented so far.

The distribution of Au deposits in different igneous rocks can be combined with similar data for Cu, Sn and W (Fig. 10.3). The same broad classes of igneous rocks used previously will equate to rocks of different SiO_2 contents. Gold deposits are evenly spread throughout mafic to felsic igneous rocks, from 50-80 wt % SiO_2. Copper deposits show a similar spread to gold. Tungsten deposits span the intermediate to felsic rock range. Tin is incompatible in silicate melts and, as expected, tin deposits are restricted to more felsic rocks. Amongst these four elements, a strong case can be made (and commonly is made) that magmatic processes partition Sn into the more differentiated felsic silicate melts, and that this is important in forming tin deposits. The link between Sn and magma type is in strong contrast to the lack of any specific magma type for Au and Cu. These data give no indication that magmatic processes can provide the extreme partitioning necessary to form gold-only deposits.

Most of these gold deposits in igneous rocks have auriferous quartz veins, which indicate that the host rock was solid at the time veins formed

through hydraulic fracturing. The introduction of the gold and silica was therefore after the host igneous rock was solid, and this implies that the auriferous fluids came from outside the immediate host; whether that fluid came from centimetres away or kilometres afield is indeterminate. Therefore, the source of the fluid being from sedimentary or igneous rocks is also indeterminate.

10.3.1   Intrusion-related Gold Deposits

Intrusion-related gold deposits are characterised by auriferous quartz veins in or adjacent to granitic rocks and porphyries, and with elements including Au, Ag, As, W, Te, Bi, Mo and minor typically uneconomic base metals, i.e. mostly gold-only. Many are formed at shallow crustal depths less than 5 km and associated with moderately reduced I-type intrusive rocks. Several examples have endowments of 100 t Au, with examples in NW Canada and Alaska (Fort Knox, Pogo), Kidston in NE Queensland Australia, and Vasilkovskoe in Kazakhstan.

The sheeted quartz veins and breccias in intrusion-related gold deposits evolved after the igneous rocks had crystallised, and so indicate that the gold was sourced from some undefined distance outside the immediate host rock. Fluid inclusions indicate that some of these deposits have CO_2-bearing fluids at depth and more saline fluids higher in the ore systems. Given the low solubility of CO_2 in granitic magmas at these low pressures and shallow depth, the CO_2-bearing fluids are unlikely to have evolved from felsic magmas, suggesting that there may be an unrecognised source of the fluid. The localisation of the auriferous veins in specific igneous rocks is more likely to reflect rheological properties that facilitate hydraulic fracture. The experimental work involving granite magma suggests that shallow, I-type intrusions are not good candidates to generate magmatic gold deposits although they may supply heat.

As a class of gold deposit, the intrusion-related deposits are defined by characteristics shared by many other types of gold deposits and are thus non-diagnostic. For example, the supposedly defining element suite is very widespread in gold-only deposits, and proximity to various granitic rocks is common in many provinces such as the Abitibi of eastern Canada, Zimbabwe, and Barberton. These deposits are difficult to distinguish from orogenic gold deposits, in part because neither type is well defined, and they share overlapping non-diagnostic features. The intrusion-related classification of deposits is thus impractical to use because the uncertainties can lead to unproductive controversy without adding to genetic understanding. The intrusion-related class may be useful for marketing but not helpful for understanding how gold deposits form or what might be causing any enrichment through partitioning of gold.

10.4   Gold Provinces with Abundant Granitic Rocks

Granitic rocks are common in many gold provinces, and studies of the gold deposits that are adjacent to granites inevitably leads to suggestions that the particular granite may be the source of the gold. The focus here is on granitic rocks (granite *sensu lato*) but the discussion about goldfields, nearby intrusive rocks and magmatic gold-forming processes extends to intermediate and mafic intrusions.

10.4.1   Gold and Granites of the Abitibi Province, Canada

The 1993 study of gold in various granite types in the Abitibi Province of Canada is an example of inspired project design by the research team of Saskatchewan, led by Rob Kerrich. Its aim was to resolve a significant controversy about the role of magmatic processes in forming the Abitibi goldfields (Feng et al., 1993). The province is a major source of gold from Archean greenstone belts, in an area intruded by numerous granitic plutons and batholiths (Fig. 4.13) and, at the deposit scale, various granitic rocks are proximal to gold. The study addressed the controversy in the scientific community about the source of gold, and whether magmatic processes were essential in deposit formation.

The project aimed to examine the gold contents of different unaltered granitic rocks to identify any with higher background gold concentrations. Sampling was based around the publicly available maps and published subdivision of the granitic rocks based upon tectonic setting:

- Pre-arc subduction and spreading
- Syn-tectonic and metamorphic collision
- Post-collisional strike-slip faulting
- Late collision with under-thrusting.

Samples were inspected for any later alteration that may have redistributed gold after each igneous stage. Geochemical analyses used state-of-the-art analytical facilities in Saskatchewan (Table 10.1).

Despite the close spatial association between granitic rocks and gold deposits in the Abitibi Province, none of the unaltered granite types had gold levels much above the crustal average of 1-2 ppb, and some were appreciably lower. There is no evidence in this study that magmatic processes had concentrated gold. Instead, the

Table 10.1 Unaltered granitic rocks and their gold, Abitibi Province, Canada.

Granitic type	#Samples	Gold (ppb)	Std Dev.
Tonalite-trondhjemite granodiorite	10	2.1	1.6
Monzodiorite	11	0.74	1.03
Syenite- monzonite—granite	5	<0.1	n/a
Syenite-shoshonite	9	1.73	3.98
Monzodiorite-syenite	10	1.22	1.46
Garnet muscovite granite	7	0.12	0.11

Source: Feng et al. (1993)

elevated gold concentrations recorded in this and other studies in granitic rocks near gold deposits were attributed to secondary gold enrichment rather than being original features due to igneous processes. Resolution of the controversy relied on the broad scale of the study, which incorporated multiple samples that were each examined for signs of alteration.

10.4.2   Gold in Archean Granitic Rocks in Zimbabwe

Many of the 6000 (some estimates suggest 20,000) gold deposits in the Zimbabwe Craton are in greenstone belts but proximal to granitic rocks, and over 600 are within granite bodies that are 100s km^2 in area. Gold deposits are essentially absent from the central parts of the large granite masses, but their number and size increase exponentially within a kilometre of the granite contacts with the rocks of the greenstone belt. Although 80 percent of the exposed Zimbabwe Craton is granitic, this area has produced only 5 percent of the gold, or 100 t Au out of the craton total of almost 2000 t Au.

10.4.3   Gold in Archean Granitic Rocks at Barberton, South Africa

The Barberton area of eastern South Africa comprises extensive greenstone belts and granitic rocks of Archean age. One of the early tests of any geochemical links between gold and granite involved the analysis of numerous granitic rocks in the age range of 3550 to 3000 Ma, for major elements and gold. Gold concentrations were low

and near the crustal background level and ranged from 0.3 to 7.8 ppb with means of 1.2 ppb (arithmetic) and 1.0 ppb (geometric); there was no significant change in gold content with granite age.

The rocks were classified by their K/Na ratio, using this as a proxy for granitic type (i.e. tonalite to granite *sensu stricto*). When gold was plotted on the K/Na diagram there was no correlation between granitic type and gold concentration (Fig. 10.4), and thus no correlation between Au concentration and the igneous mechanism that caused the variation in the granitic rocks.

A separate outcome of this study was that some of the granitic rocks contain minor amounts of secondary pyrite, reflecting alteration. There is a clear correlation between those granitic rocks with pyrite and higher gold concentrations. The conclusion is that magmatic processes that dictated the variations in the granitic suite had not influenced gold concentration and gold concentrations are uniformly low in granitic rocks of this province except where there has been later alteration. The study also highlighted the difficulty some have had when sampling granitic rocks near gold deposits, in determining whether results reflect magmatic concentrations or are due to alteration, or both.

10.4.4   Gold in Felsic to Intermediate Intrusions in the Northeast Queensland Charters Towers Province

The Northeast Queensland Charters Towers Gold Province in Australia contains several large goldfields in Paleozoic igneous rocks, and others

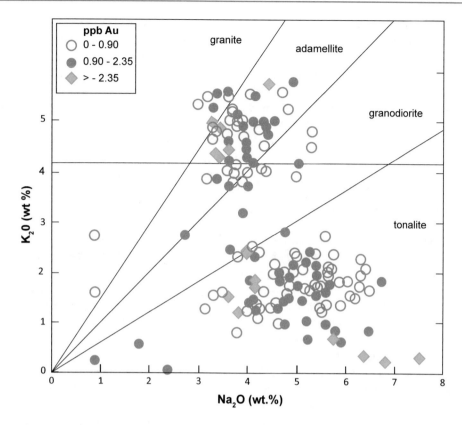

Fig. 10.4 Gold concentrations in granitic rocks of the Barberton Greenstone Belt margins. All gold concentrations are near crustal background, except where there is pyrite present, indicating alteration. There is no correlation between gold concentration and granitic type. Monzogranite is the term now used for adamellite. Data from the study of Saager and Meyer 1984, reproduced with permission.

in metasedimentary sequences (Fig. 4.11). These have been previously classified as epithermal, breccia-pipe-related, and granite-hosted, and the geochemistry of ores and altered rocks includes some elements and isotopes that match the magmatic rocks. Although considerable effort has been devoted to subdividing these deposits, based on textures and other high-level features, no attempt has been made to identify an underlying geological cause for the entire gold province. The auriferous quartz veins indicate gold introduction after the igneous host rocks were solid, and therefore the fluid source was external to the immediate host rock. Much of the geochemistry reflects elements in the igneous rock and then incorporated into the alteration zone, i.e. provided by the local host rock and not reflecting the auriferous fluid.

Previously, a magmatic origin has been inferred for these deposits based on igneous rocks being important host, and because there are some aspects of the geochemistry of the ores and alteration zones matching the igneous rocks. As discussed, neither line of evidence is diagnostic.

The preferred explanation here is that the gold deposits in igneous rocks of the Northeast Queensland Province are localised in their host rocks because of rheology (fracturing, tensile strength, rock contrasts) and host rock geochemistry, and noting also that the source of both gold and fluid was not within the host rocks. There is no evidence that magmatic processes were essential in concentrating the gold, segregating the base metals, and forming these deposits. Subdividing each deposit based on its high-level features (e.g. breccia-hosted, epithermal, sheeted veins) is useful but can also obscure issues such as why there is a gold province at all and how each deposit has formed. Geochemical including isotopic signatures of the ores and alteration using immobile elements are expected to reflect the magmatic host rocks and are not the appropriate data to constrain either gold, silica or fluid.

10.4.5 Victoria Granite Types and Mineralogical Domains

The Victorian Gold Province (see Section 4.3.8) has produced 2600 t Au mostly from Paleozoic metasedimentary rocks. The province comprises 20 percent Paleozoic granitic rocks that have a total gold production of 10 t. With 7000 mines and workings in the province, it is almost unavoidable that many of those gold deposits will be near and even adjacent to a granite.

Detailed syntheses of the various Victorian granite types and, separately, of the mineralogical domains depicting hypogene minerals and the geochemical characteristics of the sulfide assemblages in gold deposits, mean that any granitic types with a role in the formation of gold deposits should emerge. Also, any role of the igneous rocks in the various suites of ore minerals can be tested.

The granitic rocks across the province have been subdivided according to small age differences (e.g. 395 Ma; 365 Ma), peraluminosity (S-type granites: $Al_2O_3 > CaO + Na_2O + K_2O$) and the presence of any Proterozoic crust at shallow depth.

The scale of the mineralogical domains is an order of magnitude larger than the granitic bodies especially considering that the batholiths are comprised of multiple, smaller plutons, rather than having been a single large uniform magma batch. The oldest Paleozoic granitic rocks are peripheral to, and east and west of, the Victorian Gold Province, and the dividing line between 395 Ma (syn-tectonic, Late Silurian – Early Devonian) and 365 Ma (post-tectonic, Middle Devonian – Early Carboniferous) granitic rocks trends NE-SW, at a high angle to the north – south stratigraphy, major fault zones and the mineralogical domains. The line demarcating areas of I-type granite and the more aluminous S-type granite is also at a high angle to mineralogical domain boundaries (Fig. 10.5). Similarly, the outline of areas overlying deeper Proterozoic crust does not relate to the mineralogical domains but corresponds broadly to the younger granitic rocks of both I- and S-type character. Thus, there is good indication from the compiled map that variations in granite types do not relate to variations in the gold deposits across the province.

10.4.6 Prospectivity of Granites in Some Gold Provinces

In many provinces with numerous batholiths, the granitic rocks are under-represented with respect to gold production (Table. 10.2). For the Yilgarn and Zimbabwe Cratons, most of the gold has been produced from the greenstone belts, and the prospectivity favours these belts compared to the granites. For the Victorian Gold Province, a miniscule total of 10 t Au has come from granitic rocks. This means there has been on average 50 times as much gold produced from one km^2 of non-granitic rocks compared to the same area of granite.

These figures clearly do not convey the full global picture because there are important gold deposits that are predominantly in granitic rocks. Charters Towers goldfield, in NE Queensland, Australia, for example, has produced 250 t Au, and a study of gold prospectivity in granites within a small area around Charters Towers would give a different picture compared to Table 10.2. Part of the differences will relate to the rocks that surround the granitic rocks be they metasedimentary (Victoria), greenstone belts (Zimbabwe and Yilgarn) or components of an igneous province (Charters Towers).

10.5 Sulfide Melt and Magmatic Gold Deposits

Sulfide melts arise when high-temperature, reduced, mantle-derived mafic to ultramafic magmas exceed their sulfur saturation levels and

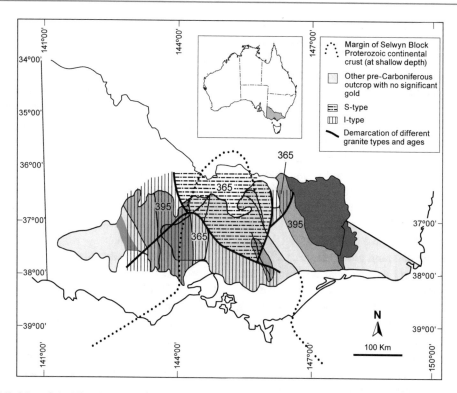

Fig. 10.5 Map of the Victorian Gold Province showing mineralogical domains (in various colours) based on the suites of hypogene minerals and the geochemical characteristics of the sulfide assemblages in gold deposits (see Fig. 4.10 for more detail). Also shown are the various areas of dominantly I-type, and of S-type, granitic rocks, ~395 Ma and ~365 Ma granitic rocks, and an area underlain by shallow Proterozoic crust. There appears to be negligible correlation between igneous rock types and features in gold deposits.

Table 10.2 Relative prospectivity of granitic rocks for gold.

	Area		Gold production		
	Granite (%)	Rest (%)	Granite (%)	Rest (%)	Relative prospectivity
Yilgarn	70	30	3	97	1:80
Zimbabwe	80	20	10	90	1:40
Victoria	20	80	0.50	100	1:50
Yilgarn example - Rest:Granite = 1:1/(30/70 × 3/97) = 1:80					

generate separate sulfide melts. In this situation, Au, Cu, Ni, platinum group elements, and some metalloids or semi-metals such as As, Sb and Te partition strongly into the sulfide liquid and thus can be removed from the silicate melt as sulfide droplets. A measure of this partitioning can be described by the sulfide-silicate melt partition or distribution coefficient (D), with values over 10,000 for the ratio (gold in sulfide melt) / (gold in silicate magma). This high partition coefficient

easily meets the requirement of a mechanism to achieve extreme gold enrichment. Some other elements also have a high D value, including Ni (250), Cu (1000) and Pd (35000).

Sulfide saturation to form sulfide melts in mafic to ultramafic magmas can occur in different ways. The assimilation of sulfur-bearing country rocks into the melts can lead to saturation; and changes within the melt, such as redox conditions and degree of fractionation can achieve the same end. When a sulfide melt forms, the distribution of gold in phases of the magma depends upon the high D values, the proportion of sulfide melt in the total magma (R factor), and the effectiveness of mixing and interaction.

Examples of mafic to ultramafic intrusions in which sulfide melts have been accompanied by metal enrichment, include the Bushveld Igneous Complex of South Africa, Great Dyke of Zimbabwe, Stillwater Complex in Montana USA, Norilsk intrusion in Russia and Skaergaard intrusion in Greenland. As an example of possible sizes and grades, the Skaergaard has a Resource of 5 Moz (150 t) Au at a grade of 0.9 g/t Au with sub-economic platinum group elements.

Magmatic gold deposits can be formed from sulfide melts as these melts are extremely effective agents in gold partitioning. They are associated with specific geological settings and bear close spatial and geologic associations with the mafic to ultramafic intrusions, but negligible correlations with any of the major gold provinces. Although they can yield significant tonnages and grades of ore, these deposits are restricted in their distribution in the crust, and the gold is typically a by-product of Ni, Cu, and platinum group element production. The element association of sulfide melts is very distinctive and comprises an array of metals unlike all other gold deposit types including the gold-only group.

10.6 Broadening the Definition of What Is a Magmatic Gold Deposit

Many different criteria have been used to label a gold deposit as magmatic or perhaps granite-related, and this leads to much of the controversy over what is, and what is not, a magmatic gold deposit. The approach here is that a magmatic gold deposit is one in which the magma plays a fundamental role in formation of the deposit, i.e. essentially that gold has partitioned strongly into the magma. This strict definition is far from universally accepted; a different endmember has been defined by some in which magmatic gold deposits are in or adjacent to igneous rocks

particularly granitic rocks. However, with respect to how gold deposits form, such an all-encompassing classification lacks a theoretical basis.

Defining *magmatic gold deposits* to include many examples in which magmatic processes are not fundamental, while being a personal choice, is not particularly useful in determining their origin. This is an example where the determination of origin is of far greater utility than simple classification; this distinction impacts on exploration and mining and whether proximity to granitic rocks, and which types, might form the focus of exploration.

As a caveat, determining the origin of gold deposits is only one reason for pursuing gold geology. For other purposes such as communicating with metallurgists, it might be important to convey that a batch of ore is arriving at the mill and is mixed with granitic waste rock. However, rather than describing this as magmatic gold, it would be more effective to simply call it gold ore within granite.

Granite provides a host rock once it has solidified, been fractured, and infiltrated by an auriferous fluid. For this process, *magmatic* is not the best term to use because the magma is irrelevant to the formation of the deposit (having solidified somewhat prior). The parallel terminology of describing a vein deposit as a *sedimentary* gold deposit because it occurs within a BIF is equally sub-optimal.

It may appear as an exercise in semantics deciding whether the role of a silicate magma in partitioning gold is fundamental or whether it is the commonly associated hydrothermal fluid that is fundamental in partitioning Au. However, the distinction becomes important if it is the hydrothermal fluid that is fundamental, in which case the role of silicate magma varies amongst deposits and provinces from enhancing gold background levels in some to a negligible role in others. If the role of the silicate magma is peripheral, then exploration might focus less on igneous rock type and more on aqueous fluids regardless of where they originate.

To test this hypothesis that silicate magmas are not critical in the formation of gold-only deposits, a list could be assembled of all gold-only deposits that formed from special silicate magmas without involvement of a hydrothermal fluid. A separate list could be compiled of all gold-only deposits that have formed from a hydrothermal fluid without involvement of that special silicate magma type.

10.7   Which Are the Magmatic Gold Deposits?

There is a good case that magmatic processes involving sulfide melts are essential in the formation of **some** deposits that contain gold. Experiments and chemical fractionation theory allow forward modelling from the sulfide melt stage, with a good match to natural occurrences, including the suites of other enriched metals. However, the product of this process is not a gold-only deposit.

For gold-only deposits, at least, the case for silicate magmas playing any direct or important part in their formation is poor. Forward modelling using chemical fractionation theory and experiments does not favour an important role for silicate magmas, and the natural observations show that there were no special igneous rocks or igneous processes that could account for the enrichment, the segregation from base metals or for the compositions of ore fluids. Anywhere that gold systems invade granitic rocks, or granitic rocks intrude pre-existing gold deposits, it is highly likely that some geochemical signatures of the granitic rocks may be present in some of the gold ores; however, this does not mean that the source of the gold was the granitic rock.

Magmatic gold has also been used to implicate magmatic volatiles (i.e. aqueous fluids) in the formation of deposits. As their chemistry demonstrates that these fluids arose from outside the immediate host rocks, some may attribute them to more distant igneous rocks. However, given the unlikelihood of the gold-only fluids being derived through magmatic processes, the source appears more likely to lie in non-igneous rocks.

To the inevitable question "There must be some gold-only deposits that form from magmatic processes" a simple response could be "Why?". Perhaps the essential steps in becoming gold-only do not include or necessitate magmatic processes; magmas may just be incidental and contemporaneous. To which the comeback might be "That all seems logical but innately I believe that there must be magmatic gold deposits".

Over a century ago Walter Lindgren and contemporaries pioneered much of the thinking around epithermal and mesothermal (i.e. medium temperature) gold deposits originating from igneous rocks. Since then, there have been many advances that allow a higher standard of proof for ore genetic models including those postulating gold from magmas. These advances include melting and crystallisation experiments with granitic compositions, variations in granite types, reliable low-level analysis of gold, and the emergence of further fluid types into the economic geology lexicon (e.g. metamorphic fluids). Viable genetic models that postulate a link to granitic rocks need to be plausible at different scales across many examples, and the magma-formation and crystallisation processes should be forward modelled to show if, and to what extent, gold might be concentrated. Such modelling would be preferable to a large body of undiagnostic or circumstantial evidence.

Snapshot

- Gold is neither strongly compatible nor strongly incompatible in silicate melts.
- A process to generate magmatic gold-only deposits should replicate the five main characteristics common to all gold-only deposits.
- The scale of gold provinces is much greater than the scale of plutons and even batholiths.
- Magmatic processes do not appear capable of strong partitioning of gold into silicate melts.

(continued)

- Any segregation of gold from base metals by silicate magmas is minor.
- Auriferous quartz veins in igneous host rocks have, by necessity, formed after the hosts are solid, and this implies that the host is neither the source of the necessary fluid nor gold.
- Crustal silicate magmas appear unlikely to generate significant volumes of the low salinity, H_2O-CO_2 fluids that are found in the quartz of gold-only deposits.
- It is the rheology of igneous host rocks that determines which rocks hydraulically fracture and become likely hosts of gold deposits.
- It has not been possible to identify any essential role for silicate magmas in the formation of gold-only deposits.
- Magmatic processes partition gold strongly into sulfide melts, leading to deposits with by-product gold, but with a suite of associated elements that does not match gold-only deposits.
- Gold enrichment of an igneous rock in the vicinity of a deposit is more likely to represent distal alteration from an external source rather than an internal source within that igneous rock.
- Igneous rock textures, such as porphyritic, even grained and seriate (variable grain size), depend on the local cooling histories of the magmas which take place on a scale far too small to be fundamental to the formation of major gold deposits. These textures may be indirectly useful if they relate to local rheology contrasts and fracturing.

Bibliography

Baker T, Pollard PJ, Mustard R, Mark G, Graham JL (2005) A comparison of granite-related tin, tungsten, and gold-bismuth deposits implications for exploration. Soc Econ Geol Newslett 61:5–17

Clemens JD (2012) Granitic magmatism, from source to emplacement: a personal view. Appl Earth Sci 212: 107–136. https://doi.org/10.1179/1743275813Y. 0000000023

Feng R, Fan J, Kerrich R (1993) Noble metal abundances and characteristics of six granitic magma series, Archean Abitibi Belt, Pontiac Subprovince Relationships to metallogeny and overprinting of mesothermal gold deposits. Econ Geol 88: 1376–1401

Holwell DA, Keays RR (2014) The formation of low-volume, high-tenor magmatic PGE-Au sulfide mineralization in closed systems: evidence from precious and base metal geochemistry of the Platinova Reef, Skaergaard Intrusion, East Greenland. Econ Geol 109:387–406

Holwell DA, McDonald I (2010) A review of the behaviour of platinum group elements within natural magmatic sulfide ore systems. Platin Met Rev 54:26–36. https://doi.org/10.1595/147106709X480913

Keays RR, Campbell IH (1981) Precious metals in the Jimberlana Intrusion, Western Australia: implications for the genesis of platiniferous ores in layered intrusions. Econ Geol 76:1118–1141

Lindgren W (1933) Mineral deposits. McGraw-Hill, New York, p 930

Mann AG (1984) Gold mines in Archaean granitic rocks in Zimbabwe. In: Foster RP (ed) Gold '82: the geology, geochemistry and genesis of gold deposits. Balkema, Rotterdam, pp 553–568

Mason B (1966) Principles of geochemistry, 3rd edn. Wiley, New York, p 329

Mysen BO, Eggler DH, Seitz MG, Holloway JR (1976) Carbon dioxide in silicate melts and crystals. Part I. Solubility measurements. Am J Sci 276:455–479. https://doi.org/10.2475/ajs.276.4.455

Noble JA (1950) Ore mineralization in the Homestake gold mine, Lead, South Dakota. Geol Soc Am Bull. 61:221–252

Patten CGC, Pitcairn IK, Molnár F, Kolb J, Beaudoin G, Guilmette C, Peillod A (2020) Gold mobilization during metamorphic devolatilization of Archean and Paleoproterozoic metavolcanic rocks. Geology. 48: 1110–1114. https://doi.org/10.1130/G47658.1

Saager R, Meyer M (1984) Gold distribution in Archean granitoids and supracrustal rocks from southern Africa, a comparison. In: Foster RP (ed) Gold '82: the geology, geochemistry and genesis of gold deposits. Balkema, Rotterdam, pp 53–70

Tamic N, Behrens H, Holtz F (2001) The solubility of H_2O and CO_2 in rhyolitic melts in equilibrium with a mixed CO_2–H_2O fluid phase. Chem Geol. 174:333–347. https://doi.org/10.1016/S0009-2541(00)00324-7

Taylor SR, McLennan S (1985) The Continental Crust: its composition and evolution. Blackwell Scientific Publishers, Oxford

Wall VJ, Massey S, Taylor JR (2017) Thermal aureole (pluton-related) gold in Central Victoria. Aust Inst Geosci Bull 65:134–140. In (ed. JR Vearncombe) Granites2017@Benalla Symposium abstract

Wedepohl KH (1995) The composition of the continental crust. Geochim Cosmochim Acta 59:1217–1232

Wyman D, Kerrich R (1988) Lamprophyres: a source of gold. Nat News Views 332:209

Zajacz Z, Candela PA, Piccoli PM, Sanchez-Valle C, Wälle M (2013) Solubility and partitioning behavior of Au, Cu, Ag and reduced S in magmas. Geochim Cosmochim Acta 112:288–304

Zhou T, Phillips GN, Denn S, Burke S (2003) Woodcutters goldfield: gold in an Archaean granite, Kalgoorlie, Western Australia. Aust J Earth Sci 50: 553–569

Abstract

Water has some extraordinary properties, and hydrothermal fluids (literally hot waters) play a major role in many processes in the Earth's crust. At moderate to elevated temperatures, corresponding to the metamorphic and magmatic environment, fluids have considerable capacity to dissolve and transport many elements. This environment is where magmatic fluids are active and where metamorphic fluids are generated during prograde metamorphism.

Metamorphic fluids are difficult to conceptualise because their source cannot be seen, and their movement cannot be directly measured. Such fluids form as metamorphism intensifies and low temperature hydrous minerals break down and are replaced by anhydrous minerals stable at higher temperatures; the process releases aqueous fluids. This indirect evidence that vast amounts of metamorphic fluids have been generated and then migrated through the crust raises the question of how they permeate otherwise solid rock.

Metamorphic fluids are explained by way of three concepts: evidence of their migration, how they are generated through devolatilisation, and how they migrate through solid rock.

Keywords

Crustal waters · Quantifying formation and migration of metamorphic fluids · Essential concepts

The properties of water are well known in undergraduate chemistry. H_2O consists of two H atoms bonded to a central O atom with a further two lone pairs of electrons linked to that oxygen. The four pairs are arranged around the O in a quasi-tetrahedral configuration giving an asymmetric charge distribution such that the end of the molecule with two H atoms is slightly positive, and the other end with the two lone pairs of electrons is slightly negative. This polarity contributes to many of the properties of water such as it being a liquid at surface conditions because of the strong electrostatic attraction between its molecules. The polarity also means that water can dissolve many metal cations. Another property of water influencing the solubility of metals is its dielectric constant. This is the measure of the number of dipoles per unit volume and highlights the considerable ability of water to separate metal ions. The dielectric constant of water informs on its ability to dissolve many elements as ions, and for water the dielectric constant of 80 is high compared to values of $1 - 15$ for other liquids.

Water in its various forms is remarkably widespread in the Earth and of great interest in economic geology because of its role in forming ore deposits. In geological processes, meteoric waters describe the near surface water derived from rain including rivers, glaciers, lakes, and groundwater. These meteoric waters are continually contributing salts and water to the oceans. Deeper in the Earth's crust two changes take place to any water. Temperature increases with depth, and the water dissolves many elements, so it is no longer pure water. These are now referred to as hydrothermal fluids, or the term aqueous fluid may be used to stress the water-based nature of the fluid. At temperatures above 300°C there is much less distinction between the physical properties of liquid water and gaseous steam, and the term fluid covers both physical states. Upon tracking a water – steam mix to higher temperatures, the critical point of water is reached at 374°C and 218 atmospheres (a pressure equivalent to a 0.5 km rock pile) beyond which there is only one phase of H_2O, i.e. a supercritical aqueous fluid. Aqueous fluids can also contain appreciable amounts of other volatile components particularly CO_2 and S.

Many gold deposits owe their origin directly to hydrothermal fluids. There are several different origins for hydrothermal fluids in the Earth's crust, but only some of these have significant capacity to dissolve and transport gold in solution and then form gold deposits. Magmatic fluids have been discussed in the previous chapter, and metamorphic fluids are discussed in some detail here.

11.1 Types of Geologic Fluids

There are many fluid types in the Earth's crust with differing origins, chemical compositions, and physical conditions but a few endmembers cover most situations (Table 11.1; Fig. 11.1). Mixing of more than one type is possible.

11.1.1 Focus on Aqueous Fluids

For the fluids that form many ore deposits it is possible to demonstrate that they were aqueous

from the stability of hydrous minerals like micas, aqueous fluid inclusions, phase relationships between co-existing minerals, and the efficacy of partial melting to form silicate magmas whose internal molecular structure and physical properties are heavily influenced by dissolved water. To be viable for gold deposit formation a hypothesised non-aqueous fluid would need to be widespread globally and distributed throughout time, be present during periods of metamorphism and deformation, share the special chemical properties that make water so important in dissolving metals, and provide a means of segregating gold from base metals. No non-aqueous fluid has been reported that meets these criteria so non-aqueous fluids are not discussed any further with respect to formation of gold deposits.

11.1.2 Meteoric Water

Meteoric waters are essentially rainwater, and as such, they approximate pure water without salinity or other dissolved elements. They are generally oxidising which reflects the influence of the atmosphere. Over time, meteoric waters may evolve through interaction with soil and rock to dissolve and move various metals including salts, uranium, gold, and many more.

11.1.3 Seawater

Seawater is derived from minute amounts of salts brought by rivers to the ocean. Near-surface seawater is oxidised, but deep ocean water can be quite reducing. The Black Sea, for example, has a toxic low-oxygen bottom layer rich in decomposing algae and reduced organic matter.

Table 11.1 Fluids types in the Earth's crust.

	Temp (°C)	Salinity
Meteoric	Below 100	Low
Diagenetic	Up to 200	High for marine sequences
Seawater	Below 100	Moderate, 3% NaCl by mass
Magmatic	400–600	Low to high
Metamorphic (typical)	200–500	Low
Metamorphic (evaporites)	400–600	Very high

Fig. 11.1 Schematic cross-section of the upper crust showing various sources of fluids. Conceptually most types are quite distinct but there will be parts of the crust where more than one of these fluid types is present. On the left, an example of metamorphic fluid generation and migration is illustrated using the greenschist to amphibolite facies transition and a steep dipping fault zone.

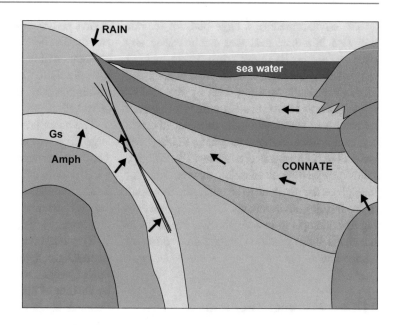

When marine sediments are buried, they may trap several percent of the seawater between mineral grains.

11.1.4   Diagenetic Water

Diagenesis covers the stage in which many rocks have space between grains or porosity which allows water, other liquids and gases to move in response to gravity and hydraulic head. Significant porosity is present in many, but not all, sedimentary sequences at 100°C but is limited above 200°C.

Diagenetic waters are the trapped liquid between clastic grains when a sediment is buried, and initially may comprise 20 percent of the sediment. This water can be fresh or saline reflecting a non-marine or marine setting, respectively. Upon burial, reaction between the interstitial waters and surrounding mineral grains commences and this reaction alters both the minerals and the water composition. One common reaction is for the water to react with grains of feldspar and rock fragments in the sediment to form clays and a more saline water.

11.1.5   Magmatic Fluid

Magmatic fluids are those evolved when a silicate magma crystallises. Rocks can melt in the crust at temperatures above 650°C and because H_2O is essential in the melting reaction, the melts so produced may contain 1 – 5 weight percent water and other volatiles. Later as the magma crystallises to form anhydrous minerals like feldspars and quartz the amount of water in the melt builds up, eventually becomes saturated, and then is released as an aqueous fluid phase with or without CO_2 and S components.

Much of our understanding of the fluids directly involved in magmatic processes comes from experimental work, some sampling such as fluid and melt inclusions trapped in crystals as the magma crystallises, sampling of volcanic gases, and from observation of the alteration effects when the fluids react with rocks.

A large igneous intrusion may release considerable amounts of magmatic fluid. A 100 km^3 melt with 5 weight percent water could release 5 km^3 of fluid although many granitic bodies comprise multiple much smaller sheets that are introduced as separate intrusive events (Clemens 2012; Tikoff et al. 2013).

The emplacement of a large body of magma will also introduce heat that can mobilise any other waters in the upper crust. These may be groundwater, seawater, basinal brine or river and lake water. These waters driven by the heat of magmas are just that (i.e. other water types driven by heat), and technically quite different in origin and composition from the magmatic fluids which are evolved as a magma crystallises. The different fluids may be difficult to isolate or even differentiate given the possibility for mixing and overlap in ore-forming environments. Definition broadening that attempts to include all these scenarios as magmatic fluids confuses several quite different endmembers.

11.1.6 Metamorphic Fluid

Metamorphic fluids and diagenetic waters, as conceptual endmembers, are generally different with respect to porosity of the rocks they inhabit since sedimentary rocks have higher initial intergranular porosity than metamorphic rocks. Definition broadening to overlap or merge the two has largely been avoided, in part aided by the hydrocarbon industry being so focused historically on the diagenetic stage.

Metamorphic fluids are a major source of water in the crust and their vast volumes and ore-forming potential has only been appreciated in the last half century. These fluids are derived from hydrous minerals as they are heated during metamorphism to then break down to form anhydrous minerals and release aqueous fluid. The hydrous minerals include micas, chlorite and clay minerals that are either clastic particles in sedimentary rocks or the products of early diagenetic reactions. Importantly, the source of the metamorphic water is the hydroxl (OH) component of minerals such as muscovite which is $KAl_3Si_3O_{10}(OH)_2$. The bound OH in minerals is not affected by rocks being consolidated, losing their porosity and evolving diagenetic waters; in metamorphism the OH is only released when the hydrous mineral, here muscovite, breaks down at elevated temperatures. This breakdown during metamorphism almost always involves further minerals, and any carbonate and sulfide minerals participating in the reactions will contribute CO_2 and S to the metamorphic fluid. The process is then referred to as devolatilisation rather than the more specific dehydration.

The potential volume of metamorphic fluids is large for even a modest sequence based on having 1 to 5 percent volatiles over 10,000s km^2 area and 1 km or more thickness. This volume of fluid is at least an order of magnitude greater than that derived from most plutons during crustal processes.

Two characteristics of the metamorphic environment greatly affect fluids and their migration. Porosity refers to open spaces in these rocks and by the time of lithification, porosity is nearly zero. Permeability is the ability to allow fluid to flow through the rocks and this is also diminishingly low for most of the time during metamorphism. Transient changes to the static conditions of low permeability and porosity are important.

Major advances have helped to understand three aspects of metamorphic fluids and how they apply to gold geology:

- Evidence that metamorphic fluids have moved in the crust—sometimes for several kilometres.
- How dehydration (or more generally devolatilisation) forms metamorphic fluids
- Mechanisms for fluid migration in the metamorphic environment.

The integration of each of these concepts into gold geology has been relatively recent; they were rarely discussed in the 1970s and rarely taught in the 1980s so that understanding of them still has an institutional and generational correlation.

11.1.7 How the Metamorphic Environment Is Different from Diagenesis

Diagenesis describes processes at low temperatures after burial of sedimentary sequences.

Maximum temperatures are a little over 200°C, but temperature alone is not ideal for separating diagenesis from the metamorphic environment. For understanding fluids in the crust, it is rock permeability and porosity that distinguishes the behaviour of diagenesis and metamorphism, and hence a major contrast exists between the environments of oil and gas formation versus gold deposit formation.

Upon burial of a clastic sedimentary sequence there is likely to be space (porosity) between clasts, and this space may be filled with marine or non-marine waters (Fig. 11.2). There may be connectivity so that the waters can migrate or flow through the rock mass (permeability). The environment of oil and gas accumulation is one example of diagenesis, and here faults may be fluid barriers and flow can be through major homogeneous permeable units such as sandstone aquifer beds. Giant oilfields require thick and rich hydrocarbon source beds, thick permeable aquifers, and large-volume traps. The formation of goldfields can be described in terms of source, flow pathways and a sink, but each of these is fundamentally different from the oil and gas description. Part of the difference arises because the metamorphic environment is generally one of negligible permanent porosity or permeability.

The devolatilisation model described here is not the same as a multi-stage leaching process of water being generated and then pervading a sequence to 'collect' gold (Fig. 11.3). The multi-stage leaching process requires a fluid pervading a complete rock mass to dissolve gold on a grain-by-grain scale and there is no evidence of fluids doing this at the metamorphic stage; the process may be more applicable to diagenesis during which the permeability allows this complete infiltration of the rock. The devolatilisation model does not require the fluid to migrate to dissolve gold; instead, *the gold is in the metamorphic fluid from the moment it forms* just like the CO_2 and S. The analogy with partial melting to form granite is instructive; a granitic melt is formed with Si, Al, K, Na, Rb, Th and many other elements from the outset—it does not have to migrate and then collect its various components from somewhere else.

11.2   Aqueous Fluids in the Metamorphic Environment

The discussion here is restricted to the burial stage beyond diagenesis where igneous and sedimentary sequences are lithified, and porosity is essentially zero. The more generic term *fluid* is used to acknowledge that supercritical conditions for water are relevant to the metamorphic environment. In this subsurface environment of negligible (static) porosity and elevated P and T, geological processes follow the principles of metamorphic geology. This is referred to as the

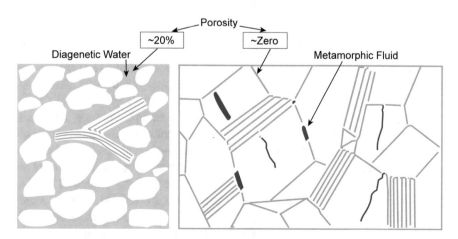

Fig. 11.2 Contrast of: (**left**) diagenetic stage in which water is trapped in pore spaces; (**right**) metamorphic stage in which there is negligible porosity, but the fluids utilise transient microcracks.

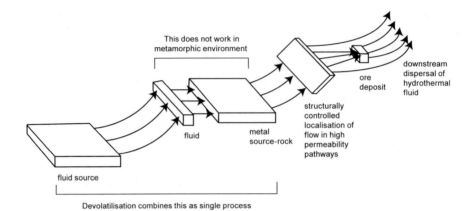

Fig. 11.3 This full multi-stage process of fluid formation, migration and then metal leaching may be relevant to the diagenetic environment for base metals but not for metamorphic rocks with negligible permeability. Metamorphic devolatilisation produces a fluid that is auriferous from its conception, and as such devolatilisation compresses the first three of these stages into one process.

metamorphic environment, but the terminology makes no assumptions as to whether the immediate rocks were originally sedimentary or igneous, and solid or molten. It is not appropriate to consider this as simply an igneous environment because an igneous rock is nearby, or to treat it as a sedimentary environment because there is a sedimentary rock nearby.

The behaviour of aqueous fluid in the metamorphic environment embraces some of the more difficult geological concepts to understand as all rocks appear to be solid and impermeable, and lacking in porosity or open cracks. Metamorphic fluids are discussed here in three parts. Concept A provides the evidence that metamorphic fluids have migrated; Concept B outlines the process of devolatilisation that generates metamorphic fluids; and Concept C discusses how these fluids migrate through solid rock masses.

11.2.1 Concept A: Evidence that Metamorphic Fluids Move in the Crust

The notion that fluids must have moved through rocks of negligible porosity and permeability is readily demonstrated in contact metamorphic aureoles in which there has been significant loss of aqueous fluid. In a clastic sequence including shale (rich in clay minerals equivalent to a pelitic bulk composition), it is common with increasing metamorphic grade to record a progression from shale to spotted slate to biotite zone to a cordierite zone and potentially to a cordierite – K-feldspar zone. These rocks are at metamorphic grades above the interval of diagenesis, and all have negligible porosity. In mineral terms, this is a progression from rocks containing abundant minerals with bound water molecules to rocks comprising dominantly anhydrous minerals. As this process progresses, the aqueous fluid needs to escape.

In a contact aureole of Paleozoic clastic metasedimentary rocks, cordierite hornfels is formed adjacent to the Strathbogie granite of Central Victoria (Fig. 11.4). Additional hornfels surrounds this granite as an aureole of 2 – 3 km width. As all the rocks have negligible porosity, the only water in the rocks is that which is bound as (OH) molecules in minerals. Estimates can be made of the bound water in pelitic rocks of the various zones by combining the modal mineral abundances and the theoretical H_2O content of each mineral (Table 11.2). This bound water is around 5% in shale-rich rocks outside the contact aureole and in the spotted zone, and 1% and below in the biotite and cordierite zones.

Fig. 11.4 Cordierite-bearing hornfels from within 100 m laterally of the Strathbogie granite. Despite being heated beyond 500°C by the granite and having lost over 1 % H_2O from hydrous minerals, the sedimentary texture is still preserved. The upper third of the photo is sand and clay laminae, there is a mass of sand above a scour in the centre, and then the lower third is of cross bedded sands. Photo by Olivia Page with permission.

Table 11.2 Volume of fluids in a contact aureole.

Approximate volatile content of various parts of a contact metamorphic aureole		
Minerals	H_2O content in mineral: %	
K-feldspar	0	
Cordierite	0	
Quartz	0	
Biotite	2	
Muscovite	4	
Chlorite	10	
Zone/Assemblage	Distance from granite	Inferred rock H_2O
Cordierite - K-feldspar zone	0–100 m	Less than 0.5%
K-feldspar - cordierite - biotite - quartz		
Cordierite - muscovite zone	Up to 1000 m	Less than 1%
Cordierite - muscovite - biotite - quartz		
Biotite zone	1000–2300 m	1%
Biotite - muscovite - chlorite - quartz		
Spotted zone	2000–3000 m	5%
Muscovite - chlorite - quartz		
Shale - outside aureole	Beyond 3000 m	5%
Muscovite - chlorite - quartz		

Therefore, these shales have lost 1-5% H_2O during contact metamorphism yet still preserve delicate bedding and cross-bedding features (Fig. 11.4).

In regional metamorphic terranes rocks at greenschist facies may contain 5% volatiles (predominantly H_2O and CO_2 in clays, chlorite and calcite) that are then lost during the progression from the greenschist facies to amphibolite and granulite facies grades (potentially an increase from 300°C to 800°C). For a granulite facies mineral assemblage such as sillimanite – cordierite – garnet – K-feldspar – quartz, the bound water and CO_2 are essentially zero implying that virtually all the 5% of volatiles have been lost from the surrounding sequence during prograde metamorphism. For a terrain of 100 km by 100 km by 1 km thickness, equivalent to a volume of 10,000 km^3 of rock, the release of 1% as water is potentially 100 km^3 of water or 10^{14} litres.

These calculations are not applicable to all rock types. For example, an anhydrous basalt or a quartz-rich sandstone would not generate fluid on the scale of a shale. However, if the basalt was strongly altered to a chlorite – carbonate assemblage prior to its burial to depths where metamorphism commences, there would be considerable scope for metamorphic fluid generation.

The siting of this bound H_2O is in every single hydrous mineral grain throughout the lower metamorphic grade zones, so it is uniformly distributed on a millimetric scale through huge rock volumes. Therefore, for the described contact aureole around the Strathbogie granite, every single cubic centimetre within 10s of cubic kilometres of pelitic rock has lost a few percent of its mass as H_2O implying fluid migration of at least 100s metres to kilometres. The distance involved in the volatile loss is even greater for regional metamorphic sequences. For all the hydrous minerals originally near the centre of that 1 km thick slab, the aqueous fluid released during metamorphism must have moved at least 0.5 km to now be outside the confines of that metamorphic slab, but this loss has happened in a lithified rock with no apparent signs of permeability.

11.2.2 Concept B: How Does a Fluid Form Through Metamorphism— Devolatilisation

The production of metamorphic fluid through devolatilisation reactions is a key part of the gold forming process. Devolatilisation generates a large volume of fluid with its composition dictated by the minerals involved in the reaction; and that fluid is ultimately coming from every grain on a millimetric scale throughout vast rock volumes.

A rock with a mix of minerals at low metamorphic grade might have hydrous minerals, carbonate minerals and minor amounts of sulfide as pyrite. If metamorphosed to granulite facies of 800°C or more, much of the H_2O, CO_2 and S will have been lost through dehydration, decarbonation and desulfidation reactions (devolatilisation). These different minerals do not break down independently, but instead, a mix of hydrous minerals and carbonates may become involved in a mutual devolatilisation reaction that generates a H_2O-CO_2 fluid with S.

An igneous analogy for this devolatilisation involving several minerals is provided in the melting of crustal rocks. A mix of quartz, K-feldspar, plagioclase, and muscovite will begin to melt around 700°C despite none of those minerals melting on their own until over 1200°C. When this mix of four minerals melts, the initial product is a magma of granitic composition, that is a composition that is different from any of the mineral components but controlled by them collectively. The silicate melts so formed by this process have a remarkably consistent bulk rock composition globally and through time (called granite). While all four minerals remain present in the mix their proportions do not affect the composition of the initial melt. The proportions of the minerals become important if one mineral is consumed because then the production of melt by that reaction terminates. Upon further heating a new melting reaction involving the remaining minerals may begin and it would generate a different melt composition. The melting process is highly effective at extracting selected elements. Taking an element like Rb that substitutes for K and is incompatible, it would partition into the initial melt: the Rb would be coming from every K-feldspar and muscovite grain throughout a large volume of rock undergoing partial melting. The Rb is part of the magma from the outset; it is not scavenged by the magma at a later stage.

Like the igneous analogue, the breakdown of a group of hydrous minerals, carbonates and sulfides in a metamorphic reaction generates a mixed volatile phase with a composition dictated by the specific mineral reactions. As well as having H_2O, CO_2 and S, the fluid will be of low salinity in the absence of any Cl in most minerals; but noting that where there are evaporites, the salinity could be high. Depending upon what metals are soluble in the metamorphic fluid, there is a high potential for segregation of various elements by partitioning between the source and the fluid. This applies particularly to the rare trace elements that can partition strongly into the fluid. Importantly, metals are part of the metamorphic fluid from its outset, and *this would include gold*; they are not scavenged by the fluid at a later time from a different site.

11.2.3   Concept C: How Do Metamorphic Fluids Migrate Through Impermeable Rock

The migration of aqueous fluid through metamorphic rocks is best described in three stages; one is the source area in which the fluid is generated, the second involves channels that move fluid for several km through the crust, and the third is the deposit site or sink area (Fig. 11.5).

The initial migration of this metamorphic fluid away from individual grains and the source area is assumed to be along grain boundaries and microcracks (Fig. 11.6) until these can amalgamate into larger cracks and fault zones (Fig. 11.5). There are no permanent open channelways on this grain scale and no permanent permeability or porosity in these metamorphic rocks but deformation involving small strain events such as earthquakes

combined with high fluid pressure may generate transient permeability. This source process of fluid generation and initial migration is completely pervasive and rock-buffered, i.e. the mineral assemblage dictates the composition of the few percent by volume of fluid. This fluid movement causes some continuous healing of the micro-cracks in a process that works against further fluid migration, until further strain such as from episodic earthquakes restarts the process.

The amalgamation of smaller cracks into larger faults (shear zones) provides the channelways for large scale fluid migration through the crust (Fig. 11.7). Such fault zones have been mapped by seismic reflection studies including across whole gold provinces; and their geometry can focus metamorphic fluids through several km vertically and laterally into confined channels and eventually deposits. These shear zones are centimetres to many metres thick, may be many tens of km in length and, in the case of goldfields, are inferred to have guided fluids vertically for 3-10 km from around 500°C to depositional sites of 350°C. This stage of migration is inferred to be fluid-buffered in that the fluid flow is constrained within the faults and has little interaction with the wallrock; this means a high fluid-to-rock ratio within the fault. The vein structures

Fig. 11.5 Cross-section through the crust identifying a source region for metamorphic devolatilisation, channels for transporting metamorphic fluid laterally and upward, and a site of structural complexity that becomes a deposit, given optimal physical and chemical conditions.

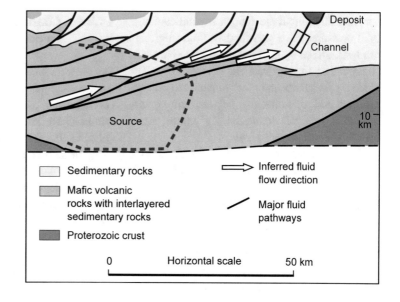

Fig. 11.6 Grain-scale depiction of the initial formation of metamorphic fluid at grain contacts which will be followed by its migration along cracks and grain boundaries. Modified from Phillips and Powell 2010.

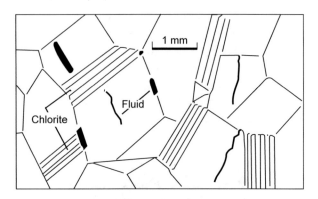

Fig. 11.7 Schematic cross-section through the upper crust with faults (shear zones) as the pathway for metamorphic fluids moving laterally and upwards into areas that become deposits.

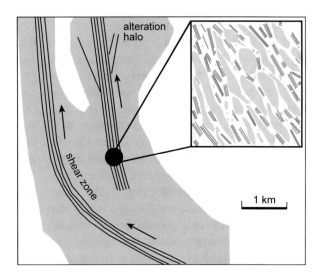

here and throughout the deposit site indicate high fluid pressures have prevailed so fluid migration has been driven by differences in effective mean stress rather than permanent permeability. The tendency for quartz and other minerals to precipitate out and seal the channels as temperature decreases can be countered by minor strain including earthquake activity.

A third stage of fluid migration is into a focused area to become an orebody (Fig. 11.8); and here the quartz vein textures indicate fluid ingress under high fluid pressure. In contrast with the second stage, the deposit area is characterised by high fluid-to-rock ratio in and adjacent to the faults with decreasing fluid-to-rock ratios towards distal parts. Extensive fluid – wallrock interaction (alteration) is important in gold deposition (see Chapter 12).

11.2.4   Channelways and Gold Deposits

On the province or craton-scale, faults may be semi-continuous for 100s km and are important economically as they have numerous deposits within 5-10 km on either side of the fault (Fig. 11.9). The individual goldfields situated along these trends may be 5-10 km long, 1-2 km wide and with mineralisation continuous to a kilometre depth indicating the scale of auriferous fluid migration. Important features coinciding with the linear trends and reflecting the fluid infiltration are widespread quartz veining, carbonate and pyrite bearing alteration assemblages, and higher strain (i.e. more deformed rocks).

On a local scale in and around orebodies, the sites favourable for fluid and hence gold will

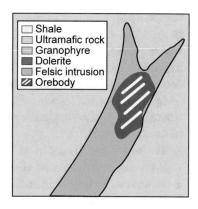
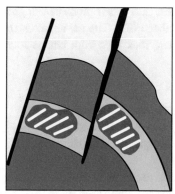

Legend:
- ☐ Shale
- ☐ Ultramafic rock
- ☐ Granophyre
- ◼ Dolerite
- ☐ Felsic intrusion
- ▨ Orebody

Fig. 11.8 Two examples in which adjacent rock types have differing rheology. The competent unit (felsic intrusion – left; granophyre – right) has fractured under high fluid pressure to allow ingress of auriferous fluid and formation of quartz stockwork mineralisation. The examples can be on scales from centimetres to 100s metres. Although these auriferous quartz veins are within igneous rocks, the auriferous fluid has been derived from well outside the host rock. Modified from Phillips and Powell 2010; originally produced with D Groves.

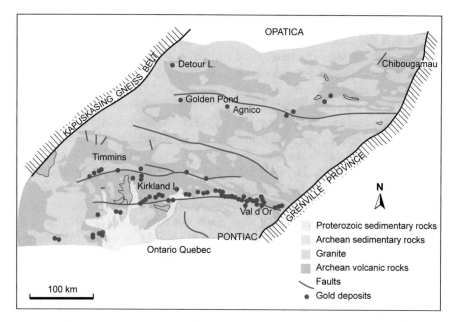

Fig. 11.9 Map of the Archean Abitibi area of eastern Canada with major east – west faults along which many large gold deposits are located. These faults and particularly many smaller associated faults are inferred channelways for auriferous fluids migrating from depth. For context, Toronto is 500 km south of Timmins.

include faults, breccias and heterogeneities in rock properties created by bedding planes, unconformities, dykes and small plutons. Even small changes in rock properties can translate to substantial focusing of fluid into plutons in metasedimentary sequences or even into one igneous variant within a larger intrusion (Fig. 11.8). Under directed stress, more competent and less competent rocks in a layered sequence will behave differently to create areas of low mean stress into which fluids can migrate (Ramsay and Huber 1987). The net result can be an orebody in one rock type with barren rocks adjacent.

11.2.5   Association of Metamorphic Fluids, Igneous Systems, and Magmatic Fluids

A relationship is expected between igneous activity and metamorphic fluids as both are the products of heat in the crust. Most devolatilisation and partial melting reactions are endothermic so their progress will coincide with the input of heat energy during igneous and metamorphic processes. There is an inevitability that complex geological environments of igneous, metamorphic, and metasedimentary rocks may result from major thermal events in the crust with a complex mix of fluid and rock types.

In the sub-volcanic system, the passage of magmas will induce contact metamorphism of wallrocks, generate metamorphic fluids (from the breakdown of hydrous and carbonate mineral assemblages), possibly lead to high fluid pressure and hydraulic fracture depending upon depth, and enable physical mixing with any fluid released by crystallisation of the magma (Fig. 11.10). It will be difficult to identify or isolate the metamorphic fluid or any component thereof. The volume of metamorphic fluid may be modest depending upon the country rocks, scale of the igneous system and local geometry.

In the plutonic environment such as adjacent to large granitic bodies metamorphic fluids are generated through contact metamorphism of what becomes the contact aureole, and considerable volumes of fluid are involved (Fig. 11.10). The metamorphic fluid formed in these plutonic environments is virtually never identified as such in field and geochemical studies which might suggest that it usually mixes intimately with the magmatic fluids released upon crystallisation. If this is the case, then studies of magmatic fluids around these plutons and batholith is usually a study of a magmatic – metamorphic fluid mix without any easy ways to discriminate the

contributions. This lack of discrimination applies to sampling volcanic gases. A more productive way to predict magmatic fluid compositions may be through forward modelling of the melting process using thermodynamics and involving quartz, feldspars, micas and assorted other minerals (this would be a similar approach to that used for metamorphic fluids).

The broader relationship of igneous activity with plumes, hot spots and rifts means that all these tectonic settings will include considerable metamorphic fluid where altered igneous rocks and metasedimentary rocks with volatile-bearing minerals are involved.

At a deeper level in the crust of amphibolite and granulite facies conditions, it becomes difficult to sustain a separate aqueous fluid and instead any such fluid would be consumed in the production of additional silicate magma. This limits the options for hydrothermal fluid flow, and indeed any hydrothermal ore deposit formation, at the higher metamorphic grades once partial melting has commenced.

11.2.6   Quantifying the Metamorphic Devolatilisation

For complex mineral assemblages such as those in altered basaltic rocks, thermodynamic calculations provide an indication of volumes of fluid lost through devolatilisation, the composition of such fluids, and the temperature intervals over which devolatilisation occurs (Elmer et al. 2006). This is achieved by taking a representative altered mafic rock composition, assembling a thermodynamic dataset for its minerals and potential fluid, and calculating stability fields for many assemblages (Fig. 11.11). Temperature is a key parameter reflecting progress of metamorphic heating, and the evolved fluid composition is reflected by a ratio of H_2O to CO_2.

Metamorphic fluid is evolved quite unevenly as the temperature rises. At low temperatures equivalent to the greenschist facies, mineral assemblages include actinolite, chlorite, albite, clinozoisite, albite and quartz. At higher temperatures, the amphibolite facies mineral

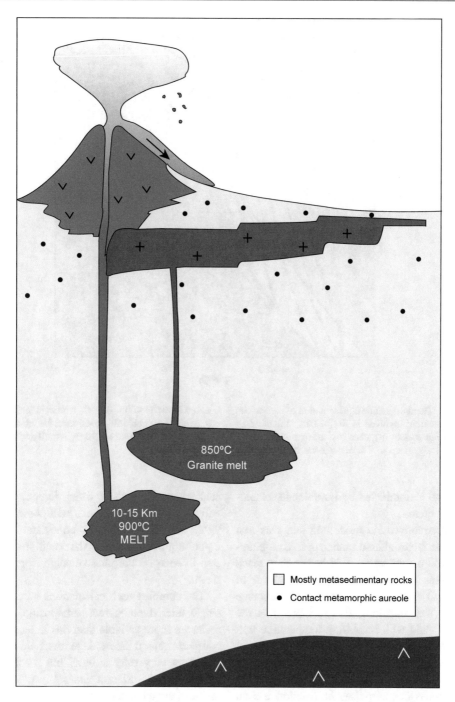

Fig. 11.10 Cross-section of a volcanic pipe and pancake-shaped granitic batholith. Country rocks adjacent to the granitic rocks and the volcanic pile will be contact metamorphosed with potential to generate metamorphic fluids from the breakdown of hydrous and carbonate-bearing minerals. It may be difficult to recognise and isolate these metamorphic fluids because this is a complex environment in which magmatic fluids may be released with volcanic activity, magmatic fluids evolved from the crystallisation of the granite, and meteoric waters percolating down from the surface.

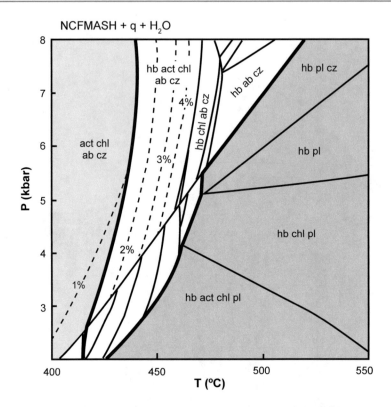

Fig. 11.11 Thermodynamic representation of the various minerals and fluid involved in the metamorphism of an altered basaltic rock from greenschist to amphibolite facies grade. At 4 kbars, up to 4% of aqueous fluid could be released between 400 – 450°C. Abbreviations: ab-albite, act-actinolite, chl-chlorite, cz-clinozoisite, hb-hornblende, pl-plagioclase. Prepared by Roger Powell and reproduced with permission.

assemblage is dominated by hornblende, plagioclase, and quartz.

The thermodynamic modelling suggests that greenschist facies altered mafic rocks can generate 2 – 5 % metamorphic fluid before they reach 500°C. This is fluid emanating from release of bound H_2O and CO_2 in hydrous and carbonate minerals. The modelling also indicates that the release of fluid with increasing temperature will be as one or more pulses coinciding with different sets of reacting minerals rather than continuously and evenly liberated over 100s degrees.

This behaviour as pulses of reaction has an analogy in the simple H_2O (water and steam) system. Heating progressively from room temperature to 99°C involves water but heating from 101°C upwards is totally involving steam. All the reaction from water to steam has occurred at 100°C essentially as a pulse. Also, with a steady

heat source the rate at which temperature rises will be quite uneven and will slow between 99°C and 101°C due to considerable energy input required to make the small temperature step to energise the phase change from water to steam.

The complex rock calculations here are for a single altered mafic rock composition whereas rocks are more variable than this in nature. For a constant mineral assemblage, the fluid composition may vary only a little, but reactions will terminate at different stages as individual minerals are consumed.

The same thermodynamic dataset and same altered basalt starting material can be used to monitor the proportion of CO_2 in the aqueous fluid for the interval of 400 – 450°C in which the bulk of the metamorphic fluid is generated (Fig. 11.12).

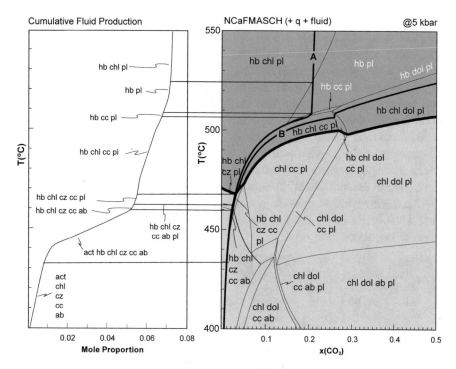

Fig. 11.12 Thermodynamic representation of the various minerals involved in the metamorphism of an altered basaltic rock from greenschist to amphibolite facies grade with figures showing: **(left)** the volume of fluid released as temperature rises; **(right)** the proportion (mole CO_2/ [CO_2+H_2O]) of CO_2 in the aqueous fluid released by devolatilisation during heating from 400°C to 550°C. Abbreviations: ab-albite, act-actinolite, cc-calcite, chl-chlorite, cz-clinozoisite, dol-dolomite, hb-hornblende, pl-plagioclase. Prepared by Roger Powell and reproduced with permission.

The thermodynamic modelling suggests that a significant amount of aqueous fluid will be released across the greenschist to amphibolite facies transition, i.e. actinolite – chlorite – albite to hornblende – plagioclase around 450°C. As this is a common set of minerals, a similar fluid could be generated both globally and through geological time in different terranes. Given the typical absence of evaporites, the fluid will contain negligible salinity. The main fluid will comprise essential H_2O and CO_2 with an intermediate composition buffered by mineral equilibria; it will not resemble the product of physical mixing of separate H_2O-rich and CO_2-rich fluids. The timing of the fluid release will coincide with thermal input into the surrounding crust and may be accompanied by intrusions and volcanism but without any specific correlation to the type of magmas involved in supplying the heat.

The actual fluid sampled in studies such as through fluid inclusions will be more complex because it is the combination of slightly different metamorphic fluids formed from several different mafic rocks at different times, depths, and temperatures.

11.3   An Unreasonably Effective Answer Explaining Auriferous Fluids

Forward modelling using thermodynamics to predict the composition of metamorphic fluids provides an uncanny match for three of the main characteristics of gold-only deposits—scale of provinces, timing, and fluid composition. The scale of a gold province and the scale of a metamorphic terrain are comparable (and neither matches the scale of individual igneous intrusions). The timing of gold-only deposits is synchronous with metamorphic processes, heat production and with igneous activity. The $H_2O - CO_2$ low salinity gold-only fluid matches the predicted metamorphic fluid arising from the greenschist to amphibolite facies transition in altered basaltic rocks (and potentially other rock units). Two questions are being addressed simultaneously with a single *unreasonably effective* answer. First, where are the fluids in nature predicted by the thermodynamic modelling, and second, what is the origin of the unusual fluids that are measured in gold-only deposits.

Snapshot

- Water has some special characteristics making it essential in many geological processes.
- Deeper crustal waters are usually referred to as fluids to recognise their complex composition and potential to be supercritical—this is when liquid and vapour properties are indistinguishable.
- Aqueous fluids in the crust have variable salinity, redox condition, and temperature ranges.
- The fluid regime in active metamorphic environments undergoing deformation is different from the processes of diagenesis described in the oil and gas literature.
- Metamorphic fluid is vast in volume and migrates for kilometres through rock masses of negligible porosity.
- Every mineral grain within large volumes is involved in metamorphic devolatilisation.
- Devolatilisation can yield H_2O, CO_2, and S bearing fluids; but Cl is usually negligible reflecting the low Cl content of the source rocks.
- Compositions of metamorphic fluids are based upon the mineral assemblage involved in a reaction.

- Metals including gold are in the metamorphic fluid from the moment it forms.
- Gold deposits are spatially related to faults, and structural complexity is the rule.

Bibliography

Chauvet A (2019) Structural control of ore deposits: the role of pre-existing structures on the formation of mineralised vein systems. Fortschr Mineral 9:56. https://doi.org/10.3390/min9010056

Clemens JD (2012) Granitic magmatism, from source to emplacement: a personal view. Appl Earth Sci 121: 107–136

Cox SF (2005) Coupling between deformation, fluid pressures, and fluid flow in ore-producing hydrothermal systems at depth in the crust. Economic Geology, 100th Anniversary volume, pp. 39–75

Elmer FL, White RW, Powell R (2006) Devolatilization of metabasic rocks during greenschist–amphibolite facies metamorphism. J Metamorph Geol 24:497–513

Ferry JM (1980) A case study of the amount and distribution of heat and fluid during metamorphism. Contrib Mineral Petrol 71:373–385

Ferry JM (1981) Petrology of graphitic sulfide-rich schists from south-Central Maine: an example of desulfidation during prograde regional metamorphism. Am Mineral 66:908–930

Ferry JM (2016) Fluids in the crust during regional metamorphism: forty years in the Waterville limestone. Am Mineral 101:500–517. https://doi.org/10.2138/am-2016-5118

Greenwood HJ (1975) Buffering of pore fluids by metamorphic reactions. Am J Sci 275:573–593. https://doi.org/10.2475/ajs.275.5.573

Kerrich R, Allison I (1978) Vein geometry and hydrostatics during Yellowknife mineralisation. Can J Earth Sci 15:1653–1660

Norris RJ, Henley RW (1976) Dewatering of a metamorphic pile. Geology 4:333–336

Peters SG (1993) Nomenclature, concepts and classification of oreshoots in vein deposits. Ore Geol Rev 8: 3–22

Phillips WJ (1976) Hydraulic fracturing and mineralization. J Geol Soc London 128:337–359

Phillips GN, Powell R (2010) Formation of gold deposits: a metamorphic devolatilization model. J Metamorph Geol 28:689–718

Powell R, Will TM, Phillips GN (1991) Metamorphism of Archaean greenstone belts: calculated fluid compositions and implications for gold mineralization. J Metamorph Geol 9:141–150

Ramsay JG, Huber M (1987) The techniques of modern structural geology, Strain analysis, vol 1. Academic Press, London

Ridley JR (1993) The relationship between mean rock stress and fluid flow in the crust: with reference to vein- and lode style deposits. Ore Geol Rev 8:23–37

Robb L (2005) Introduction to ore-forming processes. Blackwell Publishing, Oxford, p 373

Tikoff B, Blenkinsop T, Kruckenberg SC, Morgan S, Newman J, Wojtal S (2013) A perspective on the emergence of modern structural geology: celebrating the feedbacks between historical-based and process-based approaches. Geol Soc Am Spec Pap 500:5–119. https://doi.org/10.1130/2013.2500(03)

Vearncombe JR (1998) Shear zones, fault networks, and Archean gold. Geology 26:855–858

Alteration in Gold-Only Deposits

12

Abstract

As metamorphic fluids depart from their source they move to areas of lower temperature, lower pressure and into country rocks with which they are no longer in equilibrium. Reaction between rocks and fluid changes the rocks and this produces an alteration zone or halo. The reaction also modifies the fluid with the potential to cause gold deposition.

Keywords

Alteration · Fluid—rock interaction · Element mobility · Buffering

Alteration is a key component of many ore deposits, is critical during the ore-forming process, and contains a wealth of information of use in science and in exploration. Once established that the alteration halo is related to the ore-forming event, the alteration mineral assemblages can constrain the conditions such as temperature and pressure, likely ore fluid composition, changes in composition of rock and fluid, and effects on metal concentration in the fluid. For gold exploration, the alteration halo presents a much larger target than looking for economic levels of gold alone (Fig. 12.1).

Numerous terms have been used for alteration including fluid – wallrock interaction, metasomatism and non-isochemical metamorphism. Usage varies between economic geologists and petrologists, and whether the purpose is to use

alteration to find ore, or to understand the science of ore formation, or both. Most metamorphism is non-isochemical to some degree involving loss and gain of the more mobile elements including volatiles like H_2O and CO_2.

Alteration is a consequence of a hydrothermal fluid interacting with a country rock. This leads to the fluid adding (or removing) one or more elements to the rock, and the net result is a changed fluid and an altered rock. Changing of the fluid composition has the potential to deposit gold (Chapter 14); whereas altering the country rock leaves a record of information about the ore-forming process and nature of the fluid. The degree of rock alteration in part reflects distance from the fluid channelway and this variable may produce distinct alteration zones informally described as distal and proximal alteration (Fig. 12.2).

12.1 Alteration Whole Rock Geochemistry

The alteration geochemistry of the altered rock, as opposed to alteration mineralogy, is a function of elements added during wallrock alteration by the gold-bearing fluid, and elements already present in the country rock. Whether an element is transported in the fluid or is immobile is influenced by the size and charge of that element and its stable ions. The fluid temperature and its redox conditions are also important in

© The Author(s), under exclusive license to Springer Nature Singapore Pte Ltd. 2022
N. Phillips, *Formation of Gold Deposits*, Modern Approaches in Solid Earth Sciences 21,
https://doi.org/10.1007/978-981-16-3081-1_12

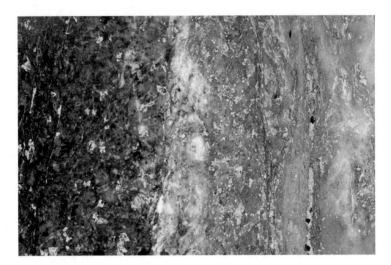

Fig. 12.1 Cross-section of a small-scale alteration halo in dolerite from the Mt Charlotte mine, Kalgoorlie goldfield, Western Australia. The dark green area (left) is the distal chlorite alteration zone, and the far right is an auriferous quartz vein bound by a high strain interval of elongate white mica and pyrite. Between the chlorite zone and vein is the proximal carbonate alteration zone that despite being bleached and with original igneous and metamorphic minerals replaced is mostly low strain and preserving the igneous texture. Width approximately 20 cm.

determining element transport. Any resulting element pattern within a gold deposit reflects the source of the fluid, the fluid transport pathway, country rocks and the minerals present near the site of deposition, and the temperature and redox conditions prevailing at deposition.

In gold-only deposits, the alteration halo typically records the addition of some distinctive elements that are negligible in the country rocks (Fig. 12.3). Such elements consistently added from solution include:

- Au, Ag, As, Sb, S, W, B, Se, Te, Hg.

Further elements such as Mo, Bi and Pb are at elevated levels in gold-only deposits of specific regions. In the Yana–Kolyma belt and Okhotsk–Chukotka arc to the NW of the Okhotsk Sea (see Fig. 4.16), Mo is widespread in gold-only deposits along with Sn (the latter is rare in gold deposits globally). In Victoria (see Fig. 4.10) there are specific mineralogical domains identified by the elevated Ag and Pb in ores

Fig. 12.2 The boundary of two alteration zones with dark green chlorite on the left (distal alteration), cream ankerite on the right (proximal alteration), and preservation of the igneous texture of 2-4 mm plagioclase laths throughout. From the Mt Charlotte mine, Kalgoorlie goldfield.

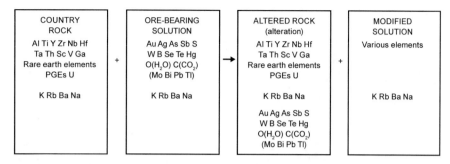

Fig. 12.3 Schematic figure for alteration geochemistry in a gold-only system showing immobile elements contributed by the country rock, and the main mobile elements introduced by the auriferous fluid. The resulting altered rocks are a combination of both sets of elements. The set of K, Rb, Ba and Na is different in that these are a group of highly mobile elements that may be contributed by country rock and by fluid.

from gold deposits throughout that domain. As these domains are sharply defined, fault-bound and 100 km by 10s km dimensions, the cause of Ag and Pb being elevated is unlikely to be related to the immediate surrounds of a deposit such as a local host rock or specific pluton.

Other elements commonly contributed to the altered rocks from the ore solution include O, H and C (in H_2O, CO_2), K, Rb, Ba and Na though these may already be present in some country rocks. The distribution of H_2O, CO_2, K and S have a major role in determining the mineralogy of the alteration halo. For the greenschist facies the generalised distribution of the major components contributed by the solution is:

- CO_2 is added in the broadest envelope around the ore zone.
- K, S and Au are abundant in and adjacent to the ore zone but added throughout the whole alteration halo in much lower concentrations.

H_2O may be added or removed depending upon the country rock and prevailing conditions.

The other source of elements in the gold-only alteration halo is from the country rock. The geochemistry of host rocks is quite variable so this part of the contribution to alteration geochemistry varies. For example, once altered, an igneous rock is likely to retain a distinctive signature indicating its original nature; similarly, a suite of sedimentary rocks may retain evidence of the precursor history. Immobile elements that are typically inherited from the country rocks because they are tightly bound and not taken up in the solution include:

- Al, Ti, Y, Zr, Nb, Hf, Ta, Th, Sc, V, Ga.
- Immobile rare earth elements, i.e. La, Ce, Nd, Sm, Eu, Gd, Tb, Dy, Ho, Er, Tm, Yb, Lu (noting that the rare earth element Eu is mobile in some environments).

Further elements that would not normally be considered immobile are contributed from country rock to the alteration halo because they are not particularly soluble in the gold-only ore solution. These include Sr, Cu and Zn, and, in most provinces, Pb.

12.2   Alteration Isotope Geochemistry

Many of these elements in the alteration halo and veins have multiple isotopes that may provide more information about the fluid source and its evolution. For example, oxygen has a specified number of protons (8) but can have different numbers of neutrons (8 or 10 in nature). These two isotopes of oxygen differ in their mass (approximately 16 : 18) and geological processes can favour (i.e. fractionate) one isotope against the other based on these mass differences. Several lighter elements like H, B, C, N, O and S have more than one stable isotope which can

Table 12.1 Isotopes in gold-only systems.

Atomic #			In auriferous fluid
1	^{1}H	^{2}H	Yes
5	^{10}B	^{11}B	Yes
6	^{12}C	^{13}C	Yes
7	^{14}N	^{15}N	Yes
8	^{16}O	^{18}O	Yes
16	^{32}S	^{34}S	Yes
19–20	^{40}K	^{40}Ar	K commonly yes
37–38	^{87}Rb	^{87}Sr	Rb commonly yes
62–60	^{146}Sm	^{146}Nd	No
71–72	^{176}Lu	^{176}Hf	No
75–76	^{187}Re	^{187}Os	No
90–82	^{232}Th	^{208}Pb	Th no, Pb rarely
92–82	^{235}U	^{207}Pb	U no, Pb rarely
92–82	^{238}U	^{206}Pb	U no, Pb rarely

fractionate in geological processes based on mass differences. Another group of elements have one or more isotopes that are radioactive and decay over time, and this decay changes the ratio between different isotopes of the radioactive element and its decay product element (Table 12.1). Analytical equipment today can easily measure the small isotopic differences, and this has led to a whole research direction of measuring as many stable and radiogenic isotopic systems in gold deposits as possible with the aim of determining where the gold, the ore solution, and other components have come from.

The fractionation of stable isotopes such as H and O is diagnostic of different water types, and C and S isotopes can reflect different source regions. Having definitive answers from these isotope systems would mean being a long way towards answering how gold deposits form. However, the reality of the various isotope systems is far from being a panacea. Rarely do the stable isotopes have diagnostic signatures that indicate a unique source of gold, although they are useful in eliminating possibilities. Some, such as isotopes of H, O and S, provide interesting constraints on the processes nearer the gold depositional site.

The elements involved in radioactive decay (Table 12.1) are mainly being sourced from the country rock in gold-only systems and unlikely to be introduced by the ore solution regardless of whether they occur in the alteration halo or as new minerals in veins. None of Sr, Nd, Sm, Lu, Hf, Re, Os, Pb, Th and U are systematically introduced in the auriferous solutions. Although their decay properties are informative for absolute dating, these elements are not especially useful in tracing the origin of gold or the solution. The ideal approach would be to apply isotopic studies directly to gold; however, with only one naturally occurring isotope, ^{197}Au does not offer the opportunity to monitor any isotopic variations. That so many of the isotopic systems do not indicate the source of auriferous fluids is not unexpected given their ionic bonding characteristics. Once the gold-only fluids were found to be low salinity in the late 1970s, it was clear that radiogenic isotope systems may have limited use as geochemical tracers of the auriferous fluid components in gold-only systems.

A second use of radioactive elements is to date rock units where the temporal relationship to gold introduction has been established by mapping, such as a barren dyke cross cutting mineralisation. In many deposits, however, several geological events are synchronous or nearly so, and thus difficult to reliably separate with absolute radiometric dating. An example might be metamorphism, partial melting, a period of deformation, magmatism and gold mineralisation being coeval. The similar time of these events is not unexpected as each is impacted by thermal events in the crust, and possibly the same thermal event for all.

12.3   Alteration Mineralogy

The alteration mineralogy is a function of the metamorphic temperature, extent of the alteration, country rock geochemistry and fluid composition. Increasing metamorphic grade can be related to distinctive mineral abundance trends (from greenschist facies to amphibolite and granulite facies):

- Chlorite (greenschist) to amphiboles to pyroxenes (granulite).
- Abundant carbonates especially ankerite, to less carbonate minerals.
- Muscovite to biotite to K-feldspar.
- Abundant pyrite to increasing pyrrhotite, and overall, less sulfide minerals.
- Arsenopyrite (FeAsS) to loellingite ($FeAs_2$).

The extent of the alteration is based on a premise that rocks close to the solution channelway are usually more strongly modified and rocks farther away less so. This can be reflected in qualitative terms of proximal and distal alteration (Table 12.2). For the greenschist facies the mineral pattern mimics the geochemical trends:

- Carbonate minerals are abundant in the broadest envelope around the ore zone.
- K-minerals and sulfides are abundant in and adjacent to the ore zone but distributed throughout the whole alteration halo in much lower concentrations.
- The scale of the proximal and distal alteration zones can be millimetres around small veins,

to kilometres around major goldfields, reflecting the volumes of hydrothermal fluid involved.

One or more distinctive minerals may form the basis of alteration mapping, but it is full coexisting mineral assemblages that are required for optimal thermodynamic analysis of the conditions of formation.

The influence of country rock geochemistry on alteration mineralogy is illustrated with K. Three K-bearing minerals (K-feldspar, muscovite, biotite) commonly occur in alteration haloes and all these minerals have essential Al in their structure. As Al is immobile and not provided by the fluid, it must be derived from the country rock. In a gold deposit in BIF made solely of interlayered quartz (SiO_2) and magnetite (Fe_3O_4) there is no Al; therefore, the absence of any Al in the rock and in the solution means that no K-mineral can form. Without any K-bearing minerals there is no K added during this alteration, but the lack of K addition does not mean that the solution was K-poor. The BIF example highlights the risk of directly correlating elements added during the alteration process with elements carried in solution.

12.4   Fluid Buffering and Infiltration, and Rock Buffering

Two endmembers are useful in understanding the interplay of the country rock and the fluid in the metamorphic environment. The progress of a fluid from its source through to its interaction with rocks at a gold depositional site can involve stages of either fluid buffering or rock buffering conditions.

Table 12.2 Alteration minerals at different metamorphic grades.

Metamorphic grade	Distal	Proximal and Ore zone	Ore zone sulfide minerals
Mid-greenschist	Chlorite, calcite	Ankerite, muscovite	Pyrite, aspy
Upper greenschist	Chlorite, calcite	Ankerite, biotite	Pyrite, pyrrhotite, aspy
Amphibolite	Chlorite, calcite, biotite	Amphibole, plagioclase, Biotite	Pyrrhotite, pyrite, aspy
Amphibolite	Amphibole, biotite, garnet	Cordierite, K-feldspar	Pyrrhotite, loellingite, aspy
			Aspy = arsenopyrite

Fluid buffering or infiltration occurs when a large volume of fluid interacts with a relatively small volume of rock; here the fluid can strongly alter the country rock, and the country rock may have less effect on the fluid. When there is only a small amount of fluid in the country rock, the rock composition determines the composition of the fluid, and the fluid has little capacity to alter the country rock. A complete alteration halo around a gold deposit may contain a transition from distal, mostly rock buffered conditions to proximal, infiltration dominated conditions of much higher fluid-to-rock ratios and stronger gold mineralisation.

the magmatic rocks have provided the auriferous fluid (and similarly for sedimentary rocks).

- The elements used in radiometric studies are chemically dissimilar to gold, not likely to be consistently transported under the same conditions as gold, and inappropriate as tracers.
- Not all elements carried in solution will necessarily be added during wallrock alteration.

Snapshot

- The reaction of fluid and country rock modifies the fluid, and yields altered country rock referred to as an alteration zone.
- The alteration zone reflects both the country rock and the fluid.
- Alteration zones in igneous host rocks are expected to have geochemical signatures that partly reflect those magmatic rocks, but this does not mean that

Bibliography

Mueller AG, Groves DI (1991) The classification of Western Australian greenstone-hosted gold deposits according to wallrock alteration mineral assemblages. Ore Geol Rev 6:291–331

Neall FB, Phillips GN (1987) Fluid–wallrock interaction around Archean hydrothermal gold deposits: a thermodynamic model. Econ Geol 82:1679–1694

Rice JM, Ferry JM (1982) Buffering, infiltration, and the control of intensive variables during metamorphism. In (Ferry J M ed) characterization of metamorphism through mineral equilibria. Rev Mineral (Mineral Soc Am) 10:263–326

Case Study: The Formation of a Giant Goldfield: Kalgoorlie, Western Australia

13

Abstract

The size of some of the world's largest goldfields raises the question of why they are so much larger than all the goldfields around them. By 1980 the Yilgarn Craton of Western Australia had produced 70 Moz Au (2200 t) since discovery around 1880 from 2000 mines and workings; more than a half of that total had come from the Kalgoorlie goldfield. Four decades later the known endowment of the Kalgoorlie goldfield is 2800 t.

Why is Kalgoorlie so much larger than its surrounds? Are there unique aspects of the geology found at Kalgoorlie and not elsewhere? What are the indicators of another Kalgoorlie or even another more modest 10 Moz (600 t) goldfield? Where and how does the Kalgoorlie goldfield terminate both laterally and with depth? What do the findings at Kalgoorlie mean for other large goldfields globally?

Keywords

Kalgoorlie goldfield · Structures · Alteration · Host rocks · Critical factors

Kalgoorlie is the largest goldfield in an Archean greenstone belt and its complex geology illustrates both the features of commonality within and between goldfields, and much of the diversity. The goldfield is 10 km in length and comprises numerous orebodies that each have slight differences. To explain how the Kalgoorlie goldfield formed it is necessary to address both the commonality and the diversity of its geology.

The goldfield has produced almost 70 Moz (2200 t) Au since its discovery in 1893; and has produced continuously since then from various mines and gold-only orebodies. At its century in 1993 its all-time production was quoted as 40 Moz (1250 t Au) with Reserves of 3 Moz—a classic example of brownfield exploration success following its rejuvenation in the 1980s. Most of the gold came from underground mining until the 1980s when modern open pit mining took over to establish the 5 km by 2 km open pit to 600 m depth (the Superpit).

Common features of all parts of the goldfield are the enrichment of gold to form the various ores, the segregation from base metals, epigenetic timing that is late in the deformational history, and its H_2O-CO_2 rich ore fluid. It also occurs with 100s of smaller gold deposits in the loosely defined but well-endowed Eastern Goldfields Province of the Yilgarn Craton. In these respects, the Kalgoorlie goldfield has characteristics typical of many other gold-only deposits globally and through time. Some of its other features are not as widely shared, and of interest are any factors accounting for Kalgoorlie being so much larger than most other Archean deposits.

Weathering has affected rocks near the surface across the Kalgoorlie goldfield but given that mining commenced in the 19th century, the weathering has not played a critical role in the

Fig. 13.1 Map of the Kalgoorlie goldfield showing the main rock units, the location of the Golden Mile, the Lake View Shaft section (dot is the former shaft position), and the Oroya B-B' section.

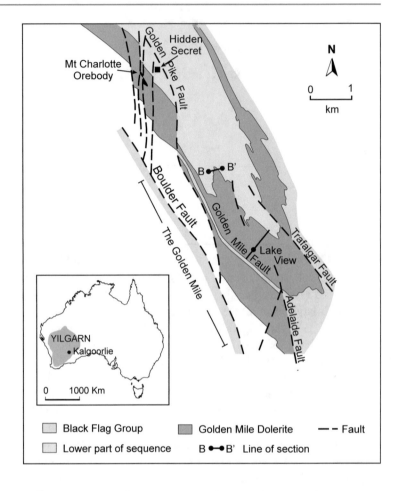

mining for much of its life in the 20th century. The oxidation has produced assemblages of iron oxides and clay minerals and it varies from 10 to 50 metres depth except within the lode structures where oxidation is locally over 100 m depth.

13.1   Stratigraphy

The stratigraphic succession of the Kalgoorlie goldfield comprises older ultramafic and mafic volcanic rocks overlain by clastic sedimentary rocks that have been metamorphosed to greenschist facies. Between the ultramafic – mafic succession and the metasedimentary rocks, at what is a significant change from dense to less-dense rocks, the Golden Mile Dolerite (GMD) has been intruded as a 600 m thick sill

with a magmatic composition equivalent to a basalt (Fig. 13.1).

The main host rocks for gold are the Dolerite, the Paringa Basalt immediately underlying it, and carbonaceous shale either as interflow sediments near the top of the Basalt or the base of the clastic metasedimentary rocks. Subordinate deposits occur in other rocks within the goldfield.

13.2   Nature of the Main Host Rocks

Although they have formed in different ways, the GMD and Paringa Basalt both originated from magmas of low-K tholeiitic basalt composition being a common type through Earth history and a basaltic magma that has elevated Fe and Fe/Mg. The Paringa Basalt was emplaced as a series of flows in a sub-aqueous environment as demonstrated by its well-preserved pillow basalt structures. The GMD was intruded after some sedimentary rocks had been deposited upon the Basalt but while this part of the sequence was undeformed and approximately flat lying. The Dolerite appears to extend 10 km laterally, but it is not always possible to uniquely identify it from other dolerite intrusions in the district where there is drilling but no underground access.

Cooling of the 600 m GMD involved differentiation into a wide range of rock types that can be mapped as ten units from bottom (Unit 1) to top (Unit 10) based on texture, mineralogy, and geochemistry. The unit boundaries are all gradational, hence it is correctly a differentiated intrusion rather than a layered intrusion. Crystallisation commenced with chilled top and bottom contacts (Unit 1 and Unit 10), then a near-ultramafic basal interval (Unit 2-3), mafic section that is quite Fe-rich (Unit 4-7), and intermediate granophyre being the most extreme form of the differentiation (Unit 8), then a quasi-repeat of this sequence in reverse order (Unit 9).

The differentiation within the Dolerite means that there are different chemical and physical properties throughout the 600 m thick sill. Being more siliceous and feldspar-rich, Unit 8 has a lower tensile strength than earlier-crystallised units that have abundant amphibole (Unit 2-4) so the former would be expected to fracture preferentially under high fluid pressures in the metamorphic environment. Although the GMD magma had elevated Fe overall, the differentiation has led to strong Fe enrichment and high Fe/Mg in the more differentiated Units 7 and 8.

13.3 Multiple Ore Types

There is a wide range of deposit types in the Kalgoorlie goldfield reflecting different host rocks and structural settings (Fig. 13.2):

- Golden Mile style comprises steeply-dipping brittle – ductile faults in the Golden Mile Dolerite and immediately adjacent rock units. It has yielded 60 Moz Au (2000 t) and in 2020 had a further ~20 Moz (600 t) of Resources. This style has also been referred to as Fimiston style.
- Charlotte style comprises a quartz stockwork concentrated in Unit 8 GMD and has produced 6 Moz Au (200 t).
- Oroya style refers to orebodies adjacent to the carbonaceous shale and mostly between the Paringa Basalt and GMD. The Oroya Shoot is the largest example of this style and is one km long and has produced 2 Moz Au (60 t) Au at high grade. Much of this ore has been counted within the Golden Mile production.

Fig. 13.2 Schematic cross-section illustrating the various settings for mineralisation in the Kalgoorlie goldfield. Host rocks include ultramafic, mafic and metasedimentary rocks, and structures include brittle – ductile faults, quartz stockworks, bedding parallel ores and high-grade shoots with breccias. Modified from Groves and Phillips 1987.

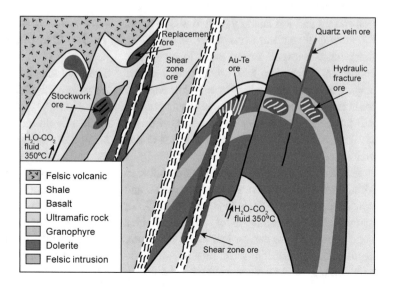

- Mt Percy style in felsic dykes within ultra-mafic rocks (10 t Au).

The alteration geochemistry and similar ore fluids suggested that the styles were likely to be different expressions of a single mineralising event, something subsequently confirmed by radiometric dating.

13.4 Geometry Including Multiple Lode Orientations

Deformation has led to a steeply dipping succession that is locally thickened by folding and structural repetition especially in the Golden Mile area where the GMD is repeated across the Golden Mile fault (see Fig. 13.1). Pelitic metasedimentary rocks have developed a weak planar fabric and basalt and dolerite have retained igneous textures with some preferred orientation of metamorphic minerals including actinolite.

The sequence is offset by many brittle – ductile faults that have an important role in localising mineralisation in the Golden Mile area. Cross sections through the heart of the Golden Mile (Eastern lode system) show where the mineralisation in these faults has been stoped (i.e. removed by underground mining) prior to 1980, and these stoped areas mark positions on major structures that were economic (Fig. 13.3). The structures can be traced from section to section and up to 2 km laterally, and to 800 m depth by mining (even more on other sections) or interpolated between stoped areas to give more continuous lines. The lack of continuity of the stoping at depth shown here does not necessarily mean that a brittle – ductile fault is absent but that it is not economic to mine in that area, or that it continues deeper that viable mining access in that vicinity.

Much of the economic viability of the Golden Mile, first as an extensive underground mine and now as a large open cut, is based on the 1000 or more mineralised faults (lodes) that result in a large volume of ore per vertical metre of mining depth. The termination of mining laterally and with depth is rarely a sharp geological boundary

but a combination of fewer and narrower mineralised faults, and patchy and lower gold grades. As the sub-parallel mineralised brittle – ductile faults decrease in number with depth, or the grade decreases it eventually becomes uneconomic to develop deeper.

Four main brittle – ductile fault orientations are recognised in plan, and the mineralisation associated with each is referred to as Main lode, Caunter lode, Easterly (or No.2) lode and Cross lode (Fig. 13.4). These have complex mutually cross-cutting relationships and anastomosing patterns that can be understood better through analysis in 3-dimensions. The four lode directions have a line of mutual intersection that plunges $50°$ towards the SE and this is an important ore shoot direction. The interpretation is that the four mineralisation orientations formed synchronously as part of NE – SW directed compression.

Some deformation persisted after the onset of the mineralisation event and was focused along faults (lodes) such that the quartz veins were contorted, pressure-solved and became milky leading to them being historically referred to as lodes rather than quartz veins; and the pyrite was pressure-solved against mica cleavage surfaces. Today many of the Golden Mile ores would be described as a central auriferous quartz vein surrounded by a mineralised alteration halo, both of which have experienced localised deformation.

13.5 Ore Fluids

The ore fluids inferred from fluid inclusion studies within the Golden Mile are mostly low salinity, H_2O-CO_2 dominant as for many other Archean gold-only deposits. The Oroya ores differ in that significant methane (CH_4) is indicated by the freezing point depression (see Table 8.1).

In the Yilgarn-wide study of ore fluids (Chapter 8), methane was detected in deposits that contained carbonaceous shales and this was attributed to reaction of the ore fluids with these shales. Another effect of interaction with the Oroya carbonaceous shales, which are V-rich, is the proliferation of vanadium-bearing minerals

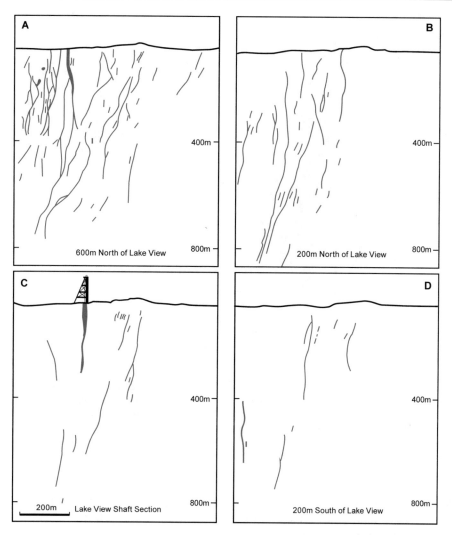

Fig. 13.3 Cross-sections through the Golden Mile showing ore zones in red that have been removed by underground mining. A-D are arranged north to south and span 800 m perpendicular to stratigraphy. The vertical distances are approximate depths below the ground surface. The ore zones (brittle – ductile faults and referred to as lodes) and surrounding geology were mapped up until the mid-1980s, but since the development of the Superpit most of these underground areas have been inaccessible or removed by mining. Lake View was the second shaft on the Kalgoorlie goldfield commencing in 1895 and was developed on the Lake View lode – the major lode below the shaft in Fig. 13.3c. Modified from Phillips 1986.

including V-hematite and V-tourmaline. The effect of the shales on modifying the fluids by the methane addition is probably on the scale of tens of metres only as the Paringa study found no detectable methane adjacent to non-carbonaceous sedimentary layers 100 m lower in the same mine. This scale is compatible with the distribution of the vanadium-rich minerals (Fig. 13.5).

13.6 Large Scale Alteration Haloes in the Golden Mile

Three alteration zones can be mapped around the Golden Mile within the GMD based upon the mineralogical siting of Fe. Outside this alteration halo the regional assemblage involves actinolite as the host of Fe which is as expected for the

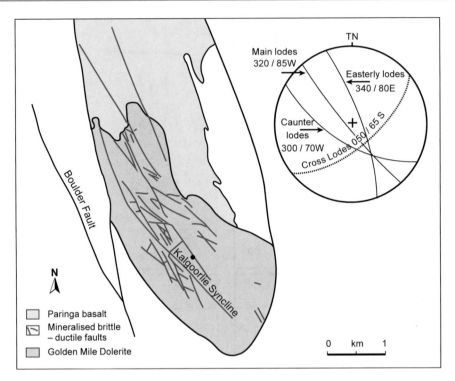

Fig. 13.4 Map of the Golden Mile showing some surface traces of mineralised faults (ore zones or lodes) at several different orientations. (**Inset**) a stereographic projection of planes of the four dominant lode orientations plotted as great circles on the lower hemisphere of a Wulff net projection. Their mutual intersection in a line plunging 50° to the SE indicates that they formed as part of a single deformation episode (rather than sequentially) and were infiltrated by the one auriferous fluid. In detail the 1000 faults within the Golden Mile are more varied in their orientation than indicated by these four generalised sets. Modified from Phillips 1986.

greenschist facies metamorphic grade and it gives the rock a very dark green colour. GMD containing the unaltered actinolite-bearing assemblage occurs in an area of 10s m on the Lake View Shaft section at 600 m depth in Units 1-6, but otherwise is absent in most underground workings.

Much of the kilometre-scale alteration features are fractal in that they are reproduced in single underground exposures (Fig. 13.6) where they can be mapped and analysed to understand the macroscopic scale in cross-section (Fig. 13.7) and in plan (Fig. 13.8).

Distal alteration around the Golden Mile is recognised by abundant chlorite with calcite imparting a moderate to dark green colour, and the absence of actinolite. This chlorite alteration zone is 5 km in length along strike, up to 2 km wide and typically of low strain with excellent preservation of textures recording the igneous crystallisation of the GMD.

The alteration mineralogy provides an important constraint on the timing of gold introduction relative to the peak of regional metamorphism. Actinolite, which formed at the peak of regional metamorphism, is overgrown by the gold-related chlorite – carbonate assemblages, and there may be little temperature difference between peak metamorphism and the gold mineralisation event. The dominant geochemical change in the formation of the chlorite alteration zone is the pervasive addition of CO_2, with only minor K, S and low but anomalous Au. The zone is interpreted as reflecting low fluid-to-rock ratios.

The carbonate alteration zone is highly variable in thickness from a few centimetres to 100s m, but laterally continuous for over 1 km and to almost the same depth. Its characteristic minerals

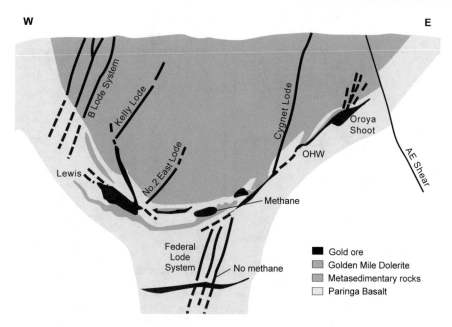

Fig. 13.5 Cross-section through the Oroya part of the Golden Mile along B-B' on Fig. 13.1. High grade gold ores including the Oroya Shoot are distributed in and adjacent to carbonaceous black shale near the base of the Golden Mile Dolerite intrusion. Fluid inclusions adjacent to the carbonaceous shale have some methane with the H_2O-CO_2; in contrast, fluid inclusions from 100 m deeper adjacent to (non-carbonaceous) interflow sedimentary rocks lack methane. Modified from Phillips et al. 2017.

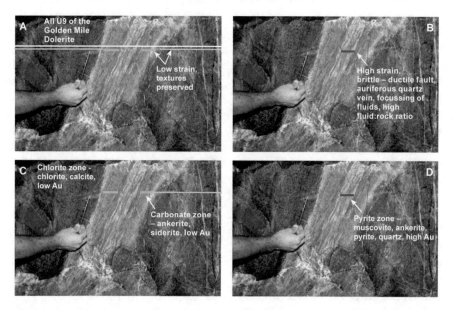

Fig. 13.6 A narrow sub-economic brittle – ductile fault (referred to historically as a lode) showing components of the gold-related alteration system in the Lake View mine, 412 east crosscut, Golden Mile, Kalgoorlie goldfield: (**a** and **b**) comparisons to document strain and fluid access; (**c** and **d**) comparisons to document the distinctive minerals reflecting alteration. Modified from Phillips et al. 2017.

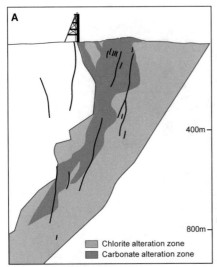
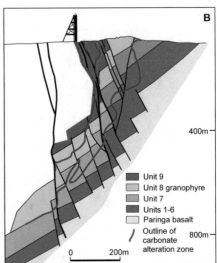

Fig. 13.7 Lake View Shaft section showing: (**a**) alteration zonation overlain on the mineralised brittle – ductile faults that have been stoped (removed through mining). These faults are in both alteration zones and cross their shared boundary; (**b**) mapped and inferred faults offsetting units of the GMD (note that these are not the same as the stoped areas in **a** though there is considerable overlap). With respect to research outcomes, it was a special day when the carbonate alteration zone was overlain on the GMD stratigraphy in **b** to reveal this spatial relationship of alteration with the Fe/Mg rich Units 7 and 8. Modified from Phillips 1986.

are carbonates particularly ankerite giving a cream to red brown colour and some hematite, and the low strain means the igneous textures are still preserved despite the complete re-organisation of the mineral assemblage. Compared to the chlorite zone, the carbonate zone involves more intense addition of CO_2, minor K, S and Au, and conditions of intermediate fluid-to-rock ratios. The geochemistry of the various alteration zones, particularly the K/Rb ratio, confirms that one auriferous fluid is implicated in the formation of all alteration zones.

The pyrite alteration zone coincides with the auriferous brittle – ductile faults (lodes) that are mined and the zone is from centimetres to several metres thick and laterally continuous for over 1 km and to almost the same depth. It is a more heterogenous interval than the other alteration zones with auriferous veins and both highly strained and unstrained pieces of altered GMD. It is characterised by common pyrite, carbonate minerals, muscovite, hematite, quartz veins and gold, and with a white to grey colour reflecting the veins and carbonate minerals. The pyrite alteration zone involves major additions of K, S and

Au with variable CO_2. The muscovite forms strongly foliated pieces of country rock intermixed with deformed quartz veins and is an important factor in focusing of strain and fluids into these brittle – ductile faults during ongoing localised deformation. The pyrite zone reflects the highest fluid-to-rock ratios. The localised addition of muscovite leads to even greater ductility, which means further strain is focussed in the same narrow alteration zone. As a result, even more gold-transporting fluid is also focussed here, and therefore gold deposition is strongly focussed too.

13.7 Alteration Geochemistry of the Golden Mile

The behaviour of various elements during alteration of the GMD can be divided into those that are essentially immobile throughout, those that are mobile throughout, and some that show evidence of mobility in the more proximal intervals of higher fluid-to-rock ratios (Table 13.1). The alteration geochemistry around the Oroya style mineralisation tends to be more complex with

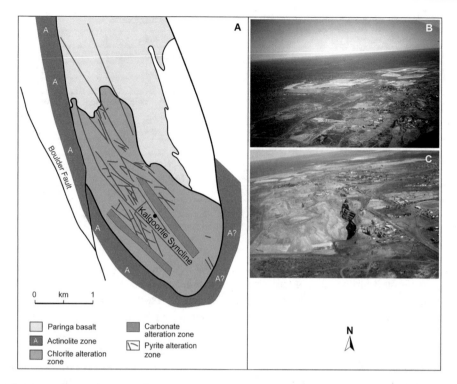

Fig. 13.8 (**a**) Map of the Golden Mile showing the extent of the chlorite alteration zone. This is effectively the extent of CO_2 addition during gold-related alteration and makes a distinctive target for exploration that seeks similar large goldfields; (**b**) aerial view over the Golden Mile at Kalgoorlie in the early 1980s prior to the Superpit with one hundred shafts that had operated in proximity and were interconnected underground; (**c**) similar view to **b** in late 1980s in the early stages of the Superpit (both looking south).

several initial country rock types (GMD, Paringa Basalt, carbonaceous shale, siltstone) that are potentially contributing components: the Oroya mineralisation also involves high inferred fluid-to-rock ratios.

13.8   Alteration Mineralogy of the Golden Mile

The alteration zones described here for the GMD in the Kalgoorlie goldfield reflect a practical scheme that can be used for underground mapping and logging core that has a theoretical petrological basis. The main differences in colour and minerals have been mapped since the early 1900s and much of these early data from long-since inaccessible areas could be integrated into mapping and thermodynamic analysis after 1980. The link between mineralogy (colour) and geochemistry is best understood by consideration of the main Fe-bearing minerals of each zone (Table 13.2) although the thermodynamic analysis is always with coexisting mineral assemblages.

The interpretation of the boundary between the chlorite and carbonate alteration zones is that it reflects increasing fluid-to-rock ratios more

Table 13.1 Element mobility during different degrees of gold-related alteration, Golden Mile.

Relatively immobile throughout alteration	Ti Al P Y Zr Ga Sc Nb La Nd Sm Eu Tb Tm Yb Lu
Mobile in proximal alteration zones	Si Fe Mn Co Ni Cu Zn V Cr Mg Ca Sr
Mobile throughout alteration zones	Au Ag As Se Sb Te Hg S K Rb Na Li Ba B W H_2O CO_2

Table 13.2 Alteration zones and mineralogy in the Golden Mile Dolerite, Kalgoorlie.

Zone	Unaltered	Chlorite	Carbonate	Pyrite
Colour	Dark green	Lighter green	Cream, bleached	Cream, pink, white
Type mineral	Actinolite	Chlorite	Carbonate	Pyrite, muscovite
Element	*Hosting minerals*			
Fe	Amphibole-titanomagnetite	Chlorite	Siderite-ankerite-hematite	Pyrite
Mg	Amphibole	Chlorite	Ankerite	Dolomite-ankerite
Ca	Amphibole	Calcite	Ankerite-calcite	Ankerite-calcite
Si	Amphibole-albite	Chlorite-albite-qz	qz	qz-muscovite
Ti	Titanomagnetite	Leucoxene	Leucoxene	Leucoxene
Al	Albite-amphibole	Chlorite-albite	Micas albite	Micas albite
Na	Albite	Albite	Albite-paragonite	Albite
K	Absent	Minor muscovite	Minor muscovite	Muscovite

Leucoxene is the term used for the mix of ilmenite, rutile, titanite and titanomagnetite (Travis et al. 1971)

proximal to the fault zones, but also reflects the chemical controls on chlorite; this mineral is more stable in more-Mg rocks such as Units 1-6 and 9 but breaks down readily to carbonate-bearing assemblages in more Fe-rich Unit 7 and 8 (Fig. 13.9).

Another source of diversity pertains to the carbonate minerals present in the alteration halo and stabilised by both CO_2 contributed by the auriferous fluid and the cations (Fe, Mg and Ca) contributed by the wallrocks. Therefore, variations in the country rock alone may be enough to vary the carbonate mineral assemblage without any variation in fluid composition.

In the Kalgoorlie goldfield at least five carbonate minerals occur in parts of the alteration halo: siderite (in Fe-rich dolerite), ankerite and calcite (dolerite and basalt), dolomite (in pyrite-bearing ore) and magnesite (ultramafic). The different minerals reflect different host rocks and do not imply any difference in the auriferous fluid. Alteration studies show that Fe, Mg and Ca are rarely added or removed in this wallrock alteration, so their concentrations in an alteration system reflect the host rock. For the GMD, ankerite is widespread throughout the alteration, but siderite is confined to the more Fe-rich rock types (e.g. Unit 7 and 8) where it is quite abundant (Fig. 13.10).

Fig. 13.9 Schematic interpretation of the alteration zonation in the Golden Mile illustrating a strong bulk rock geochemical control on the development of the carbonate alteration zone. Relative to chlorite, Fe-carbonate is much more stable in the higher Fe/Mg Unit 8 and upper Unit 7 of the GMD. Modified from Phillips 1986.

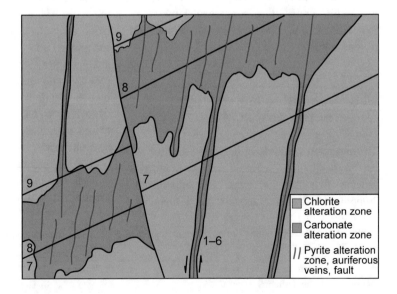

Fig. 13.10 Ca-Mg-Fe
diagram showing the bulk
rock composition of Units
7-9 as green bars, and the
tightly clustered analyses of
the composition of siderite,
ankerite and calcite in the
carbonate alteration zone as
pale pink fields. Dolomite –
ankerite compositions for
the pyrite alteration zone
are shown as pink dots and
are quite varied over 1-2
cm. The data are from the
Lake View area. Modified
from Phillips and
Brown 1987.

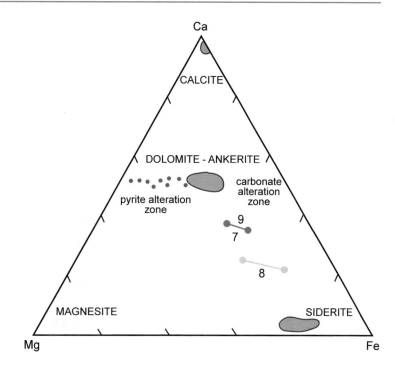

The pyrite alteration zone illustrates a further source of diversity with respect to carbonate mineral compositions (Fig. 13.10). More variation can be found in dolomite – ankerite mineral compositions within 2 cm than for ankerite compositions over 100s m in the carbonate alteration zone. The dolomite – ankerite mineral becomes much more Mg/Fe-rich approaching high-grade mineralisation with abundant pyrite. The interpretation is that the higher fluid-to-rock ratios in proximal alteration have introduced S in auriferous fluids that has combined with Fe in the country rocks to produce pyrite and leave the rest of the rock with high Mg/Fe to form dolomite.

An additional factor influencing carbonate mineral assemblages is the distribution of magnetite and hematite; the former is part of the igneous mineralogy as titano-magnetite and the addition of CO_2 in the alteration zones oxidises the auriferous fluid to the magnetite – hematite boundary (something that can be monitored with isotopic changes in sulfur). Hematite forms and imparts a local pink to red colour.

The full mineral assemblages of the various alteration zones combined with modern thermodynamic datasets can be modelled to indicate conditions during gold deposition. The distal halo is compatible with equilibrium between the country rock and a fluid of $CO_2/(CO_2+H_2O)$ of 0.1 – 0.25 at 315 – 320°C. The proximal interval reflects a fluid of 0.25. Fluid-rock buffering calculations show that, despite all the variations with host rock and proximity to faults, the alteration halo is consistent with interaction with a single fluid composition. The zoned nature of the halo reflects the volume of this fluid with which the rocks reacted, i.e. local fluid-to-rock ratios.

13.9   Factors Resulting in the Kalgoorlie Goldfield Being So Large

Despite this diversity it is still valid to summarise by saying "Nearly all the deposits of the Kalgoorlie goldfield lack economic base metals, are epigenetic and structurally-controlled and spatially associated with quartz veins and sulfide minerals."

There are several favourable features of the Golden Mile Dolerite that contribute to the size of the Kalgoorlie goldfield including:

- the host rock chemical composition such as total Fe and Fe/Mg
- the original host rock thickness
- its mechanical properties including its tensile strength
- a host sequence with rock units of rheological contrast
- the structural duplication of the GMD in the Golden Mile area
- the geometric orientation with respect to the far-field stress
- P-T conditions.

The first four factors are a consequence of the igneous processes during emplacement and crystallisation of the Dolerite; the last three factors are imposed much later at the time of metamorphism and deformation. The combination of original stratigraphy and deformation has created a structurally complex area of the goldfield with substantial repetition of faults of multiple orientations.

Any alteration halo comparable to the ~10 km^2 chlorite alteration zone surrounding the Golden Mile is a large target during a regional exploration program. Most of the features of the Golden Mile alteration are found around other goldfields including Cuiabá in Brazil where the sequence of alteration zones in mafic rocks is similar, and Natalka in Russia where the alteration halo is comparable in areal extent.

An important factor in the size of the Kalgoorlie goldfield is the great thickness of mafic rocks in the Golden Mile area and the orientation of this mass to the far field stress during the mineralisation event. This relationship is not clear using mine grid orientations but apparent once some major goldfields are oriented with respect to their far field stress at the time of mineralisation and then compared. This geometry involves a large mass of (tholeiitic) mafic material of the Paringa Basalt and GMD within less competent metasedimentary and ultramafic rocks resulting in multiple channelways in the GMD from high fluid pressure, and extensive mineralisation. Analogous patterns apply to Timmins, the largest goldfield in Canada, and the Jundee gold deposit in the northeast of the Yilgarn Craton.

Snapshot

- The large size of the Kalgoorlie goldfield is attributed to several factors at the magmatic emplacement stage, and further favourable factors during deformation and metamorphism.
- The Golden Mile Dolerite had favourable bulk rock composition, thickness, mechanical properties and contrasting rheological properties to adjacent rocks.
- The Kalgoorlie goldfield included structural repetition and had favourable P-T conditions and orientation to the far-field stress.
- Each of these favourable factors can be found in one or more other goldfields but most goldfields lack the full set of factors.
- The Kalgoorlie goldfield re-enforces the importance of structural repetition and complexity.
- A single auriferous fluid is implicated in generating many different styles of orebody.
- Changes of alteration mineralogy do not necessarily reflect changes in ore solution composition; they might equally well reflect changes in host rock or reaction progress.
- Auriferous quartz veins in metasedimentary rocks do not imply a sedimentary source of the fluid or gold; similarly, immediate igneous host rocks are unlikely to be the source of fluids or gold.

Bibliography

Boulter CA, Fotios MG, Phillips GN (1987) The Golden Mile, Kalgoorlie: a giant gold deposit localized in ductile shear zones by structurally induced infiltration of an auriferous fluid. Econ Geol 82:1661–1678

Clout JM, Cleghorn JH, Eaton PC (1990) Geology of the Kalgoorlie gold field. In: Hughes FE (ed) Geology of the mineral deposits of Australia and Papua New Guinea. Melbourne, The Australasian Institute of Mining and Metallurgy, pp 411–431

Evans KA, Phillips GN, Powell R (2006) Rock buffering of auriferous fluids in altered rocks associated with the Golden Mile-style mineralization, Kalgoorlie goldfield, Western Australia. Econ Geol 101:805–817

Groves DI, Phillips GN (1987) The genesis and tectonic control on Archaean gold deposits of the Western Australian shield – a metamorphic replacement model. Ore Geol Rev 2:287–322

Mueller AG (2020a) Structural setting of Fimiston- and Oroya-style pyrite-telluride-gold lodes, Paringa South mine, Golden Mile, Kalgoorlie: 1. Shear zone systems, porphyry dykes and deposit-scale alteration zones. Mineral Deposita 55:665–695. https://doi.org/10.1007/s00126-017-0747-3

Mueller AG (2020b) Paragonite-chloritoid alteration in the Trafalgar fault and Fimiston and Oroya-style gold lodes in the Paringa South mine, Golden Mile, Kalgoorlie: 2. Muscovite-pyrite and silica-chlorite-telluride ore deposited by two superimposed hydrothermal systems. Mineral Deposita 55:697–730. https://doi.org/10.1007/s00126-018-0813-5

Phillips GN (1986) Geology and alteration in the Golden Mile, Kalgoorlie. Econ Geol 81:779–808

Phillips GN, Brown IJ (1987) Host rock and fluid control on carbonate assemblages in the Golden Mile Dolerite, Kalgoorlie gold deposit, Australia. Can Mineral 25:265–274

Phillips GN, Groves DI, Kerrich R (1996) Factors in the formation of the giant Kalgoorlie gold deposit. Ore Geol Rev 10:295–317

Phillips GN, Hergt J, Powell R (2017) Kalgoorlie goldfield—petrology, alteration and mineralisation of the Golden Mile Dolerite. In: Phillips GN (ed) Australian ore deposits. The Australasian Institute of Mining and Metallurgy, Melbourne, pp 185–194

Ridley J, Mengler F (2000) Lithological and structural controls on the form and setting of vein stockwork orebodies at the Mount Charlotte gold deposit, Kalgoorlie. Econ Geol 95:85–98. https://doi.org/10.2113/gsecongeo.95.1.85

Travis GA, Woodall R, Bartram GD (1971) The geology of the Kalgoorlie goldfield. In: Glover JE (ed) Symposium on Archaean rocks, special publication 3. Geological Society of Australia, Canberra, pp 175–190

Vielreicher NM, Groves DI, Snee LW, Fletcher IR, McNaughton NJ (2010) Broad synchroneity of three gold mineralization styles in the Kalgoorlie gold field: SHRIMP, U-Pb, and $^{40}Ar/^{39}Ar$ geochronological evidence. Econ Geol 105:187–227

White RW, Powell R, Phillips GN (2003) A mineral equilibria study of the hydrothermal alteration of mafic greenschist facies rocks at Kalgoorlie, Western Australia. J Metamorph Geol 21:1–14

Hydrothermal Transport of Gold

14

Abstract

Gold is regarded as insoluble, and this is essentially correct in pure water. However, when other components are added to water, gold can be quite soluble at temperatures of 300°C and even near the surface at 25°C. Some of these fluids that dissolve gold can cause segregation from base metals by strong partitioning of different elements by selective solution.

Keywords

Gold solubility · Complexing agents · Ligands · Covalent bonding · Electronegativity

Part of the noble character of gold is that it is remarkably insoluble in pure water, and this has allowed coins and artefacts to last for thousands of years. Yet there is clear evidence that gold has been dissolved, migrated and reprecipitated in saline groundwaters and this migration can be important in mining of near surface deposits such as via open pits. The solubility of gold at elevated temperatures has been established and confirmed by laboratory experiments and is important in understanding the formation of gold deposits.

14.1 Chemical Properties of Gold

The chemical properties of gold are independent of how we classify and name deposits; and the properties apply universally, globally and through time. We might excise a favourite deposit type and make a case why it is different in some ways but ultimately it is the same element gold obeying the same physical and chemical behaviour. Consequently, it is not always helpful erecting barriers between different types of gold and its deposits because if all gold deposits are viewed together there might be a fuller understanding than the sum of many small parts.

Gold is a chemically stable metal in its neutral oxidation state, and this is the form of almost all gold found and mined. However, it can also exist in two more chemically active oxidation states under conditions that are relevant in the Earth's crust, these being Au^{1+} and Au^{3+}. Qualitatively, an Au^{1+} gold ion with its single positive charge to attract electrons is predicted to have a larger atomic radius than Au^{3+} as the latter has a higher charge that attracts the electrons more strongly. These predictions are correct with radii of 137 pm and 85 pm (i.e. 1.37 and 0.85 Angstroms; pm is a picometre – a billionth of a millimetre), respectively. To put the 137 pm into perspective, most metal cations (positive ions) are less than 100 pm and only the ions of alkali metals like K^+ and Rb^+ match the large size of Au^{1+}. This exceptionally large ionic size influences how Au^{1+} bonds with other elements and how it is transported in aqueous solution. As a result, Au^{1+} can be relatively mobile in aqueous fluids under conditions where Au^{1+} ions are the dominant oxidation state of gold. In general, the respective oxidation states suggest that Au^{3+} might favour more oxidising environments in nature such as on or near the land surface, whereas Au^{1+} will be stable in reducing environments more distal from any effects of atmospheric oxygen, i.e. deeper in the crust. Investigations in the chemistry laboratory have established that compounds of Au^{1+} usually adopt a linear co-ordination of two species around the gold atom, whereas compounds of Au^{3+} adopt a square planar co-ordination with four species around the gold atom. The difference in co-ordination numbers influences the number of ligands around gold ions.

14.2 Complexing Agents and Gold: Introducing Ligands

Despite gold being quite insoluble in pure water, the Au^{1+} and Au^{3+} ions can be quite soluble once combined with a complexing agent or ligand. This feature of gold being soluble when complexed is not unique but a feature of many metals and an essential part of forming major Cu, Pb, Zn and Ag ore deposits. The nature of the ligand species is important and can be quite specific to an element and an oxidation state.

For a gold complex to be important in nature it needs to meet two requirements. First the ligand needs to bond strongly with Au^{1+} or Au^{3+} to form a chemically stable and physically mobile complex. The second requirement is that the ligand needs to be present and stable in nature which eliminates many synthetic ligands that are studied in the laboratory.

Strong bonding between a ligand and a gold ion depends upon the bonding properties of both. If the gold ion favours bonding by sharing electrons (*covalent bond*) then it is important the ligand species is also predisposed to sharing electrons. Another option for a successful complex is a gold ion favouring giving up electrons (*ionic bond*) combining with a ligand species that favours accepting electrons. Much less effective will be attempted complexes between one species wanting to give or take electrons, and the other wanting to share.

The predisposition of different species for covalent or ionic bonding can be expressed by their hard / soft (also called Class A / Class B) behaviour. Softness is essentially a reflection of ease of polarisation and tendency to covalent behaviour. More specifically, hard cations bond more strongly with halides (anions) of low atomic number such as F, soft cations with halides of high atomic number like I and Br, and borderline cations show minimal preference down this column of the periodic table. A guideline is that anions and cations of similar hardness might form stable complexes together; hence hard cations combine better with hard anions, and soft cations combine better with soft anions. This suggests that Au^+, which is large with a low charge and hence the archetypal soft cation is well suited to form complexes in which it is bonded to reduced S. However, Au^+ is not suited to bond with hard anions such as F^- (Table 14.1). In contrast, Au^{3+} with a high charge and hence smaller size is harder and more predisposed to form ionic bonds with harder anions such as F^- and particularly Cl^-.

Common molecules in hydrothermal fluids that are not likely to form strong bonds with Au^{1+} are sulfate (SO_4^{2-}) and carbonate (CO_3^{2-}). Both meet the first criterion of a potentially effective ore-forming ligand by being relatively abundant in some natural waters. However, for both molecules, any bonding with Au^{1+} would be through the O atom and hence be Class A or hard in character and a poor combination with Au^{1+}.

Cyanide (CN^-) forms a stable complex with Au^{1+} with bonding through the C atom. The strong bonding between gold and cyanide has been widely utilised to dissolve gold during mineral processing since its first use on the Witwatersrand in 1890. The dissolving of gold with cyanide involves the oxidation of gold atoms and combination with cyanide to form sodium gold cyanide:

$4\ Au + 8\ NaCN + O_2 + 2\ H_2O => 4\ Na(Au(CN)_2) + 4\ NaOH$. Despite this being very effective on mine sites it is not viable in Nature because of the scarcity of cyanide in the Earth's crust.

Table 14.1 Hard and soft cations and anions as potential ligands during ore formation.

	Cation	Anion
Hard	H^+, Na^+, Ca^{2+}	OH^-, CO_3^{2-}, PO_4^{3-}, SO_4^{2-}, F^-, Cl^-
Borderline	Fe^{2+}, Cu^{2+}, Zn^{2+}, Pb^{2+}	
Soft	Ag^+, Au^+, Hg^+	H_2S, HS^-, S^{2-} [CN^- - not in Nature]

Two complexes relevant to gold deposit formation have been the subject of considerable laboratory experiments and their stability is known to elevated temperatures of several 100°C, one is Au^{3+} with Cl^- and the other is Au^{1+} with reduced S, i.e. $Au(HS)_2^-$.

At temperatures near 25°C, as applicable to the regolith, gold is soluble as chloride $AuCl_4^-$ and thiosulfate $Au(S_2O_3)^{3-}_2$ complexes (Webster and Mann 1984). Both these complexes of gold play a role in gold mobility in the near-surface environment and the migration of gold and growth of nuggets.

14.3   Gold Solubility in Saline Hydrothermal Fluids

Historically, research on metal chloride complexing received much attention during the 1960-70s because of the relevance to the porphyry copper industry especially of SW USA. The findings have been extended globally to other porphyry copper regions with elevated gold such as in the Pacific Rim. The conditions favouring solubility of gold (Au^{3+}) chloride complexes are acidic and oxidising conditions and high concentration of Cl^-, i.e. high salinity as demonstrated by their fluid inclusions with multiple daughter minerals including hematite. Elevated temperature of several 100°C increases solubility but even at 25°C gold is soluble in acidic saline groundwaters where it contributes to a viable deposit forming or modifying process. Thermodynamic calculations and laboratory experiments demonstrate that achieving levels of several ppm Au in solution is quite feasible in acidic saline solutions at elevated temperatures (Fig. 14.1).

14.4   Gold Solubility in Low Salinity Hydrothermal Fluids with Reduced Sulfur

Experiments demonstrate that gold is soluble in its Au^{1+} form complexed with reduced S as predicted from hard – soft theory (Table 14.1). The gold bonds in linear co-ordination meaning two ligands per central Au^{1+} giving $Au(HS)_2^-$ as the relevant ion complex for many geological conditions. To show the effects of redox and acidity, a diagram of $\log[fO_2]$ plotted against pH includes fields of common Fe minerals – hematite with sulfate and bisulfate in oxidising conditions (less appropriate here), and pyrite, pyrrhotite and magnetite stability linked to pH in the reduce S fields.

The experiments using $Au(HS)_2^-$ demonstrate that achieving levels of several ppm Au in solution is quite feasible at elevated temperatures (Fig. 14.2). The contours of gold in solution centre on a point of intermediate redox and acidic conditions corresponding to the field of reduced S (H_2S, HS^-, S^{2-}) and oxidised C (CO_2).

14.4.1   Maximising Gold Solubility and the Role of CO₂

It is one thing to identify a $\log[fO_2]$-pH space in which gold is soluble in realistic fluids but quite another to determine how such conditions can be achieved and maintained in nature. This becomes a matter of achieving high gold concentrations, maintaining those levels during migration of an auriferous fluid through the crust, and then having gold deposition in a constrained volume of rock. It is less useful to have substantial migration of gold that involves unfocused deposition through

Fig. 14.1 Solubility of gold in an H-O-Cl-S system as a function of fO_2 and pH. For high salinity and acid oxidising conditions (top left), Au^{3+} is soluble as a chloride complex. fO_2 is oxygen fugacity and essentially a measure of more oxidising conditions up the page, and more reducing conditions down the page. https://geoscienceaustralia.shinyapps.io/fo2-ph-contour/ © Commonwealth of Australia (Geoscience Australia) 2020.

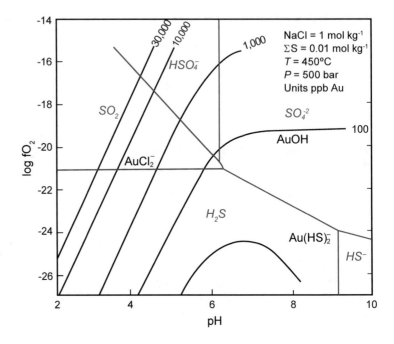

Fig. 14.2 Gold solubility contours (dashed lines) highlighting the importance of redox ($\log[fO_2]$) and acidity (pH) conditions in maximising gold in solution. Oxidising conditions are up; acid conditions are to the left. The figure is for 300°C and shows parts-per-million levels of Au in solution are achievable under realistic conditions of the greenschist facies. Abbreviations: hem, hematite; mt, magnetite; py, pyrite; po, pyrrhotite. Modified from Phillips and Powell 2010, based on Stefánsson and Seward 2004 and reproduced with permission.

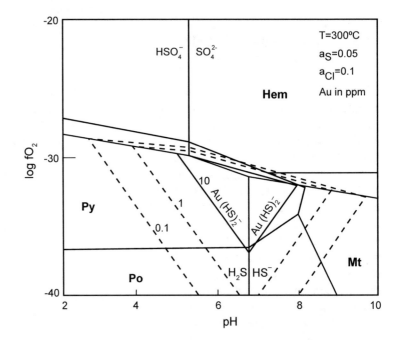

many 10s km which is hardly conducive to economic mineralisation.

The reaction of gold going into solution as the $Au(HS)_2^-$ complex can be illustrated with the equation:

$$Au + HS^- + H_2S + 0.25\,O_2 = Au(HS)_2^- + 0.5\,H_2O \quad (14.1)$$

The equilibrium constant disregarding water is

$$K = Au(HS)_2^- / [HS^-] *[H_2S]*[O_2]^{0.25}$$

Which is rearranged to give $[Au(HS)_2^-]$

$$= [HS^-] \times [H_2S] \times [O_2]^{0.25}/K$$

$$(14.2)$$

Ostensibly having elevated HS^-, H_2S and O_2 will drive Eq. (14.1) to the right and increase gold in solution but there are several conflicting forces here. If fO_2 is too high then much of the S will be as sulfate rather than the sulfide, lowering the two S components on the left of Eq. (14.1) and lowering gold in solution. Separately if fO_2 is too low it will force the equation to the left; a balance of intermediate redox is optimal.

There are conflicting forces also between the reduce S species according to the dissociation of H_2S ($H_2S = HS^- + H^+$). For reducing conditions of minimal sulfate, Total S = $H_2S + HS^-$. Maximum gold solubility from Eq. (14.2) is achieved by maximising the product $[H_2S]*[HS]$ which is achieved by having $[H_2S] = [HS^-]$. The condition of $[H_2S] = [HS^-]$ occurs at a specific pH (Fig. 14.3); and that optimal pH is dependent upon temperature.

In order to avoid haphazard gold deposition as the auriferous fluid passes through various rock sequences in the crust, effective gold transport in nature as $Au(HS)_2^-$ requires that the necessarily mildly acidic pH conditions do not vary dramatically. This requirement is facilitated when fluid flow in major structures is *fluid-dominated* (high fluid-to-rock ratios) with minimal influence of the enclosing rocks on the auriferous fluid composition.

A separate requirement of effective gold transport is the maintenance of fluid pH close to that of $[H_2S] = [HS^-]$ coinciding with maximum gold solubility; and that optimal pH varies with temperature. Fluid pH is dictated by any weak acid that is in abundance and here it is CO_2 as H_2CO_3 in aqueous fluid that is inferred to play the critical role as by far the dominant component in the fluid after H_2O. CO_2 in water is a weak acid so it can act as a buffer of pH and resist any changes to solution pH. CO_2 has a second role because its dissociation has a similar relationship to pH as does H_2S (see below).

The dissociation: $H_2CO_3 = HCO_3^- + H^+$, by involving H^+, acts as a buffer on pH in the auriferous fluid. The buffering of fluid pH, or holding pH within a narrow range, arises because a slight increase in H^+ causes the reaction to shift to the left consuming H^+ and resisting the pH change. A slight decrease in H^+ causes the dissociation reaction to shift to the right creating H^+ and again resisting the pH change. With CO_2 comprising 20-30% of the auriferous fluid, there is considerable buffering capacity to minimise any variation in pH of the fluid and minimise any loss of Au during migration through the crust.

One of the more intriguing observations of dissociation of various weak and strong acids of geological importance is how similar H_2CO_3 and H_2S are over a wide temperature range (Fig. 14.4). This means that in a fluid dominated by CO_2, the pH can be maintained very close to the ideal pH for gold solubility as $Au(HS)_2^-$ even allowing for some variation in temperature in the crust. This suggestion solves two long standing issues for gold, first is how pH in an auriferous fluid is maintained during long distance migration through the crust as temperature changes. Second, it also explains why CO_2 seems so important in auriferous fluids globally and through earth history even though it has no direct role in bonding to and complexing with Au.

The role of CO_2 in buffering of pH and in maintaining ideal conditions for gold transport represents an *unreasonably effective* explanation of two observations previously thought to be unrelated; namely "Why is there CO_2 in the fluid and alteration assemblages?" and "How is pH regulated during Au transport?".

14.4.2   Deposition of Gold

The two depositional methods of most importance in gold-only deposits are reduction and wallrock sulfidation. Both can be illustrated on the log[fO2]-pH figure below (Fig. 14.5) by shifts away from the field of maximum gold solubility.

An effective reducing agent in the crust is carbon represented in different goldfields by any

Fig. 14.3 Maximum gold solubility of Au(HS)$_2^-$ occurs at a specific mildly alkaline pH (purple line) and corresponds to an intermediate redox state (two red lines) below the hematite and sulfate dominant field. Modified from Phillips and Powell 2010.

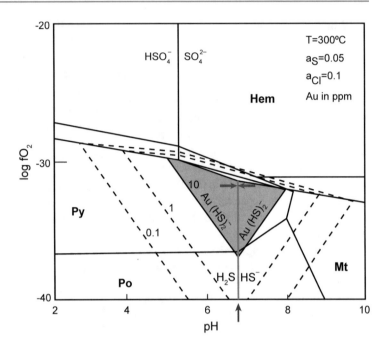

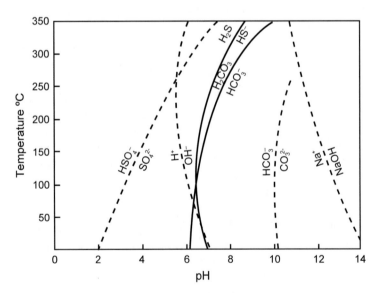

Fig. 14.4 Ionisation of some aqueous acids and bases as a function of temperature and pH. The curves represent equal activities of aqueous species and separate their various regions of predominance. The striking proximity of the H$_2$S and H$_2$CO$_3$ curves over a wide temperature range means that any fluid buffered near [H$_2$CO$_3$] = [HCO$_3^-$] will be within one pH unit of the pH of maximum gold solubility determined by [H$_2$S] = [HS$^-$]. H$_2$S data taken from Suleimenov and Seward (1997) and used with permission.

of black shale, black slate, carbonaceous shale, carbon seams and flyspeck carbon. Maximum gold solubility is within the field of oxidised

carbon (CO$_2$) so the effect of interacting with reduced C will be to move the fluid to much lower redox conditions which will lower gold

Fig. 14.5 Moving the redox conditions of the auriferous fluid from its optimal position (dot) into reducing conditions (along the arrow to lower log fO₂) leads to gold deposition. This is a common deposit forming process adjacent to carbonaceous rocks. Modified from Phillips and Powell 2010.

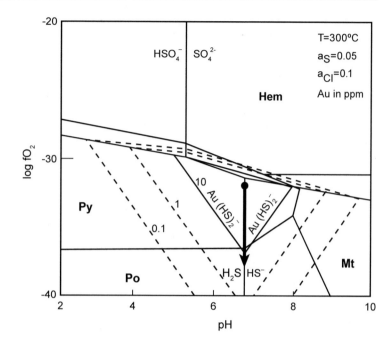

solubility levels and initiate gold deposition. The association of gold-only deposits with reduced C includes Archean Witwatsrand and greenstone gold, and Paleozoic turbidite-hosted gold and many Mesozoic and Cenozoic deposits.

Wallrock sulfidation is also an indirect but effective way to deposit gold and illustrates the importance of alteration. Reaction between the auriferous fluid and Fe-bearing country rocks can lead to the formation of pyrite (FeS_2, also known as fool's gold) and this is a common mineral in and around quartz veins in gold deposits globally. A consequence of this formation of pyrite is the removal of S from the auriferous fluid in which one of its roles was to be bonded as the gold-bisulfide complex $Au(HS)_2^-$. Lowering of the S concentration lowers Au solubility (Eq. 14.1) and leads to gold deposition with pyrite and the association of many gold-only deposits with Fe-rich rocks (Fig. 14.6). The potential Fe-rich rocks are quite varied including hematite and magnetite, and Fe-bearing silicates like chlorite.

Specific goldfields are not restricted to just the one gold precipitation method. Kalgoorlie for example (Chapter 13) illustrates the importance of reduction in Oroya-type orebodies adjacent to

carbonaceous shale, sulfidation of wallrock Fe in much of the Golden Mile Dolerite and Paringa Basalt, and possibly oxidation as magnetite reacts with CO_2 and redox conditions become more oxidising toward the magnetite – hematite boundary.

14.5 Electronegativity, and the Other Elements Expected in Gold-Only Fluids

The low salinity, sulfide-bearing nature of the auriferous fluid places limits on other elements that are in solution with gold. Some elements will be in solution because they are large, low charge and readily soluble like K, Rb, Na and Ba; these are the large ion lithophile elements. The degree to which further metals share electrons in effective covalent bonds with S can be used to predict additional elements in auriferous solutions.

Electronegativity is a measure of attraction exerted by an atom on a shared electron in a covalent bond (Zumdahl 2002). It is a number up to 4 but without units, with the lowest electronegativity being towards the bottom left of the

Fig. 14.6 Reaction of S in solution with Fe in country rocks to form pyrite (FeS$_2$) lowers the gold concentration in the fluid, as represented by the diagrammatic collapsed contours applicable to the lower S and thus lower [Au (HS)$_2^-$]. Modified from Phillips and Powell 2010.

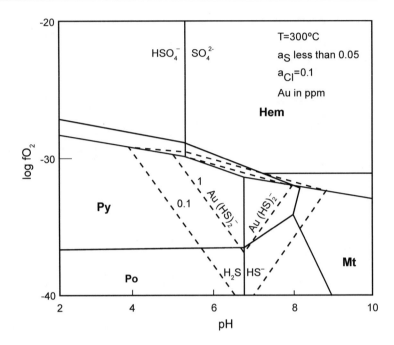

periodic table (Cs) and highest electronegativity in the top right (F). Two atoms are likely to form an effective covalent bond if they exert similar attraction on the electrons they are sharing, i.e. they have similar electronegativity. Using the electronegativity values of Pauling (Zumdahl 2002), gold at 2.54 on this scale is the highest of all metals, and close to S of 2.58 (Fig. 14.7) which explains why the complexation of Au with reduced S is so effective. Cation-forming elements with electronegativity of 2.0 and above include:

- Au, As, Se, Sb, Te, B, Hg, W, Bi and Mo.

This list is remarkable for its similarity to the elements added by the gold-only fluid during alteration (Chapter 12):

- Au, As, Se, Sb, Te, B, Hg, W with Ag, S and lesser Mo, Bi and Pb.

These elements commonly associated with gold-only fluids are not a suite usually associated with chemical or clastic sedimentary processes nor with silicate or sulfide magmas as either compatible or incompatible elements. Their association is better understood by reference to electronegativity and their chemical similarities to S and Au in aqueous fluid.

The base metals of Cu, Pb and Zn have low electronegativity, strongly favouring ionic bonding with Cl$^-$. These will not readily form covalent bonds with reduced S. Lead is one of the elements for which electronegativity has been determined for both oxidation states (2$^+$ = 1.87, and 4$^+$ = 2.33) and the lower electronegativity of Pb^{2+} is used here as it is appropriate for the reduced conditions implied by sulfide stability and gold-only fluids.

Electronegativity helps to explain the bonding with reduced S and the unusual group of elements enriched along with gold in gold-only deposits. It also addresses one of the five major characteristics of gold-only deposits being the segregation of Au from base metals Cu, Pb and Zn; this is the partitioning of Au into the auriferous fluid, but without base metals. Platinum group elements do bond with S to form bisulfide complexes but appear much less soluble than Au in gold-only fluids.

The elements associated with gold-only deposits (Au, Ag, As, Se, Sb, Te, B, Hg, W, Bi, Mo) are quite distinct from the element suites from other ore deposit types:

- Sulfide melt: Ni Cu Au PGEs (Os Ir Ru Rh Pt Pd)

H 2.20																	He
Li 0.98	Be 1.57											B 2.04	C 2.55	N 3.04	O 3.44	F 3.98	Ne
Na 0.93	Mg 1.31											Al 1.61	Si 1.90	P 2.19	S 2.58	Cl 3.16	Ar
K 0.82	Ca 1.00	Sc 1.36	Ti 1.54	V 1.63	Cr 1.66	Mn 1.55	Fe 1.83	Co 1.88	Ni 1.91	Cu 1.90	Zn 1.65	Ga 1.81	Ge 2.01	As 2.18	Se 2.55	Br 2.96	Kr 3.00
Rb 0.82	Sr 0.95	Y 1.22	Zr 1.33	Nb 1.6	Mo 2.16	Tc 1.9	Ru 2.2	Rh 2.28	Pd 2.20	Ag 1.93	Cd 1.69	In 1.78	Sn 1.96	Sb 2.05	Te 2.1	I 2.66	Xe 2.60
Cs 0.79	Ba 0.89	La 1.1	Hf 1.3	Ta 1.5	W 2.36	Re 1.9	Os 2.2	Ir 2.20	Pt 2.28	Au 2.54	Hg 2.00	Tl 1.62	Pb 1.87	Bi 2.02	Po 2.0	At 2.2	Rn 2.2

La 1.1	Ce 1.12	Pr 1.13	Nd 1.14	Pm 1.13	Sm 1.17	Eu 1.2	Gd 1.2	Tb 1.1	Dy 1.22	Ho 1.23	Er 1.24	Tm 1.25	Yb 1.1	Lu 1.27

Fig. 14.7 Electronegativity describes the power of an atom in a molecule to attract electrons to itself. Many elements with values below 2.0 involve ionic bonds. Gold is in a smaller group of elements, highlighted above in green, with electronegativity between 2.0 and 2.54; these form strong bonds with reduced S and are common around gold-only deposits. Gold and the related elements do not form strong bonds in nature with the strongly anionic elements, with electronegativity between 3.04 and 3.98, highlighted above in red. Source data based on Zumdahl 2002.

- Saline hydrothermal such as VMS deposits: Cu Zn Pb Ag Au and many others
- Saline high-T oxidising: Cu Au (Bi Co REE U)
- Silicate magma – felsic: K Na Rb Sn Th U
- Silicate magma – mafic: Ti V (Cu Zn)
- Silicate magma – ultramafic: Cr Ni Co.

14.6   Quartz and the Solubility of Silica

In many different forms, silica is the major ingredient of gold mineralisation globally. It may be virtually pure quartz in veins, laminated quartz veins with incorporated slivers of wallrock, pressure-solved veins, finer grained quartz described as cherty, colloform or with foliated wallrock and mapped as 'lode' (Kalgoorlie), or incorrectly inferred to be part of the stratigraphy. Although quartz is essentially insoluble at 25°C it becomes quite soluble at 100°C and its solubility increases systematically with temperature towards the critical point of water, and above that its solubility is retrograde, meaning that it decreases with increasing temperature (Fig. 14.8). For pressure of 20 Mpa, coinciding with about 6 km depth the solubility is over 500 ppm at 300°C.

For a fluid rising through the crust to lower temperatures there will be a tendency for quartz to deposit, explaining the formation of many quartz veins. A corollary is that descending waters moving to higher temperatures may have the capacity to dissolve and remove quartz.

14.7   The Gold-Only, Gold-plus Classification: Unreasonably Effective?

The geochemistry of gold in aqueous fluids completes the information required to forward model from the chemistry laboratory, and merge this with the inverse (backward) modelling from observations.

The classification of gold-only and gold-plus is practical to apply and is based on a simple question (does the mine have economic base metals? – Fig. 14.9) with an essentially unambiguous answer. With caution the question can be answered talking to a mine manager, looking through Resource figures in company annual reports, or reading historical descriptions of

Fig. 14.8 Solubility of quartz in pure water. As metamorphic fluids rise through the crust and cool, they have a high potential to deposit quartz veins (based on Morey et al., 1962).

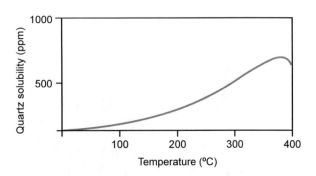

CRITICAL OBSERVATIONS

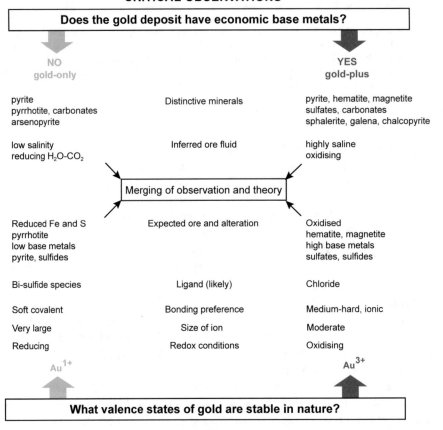

Does the gold deposit have economic base metals?

NO	YES
gold-only	gold-plus

pyrite	Distinctive minerals	pyrite, hematite, magnetite
pyrrhotite, carbonates		sulfates, carbonates
arsenopyrite		sphalerite, galena, chalcopyrite
low salinity	Inferred ore fluid	highly saline
reducing H_2O-CO_2		oxidising

Merging of observation and theory

Reduced Fe and S	Expected ore and alteration	Oxidised
pyrrhotite		hematite, magnetite
low base metals		high base metals
pyrite, sulfides		sulfates, sulfides
Bi-sulfide species	Ligand (likely)	Chloride
Soft covalent	Bonding preference	Medium-hard, ionic
Very large	Size of ion	Moderate
Reducing	Redox conditions	Oxidising
Au^{1+}		Au^{3+}

What valence states of gold are stable in nature?

THEORY

Fig. 14.9 In the Preface and early chapters of this book a division into gold-only and gold-plus deposits was made with little justification at the time but with a promise that the classification would be revisited and tested. The integration outlined above illustrates that the gold-only and gold-plus classification is practical to apply using a simple question about the presence or absence of economic base metals, and it is underpinned with sound theory in gold chemistry. Modified from Phillips and Powell 2015.

ancient gold mining operations. The classification avoids the difficulties inherent in schemes that are based on how deposits form, e.g. mesothermal (meaning formed at moderate temperatures), epithermal, magmatic-related, and orogenic. Given that one purpose of studying all gold deposits is to determine how they form, the four genetic classifications just mentioned may present a certain circular logic of setting up classifications based on how deposits form, and then determining their modes of formation.

As well as being practical, the gold-only and gold-plus classification also has a firm theoretical basis that is readily traced back to the basic inorganic chemistry of gold. Simplistically, gold deposits can form utilising either the Au^{1+} or Au^{3+} species with no overlap of these two options. They involve quite different chemical environments and quite different ore and gangue mineral associations as shown in Fig. 14.9.

There is a contrast between the longevity of the gold-only, gold-plus classification without modifications, and the regular modifications and definition broadening of other terms applied to gold deposits such as epithermal, Carlin-type and magmatic.

The idea of classifying gold deposits based on their ore forming fluids is not new. As early as 1950, Noble suggested that "The question is raised whether the widely accepted depth-zone classification of ore deposits should not be replaced by a classification based on composition of the ore-forming fluids". When that statement was made the science necessary to fulfil Noble's suggestion was not available; starting with Bill Fyfe in the 1970s we are much better placed today.

Snapshot

- Two oxidation states of gold are relevant in many natural processes (Au^{3+} and Au^{1+}).
- Au^{3+} is the smaller ion preferring Class A, harder, more *ionic bonding*.
- Au^{1+} is the larger Class B ion preferring *covalent bonding*.

- Cl^- is a viable complexing agent with Au^{3+} in oxidising acidic saline fluids.
- HS^- is a viable complexing agent with Au^{1+} in moderately reducing near neutral low salinity fluids.
- Base metals (Cu, Zn, Pb) form stable complexes with Cl in saline fluids.
- CO_2 has an important role in low salinity fluids by buffering the fluid pH to optimise dissolved gold concentrations.
- The gold-only, gold-plus classification has a sound theoretical basis in gold chemistry.

Bibliography

Finklestein NP, Hancock RD (1974) A new approach to the chemistry of gold. Gold Bull 7:72–77

Helgeson HC (1964) Complexing and hydrothermal ore deposition. Pergamon, Oxford

Morey GW, Fournier RO, Rowe JJ (1962) The solubility of quartz in water in the temperature interval from 25° to 300°C. Geochim Cosmochim Acta 26(10): 1029–1040. https://doi.org/10.1016/0016-7037(62)90027-3

Noble JA (1950) Ore mineralization in the Homestake gold mine, Lead, South Dakota. Geol Soc Am Bull 61:221–252

Pearson RG (1963) Hard and soft acids and bases. Am Chem Soc J 85:3533–3539

Phillips GN, Evans KE (2004) Role of CO_2 in the formation of gold deposits. Nature 429:860–863

Phillips GN, Groves DI (1983) The nature of Archaean gold-bearing fluids as deduced from gold deposits of Western Australia. Geol Soc Aust J 30:25–39

Phillips GN, Powell R (2010) Formation of gold deposits: a metamorphic devolatilization model. J Metamorph Geol 28:689–718

Phillips GN, Powell R (2015) A practical classification of gold deposits, with a theoretical basis. Ore Geol Rev 65:568–573. https://doi.org/10.1016/j.oregeorev.2014.04.006

Puddephatt RJ (1978) The chemistry of gold. Elsevier, Amsterdam

Ridley J (2013) Ore deposit geology. Cambridge University Press, New York, p 398

Seward TM (1973) Thio-complexes of gold and the transport of gold in hydrothermal ore solutions. Geochim Cosmochim Acta 37:379–399

Seward TM, Williams-Jones AE, Migdisov AA (2014) 13.2 the chemistry of metal transport and deposition by ore-forming hydrothermal fluids. In: Treatise on Geochemistry, vol 13.2, 2nd edn. Elsevier, Amsterdam, pp 29–57. https://doi.org/10.1016/B978-0-08-095975-7.01102-5

Stefánsson A, Seward TM (2004) Gold(I) complexing in aqueous sulphide solutions to 500°C and 500 bar. Geochim Cosmochim Acta 68:4121–4143. [the source of the Eh-pH figures 2–6]

Suleimenov OM, Seward TM (1997) A spectrophotometric study of hydrogen sulphide ionisation in aqueous solutions to 350°C. Geochim Cosmochim Acta 61: 5187–5198

Webster JG, Mann AW (1984) The influence of climate, geomorphology and primary geology on the supergene migration of gold and silver. J Geochem Explor. 22: 21–42. https://doi.org/10.1016/0375-6742(84)90004-9

Wood SA, Samson IM (1998) Solubility of ore minerals and complexation of ore metals in hydrothermal solutions. Rev Econ Geol 10:33–80

Yardley BWD (2005) Metal content of crustal fluids and their relationship to ore formation. Econ Geol 100: 613–632

Zumdahl SS (2002) Chemical principles. Houghton Mifflin, Boston, p 1047

Metamorphic Processes Leading to Gold-Only Deposits

15

Abstract

Metamorphic processes can explain and predict the five important characteristics of gold-only deposits outlined in Chapters 4–8 as well as the diversity and commonality between deposits. The metamorphic devolatilisation model can be summarised in terms of a source of auriferous fluid, the migration of that fluid, and then deposition of gold. Metamorphism of altered basaltic rocks and some clastic metasedimentary rocks at $400 - 500^{\circ}C$ and $10 - 15$ km depth produces a fluid dominated by H_2O-CO_2-H_2S that carries gold. This auriferous fluid is produced on a grain-by-grain scale before it amalgamates and moves via a fracture network upwards to around 5 km depth and $350^{\circ}C$. Active deformation and high fluid pressure facilitate fluid-rock interaction and quartz vein formation. The deposition of gold is primarily, but not exclusively, by reduction and wallrock sulfidation.

Keywords

Devolatilisation · Metamorphism · H_2O-CO_2-H_2S fluid · Reduction · Sulfidation

The distance of transport implied by the enrichment calculations for gold and related elements means that studies and classifications based solely on the deposit site are not adequate to answer how gold deposits form. Direct observational evidence at various scales and considerable indirect evidence need to be integrated with forward modelling from theory.

15.1 Source of Auriferous Fluid

The source of the auriferous fluid is from the metamorphism of thousands of km^3 of rock in response to elevated temperature. Some sequences such as altered basaltic rocks and greywacke have hydrous minerals, carbonates and minor sulfide that react to produce a supercritical H_2O-CO_2-H_2S fluid that dissolves nanometric gold at the time of fluid formation. This ability of the metamorphic fluid to dissolve gold as a complex with reduced S (as the gold bisulfide complex $Au(HS)_2^-$) means that devolatilisation has access to vastly more gold than alternative processes.

The breakdown of hydrous minerals to release fluid during progressive metamorphism is well-documented in both the field and via laboratory experiments. It is also established that hydrous minerals can break down more readily when combined with carbonates and sulfide. That gold is soluble in such fluids, as predicted from theory and confirmed in experiments, demonstrates that levels of gold of 1 ppm or more in such fluids are feasible. If the original rock has up to 5% volatiles and the gold source has 1-2 ppb Au it is likely that most if not all the gold can be extracted before the supply of fluid is exhausted. As this whole devolatilisation process is happening in the

© The Author(s), under exclusive license to Springer Nature Singapore Pte Ltd. 2022
N. Phillips, *Formation of Gold Deposits*, Modern Approaches in Solid Earth Sciences 21,
https://doi.org/10.1007/978-981-16-3081-1_15

metamorphic domain with no porosity, and normally no Cl-bearing minerals, the fluid will be of low salinity; this means that there is no significant transport of base metals, and hence a major partitioning and concentration of gold versus base metals is achieved right at the source of the fluid. Other elements with similar chemical properties to Au in aqueous fluids (particularly electronegativity) will be soluble as well, including Ag, B, As, Sb, Se, Te, Hg, Mo, W and Bi.

The metamorphic condition under which the devolatilisation reactions occur is estimated to be 400 – 500°C based on thermodynamic modelling but this cannot be confirmed by observation or measurements alone. This equates to approximately 10 – 15 km depth for the inferred relatively high paleo-geothermal gradients.

This metamorphic devolatilisation model postulates the leaching of ppb concentrations of gold from rocks undergoing prograde metamorphism. Such a loss is confirmed in a study in Otago, New Zealand (Pitcairn et al. 2006). These authors showed that 60% of gold was lost during progressive metamorphism to 600°C. Interestingly they found a major loss of As, Sb, Hg, and some loss for Mo and W; but no loss of the base metals Cu, Zn and Pb. Overall, the elements lost during metamorphism in Otago bear a close relationship to elements added in alteration haloes around gold deposits and predictions from electronegativity (see Fig. 14.7).

15.2   Fluid Channelways

The role of faults and other structures as pathways for auriferous fluids has revolutionised the understanding of how gold deposits form. At different scales, these structures have the capability to move fluids for kilometres through the Earth's crust and it is likely that earthquakes are important as low strain events creating transient permeability. In some cases, there will be focusing of large amounts of fluids into small volumes with the potential to form ore deposits (Fig. 15.1). Obviously, there will be many other times and situations in which focusing does not happen or even fluid dispersal occurs with no useful

mineralisation. On a local scale, structure is the critical influence on ore geometry for exploration and mine geology in most goldfields.

Metamorphic fluids are inferred to form at millimetric-scale grain boundaries, coalesce and migrate along microcracks and fracture networks and then move into small and then larger faults. This is a *rock dominated* chemical system with low fluid-to-rock ratios.

Further fluid concentration and migration is inferred along larger fault zones in the crust which can be imaged by seismic studies to 10s km depth and by mapping for 10s and 100s of km on the surface. High fluid pressure and areas of low mean rock stress play an important role in dictating the fluid flow. These faults give the access for auriferous fluids to move laterally and upward through the crust with minimal interaction with any country rocks during this migration. This has now become a *fluid dominated* chemical system with high fluid-to-rock ratios and internal buffering in the fluid to maintain the elements being transported in solution.

Near the transition from ductile to brittle conditions the high fluid pressures can cause hydraulic fracture of one or more rock types giving much greater fluid access to the country rock. Under directed stress, rock units of different structural competence in a layered sequence will behave differently to create areas of low mean stress into which fluids can migrate. The net result can be an orebody in one rock type with barren rocks adjacent, a common example of which would be a mineralised igneous intrusion within barren metasedimentary sequences (Fig. 15.2), such as the Golden Mile Dolerite in the Kalgoorlie goldfield (Chapter 13).

15.3   Deposition of Gold from Fluid

Of several gold depositional mechanisms, the two most important and widespread appear to be reduction particularly by carbonaceous material, and sulfidation of Fe-bearing country rocks. An example of a third but less universal depositional mechanism might be the oxidation of auriferous fluids through reaction with country rock

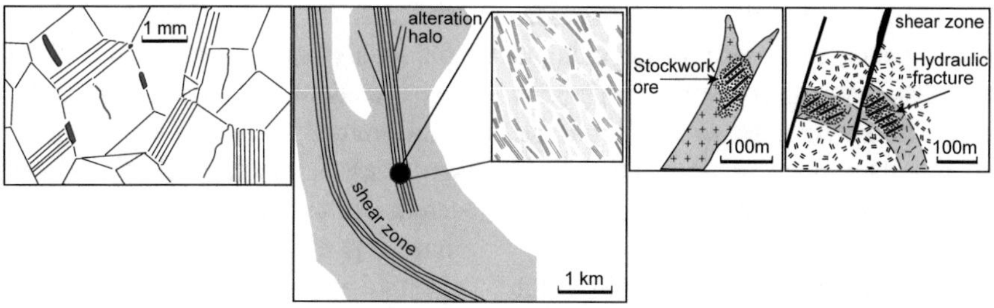

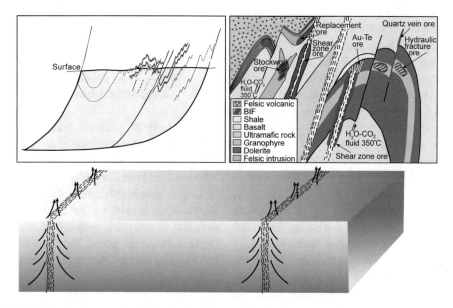

Fig. 15.1 Stages of fluid flow in the metamorphic environment: note that the four sub figures are at very different scales: (**left**) the source stage in which minute amounts of auriferous fluid develop on grain boundaries before coalescing; (**centre**) large fault zones into which the fluid at high pressure coalesces and then migrates laterally and upwards; (**two sub-figures on right**) at the gold depositional site fluid selectively infiltrates some rock types because of rheological contrasts and especially low tensile strength.

Fig. 15.2 Gold formation process (**bottom**) showing fluid generation within the block of 1000s km³ during regional metamorphism. (**upper left**) shows channelways of several km dimension leading into structural sites for ore deposits. (**upper right**) shows the complexity of orebody types in one goldfield (mineralisation is red) and includes several orebodies in igneous host rocks because of their low tensile strength leading to hydraulic fracture and auriferous fluid ingress. Other orebodies are in metasedimentary rocks parallel or oblique to bedding.

magnetite to shift the fluid to the hematite field, stabilise sulfate, and cause gold deposition by the lowering of reduced S in solution.

The composition of the auriferous fluid as determined from fluid inclusions is H_2O-CO_2-H_2S bearing, and this is predicted by thermodynamic analysis from the metamorphism of altered basaltic rocks. Some sedimentary rocks, such as immature greywackes, will produce a similar fluid.

15.4   Optimising Metamorphic Devolatilisation to Form Giant Gold Deposits

Despite a common practice to look for unusual or even unique geological features at the largest goldfields, it appears that the largest deposits have been formed by the same set of processes and conditions operative at many of the medium to small deposits. The largest goldfields appear to owe their size to having more of the favourable factors and have these much better developed and operating at larger scales compared to many smaller deposits.

The background concentration of gold in the source country rocks may directly impact the endowment of a province or the largest goldfield. All other factors being equal, background Au of 5 ppb rather than 1 ppb should be advantageous. For basaltic rocks, higher degrees of partial melting of the parent peridotites leads to greater amounts of sulfide (and therefore Au) extracted into the magma. Therefore, komatiite and plume-related magma (higher degrees of melting) have higher Au and should be better source material than mid ocean ridge basalt (low degrees of melting). The degree of melting on its own does not form gold deposits but it might contribute to some basaltic rocks being better sources of gold than others during metamorphism.

Alteration of the source rocks prior to burial is critical within the suggested metamorphic devolatilisation model. Hydrous minerals, carbonates and minor pyrite are all ingredients of the alteration assemblages that are important in the formation of auriferous fluids. Once any one of these minerals is exhausted or all the available Au is consumed by reaction it could mean the end of production of the auriferous fluid. Oxidising conditions, such as in continental basalt and red beds, might be counter-productive by precluding pyrite and mitigating against later formation of any auriferous fluid.

Sedimentary source rocks containing chlorite such as pelites and greywacke can replicate many features of metamorphosed basalt and with even higher gold concentrations being possible (Tomkins 2010). If gold is a component that is consumed early in the source reaction, then the higher concentration of gold may be less material; if other components like hydrous minerals, carbonate or pyrite are consumed early then the higher concentration of gold in solution may be critical.

The devolatilisation reaction to produce auriferous fluid is endothermic so requires thermal input across broad areas to produce gold provinces. This thermal input is achieved in tectonic settings such as the closure of large oceans, large accretionary arc complexes and continental arcs. Common favourable factors include subduction, arcs, voluminous mafic rock material and the accretion of terranes. Earthquake activity, even if minor, is part of the orogenic process. It plays an important role facilitating the upward escape of auriferous fluids.

On the goldfield scale contrasting rock types with different rheology, structural repetition, and orientation with respect to far field stress dictate fluid access. Fluid – wallrock interaction with carbonaceous rocks commonly leads to high gold grades. The country rock chemical composition controls progress of desulfidation and gold deposition, and total Fe and Fe/Mg are important.

15.5   Magmatic Gold Deposits

This summary of processes leading to gold-only deposits omits reference to magmatic gold deposits. The definition of a magmatic gold deposit being followed here is one in which the main processes to achieve enrichment is by gold partitioning into silicate magma, with the focus on *magma*. The definition of magmatic gold deposit has not been broadened to include shared thermal events, aqueous fluids that may have been evolved from magmas, host rocks that were once magmatic, or geographic proximity to an igneous rock. With the specific definition here, that is focused on silicate magma, no examples of a magmatic gold deposit have been found that can explain the enrichment, segregation, ore fluid type or scale of gold provinces. Deposits containing gold can be formed through magmatic immiscible sulfide melt processes, but they are not gold-only deposits because base metals will also be partitioned into sulfide melts.

15.6   Modifying Deposits After they Form to Make them Look Different

The devolatilisation model as described above portrays the formation of many gold deposits as a relatively uniform process with some diversity due to varying host rocks and structural geometry. The events that happen after a gold deposit has formed can have a profound effect on what a deposit looks like, its economics, and even how it is traditionally classified.

Many of the features that make various gold deposits look so different from each other do not reflect the original formation process. Instead, the differences relate to superimposed later effects. These include high grade metamorphism including partial melting, retrogression, weathering, and erosion.

To understand the fundamental processes forming a gold deposit, these later modifying effects need to be recognised and then allowed for. In many cases recognition of such effects and making of allowance are easy and almost automatic, in other cases recognition is limited by an individual's skills base and experience and can hamper progress on understanding deposit formation. Chapter 16 describes later modifying processes that occur at high temperature; Chapter 17 describes processes at low temperature.

Snapshot

- Metamorphic processes can generate auriferous fluids at every mineral grain boundary site throughout 1000s km^3 of rock mass. This is the key to converting background 1–2 ppb levels of Au into deposits of 3 Moz (100 t) or more.
- Given the large scale of this process, it is important to look well beyond the host rocks for answers about deposit formation.
- Microcracks, fracture networks of faults and then larger fault zones are the pathways along which auriferous fluids move to form economic 1–10 ppm Au deposits.
- Structure is the critical influence on ore geometry for exploration and mine geology in most goldfields.
- Gold precipitation from the fluid can occur by several mechanisms but most

notably reduction and wallrock sulfidation.
- Most gold-only deposits form around 350 ± 50 °C and 0.5–5 km depth.
- The largest goldfields reflect the optimisation of the same factors found at individual smaller deposits; it is important that multiple favourable factors have converged.
- Many of the features that make deposits look different reflect different histories after formation rather than differences in the fundamental process of formation.
- There are many inferences involved in the metamorphic devolatilisation model that cannot be observed or tested yet, but it can explain and predict the five important characteristics of gold-only deposits outlined in Chapters 4–8.
- Metamorphic devolatilisation is a model to explain the formation of many gold-only deposits and this model will inevitably evolve over time.
- *The host rock is where the process finishes, not where the fluid and gold begin.*

Bibliography

Finch EG, Tomkins AG (2017) Pyrite-pyrrhotite stability in a metamorphic aureole: implications for orogenic gold genesis. Econ Geol 112:661–674

Gaboury D (2019) Parameters for the formation of orogenic gold deposits. Appl Earth Sci 128:124–133. https://doi.org/10.1080/25726838.2019.1583310

Phillips GN, Powell R (2009) Formation of gold deposits: review and evaluation of the continuum model. Earth Sci Rev 94:1–21

Phillips GN, Powell R (2010) Formation of gold deposits: a metamorphic devolatilization model. J Metamorph Geol 28:689–718

Pitcairn IK, Teagle DAH, Craw D, Olivo GR, Kerrich R, Brewer TS (2006) Sources of metals in orogenic gold deposits: insights from the Otago and Alpine Schists, New Zealand. Econ Geol 101:1525–1546

Tomkins AG (2010) Windows of metamorphic sulfur liberation in the crust: implications for gold deposit genesis. Geochim Cosmochim Acta 74:3246–3259

Modification of Deposits at High Temperature

16

Abstract

Once a gold-only deposit has formed at a nominal temperature of 300 – 350°C, there are several ways in which it may be heated to significantly higher temperatures and undergo metamorphic changes. Although this further heating has only affected some deposits, it is an important extension to the metamorphic devolatilisation process as it has a profound effect on modifying the appearance and geological characteristics of deposits. The main gold deposit formation process via metamorphic devolatilisation can only be fully understood if these later modifying effects are recognised and allowed for.

Keywords

Modification · High temperature · Metamorphic · Big Bell/Hemlo type · Amphibolite facies · Melting

Part of producing the gold deposits, as we see them today, are processes active after the main introduction of gold. Modifying events occurring after a gold deposit has formed are the cause of much of the diversity in appearance. Such events can be either at higher or at lower temperature than gold deposit formation and each will lead to its own overprint which must be understood before determining how a deposit has formed. These modifying processes can be beneficial or detrimental to the size and gold grade of deposits.

There are different ways in nature in which an existing gold deposit can be metamorphosed after its formation. Contact metamorphism by an intruding igneous body is one possibility, another is by regional metamorphism. In Archean cratons, regional metamorphism may be an extension of the gold-forming metamorphic event in which the rising geotherms heat the gold deposit further, or it might be a later regional metamorphic event within an adjacent Proterozoic mobile zone. The contact and regional processes that generate metamorphosed gold deposits are described in this chapter with examples.

Gold deposits in higher grade domains lacking carbonate minerals have been informally grouped as Big Bell / Hemlo type after studies in the early 1980s. These studies suggested such deposits had been metamorphosed after their formation to the extent that some even included partial melting of ore and host rock.

16.1 Metamorphosed Gold Deposit in a Contact Aureole Adjacent to a Phanerozoic Granite

Gold deposits in the contact aureoles surrounding igneous intrusions are relatively common globally and through time. One example is the 2 Moz (60 t) Maldon goldfield in the Victorian gold province which was formed in Ordovician clastic metasedimentary rocks and then intruded by a Devonian granite (Fig. 16.1). The goldfield is

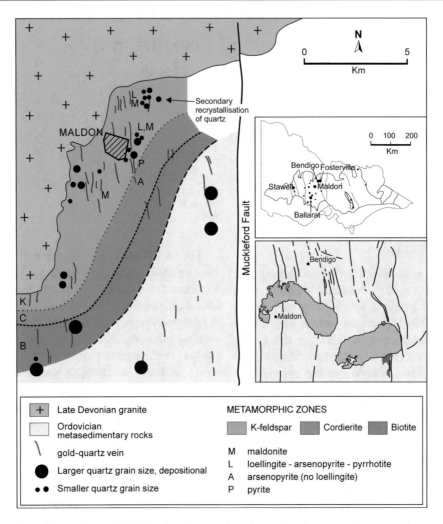

Fig. 16.1 Map of the Maldon goldfield, the discordant granite to its north, and metamorphic zones of the contact aureole (based on Hughes et al., 1997 including mapping of L Gregory). Maldonite is the mineral Au_2Bi.

approximately 10 km long, 3 km wide and consists of multiple N-S trending auriferous quartz vein sets with minor sulfide, carbonate minerals but no economic base metals, meaning it is gold-only. Eleven mines in the goldfield produced at least 1 t Au and together account for most of the production from Maldon. The vein set comprising the Maldon goldfield is semi-continuous with similar veins persisting for 90 km to the south.

The granite is strongly discordant with the N-S trend of the Maldon goldfield and truncates the

latter at its northern end (Fig. 16.2). Dykes of granite intrude the ore zones in some mines for over 100 m. The contact aureole is up to 3 km wide with an inner K-feldspar zone, then a cordierite – biotite zone, and a biotite zone; and there are auriferous quartz veins in each metamorphic zone and outside the aureole in the chlorite – muscovite slate. The contact metamorphism of the shales is essentially an isochemical heating event with loss of volatiles through breakdown of hydrous, carbonate-bearing and sulfide minerals (H_2O, CO_2, and S). Late-stage fluids, either from

Fig. 16.2 Details of the granite contact at the north of the Maldon goldfield: (**a**) view looking north along a line of mineralisation from the elevated ridge of the contact aureole. The flat land to the north is granite; (**b**) view looking south along the mineralisation with Professor Nikolai Goryachev of Magadan Russia standing on granite with hornfels a few metres behind him; (**c**) looking south from granite along the line of mineralisation and towards the low ridge of the contact aureole. The red line approximates the granite boundary; (**d**) photomicrograph of a sector-twinned, 1 mm grain of cordierite and two adjacent cordierite porphyroblasts. Between the cordierite porphyroblasts is fine-grained muscovite and biotite.

the granite or circulated by the heat of the granite, have caused some retrogression and textural changes but no measurable gold addition.

Changes in the ores approaching the granite contact include polygonisation of the vein quartz, and a smaller average grain size of this quartz. Pyrite and arsenopyrite are distal to the granite, but loellingite, arsenopyrite and pyrrhotite generally within 1.5 km of the granite.

The northern part of the Maldon goldfield proximal to the granite is vastly different to most goldfields in the Victorian gold province with respect to silicate mineral assemblages, ore minerals and the hornfels rock texture. However, despite the contrasts, the whole Maldon goldfield has the characteristics of gold-only deposits of being in a province, enrichment, segregation of gold from base metals and epigenetic timing. The differences along the 10 km length of the Maldon goldfield are explained as the result of contact metamorphism of what was a typical Victorian

gold deposit before the granite intrusion. This interpretation is supported by the continuity of exposure in the field which allows observations at progressive distances from the granite.

16.2   Gold Deposits in Higher Metamorphic Grade Domains of Archean Cratons

Metamorphosed gold deposits are widely distributed globally especially in Archean cratons both in the craton centres and around the margins (Table 16.1; note that the Yilgarn Craton appears over-represented because of a recent compilation of its deposits. Also note that for some of the goldfields in the table only parts of them are at high metamorphic grade). The largest deposit where formal study has established a pre-peak metamorphic time of formation is the Hemlo goldfield in eastern Canada (25 Moz, ~800 t

Table 16.1 Goldfields in amphibolite—granulite facies domains.

Goldfield	Location	Moz Au
Hemlo	Ontario Canada	25
Red Lake	Ontario Canada	23
Plutonic	Yilgarn margin Australia	10
Tropicana	Yilgarn margin Australia	8
Navachab	Namibia	8
Musselwhite	Ontario Canada	6
Morilla	Mali	6
Coolgardie	Yilgarn Australia	6
Bullabulling	Yilgarn Australia	4
Big Bell	Yilgarn Australia	3
Marvel Loch	Yilgarn Australia	3
Tucano	Northern Brazil	3
Crixas	Western Brazil	2
Borborema	Northeast Brazil	2
Consort	South Africa	2
Davyhurst	Yilgarn Australia	2
Challenger	Gawler Craton Australia	2
Maldon	Australia – SE	2
Renco	Zimbabwe	1

Parts of goldfields may be of lower metamorphic grade

Au). There are fewer gold deposits in granulite facies domains possibly reflecting that such domains are less common in Archean cratons.

16.2.1 Hemlo and Renco Goldfields

The Hemlo goldfield in the west of the Abitibi gold province is within a greenstone belt distant to the historically important goldfields such as Timmins and Kirkland Lake. Its discovery in the early 1980s was probably the most important gold discovery of the decade being adjacent to a major highway, in a belt not known for its gold, and including unusual host rocks of amphibolite facies grade. The sequence in the goldfield includes felsic and mafic igneous rocks and clastic metasedimentary rocks that are folded and juxtaposed across the Hemlo Fault. The mineralisation occurs as a planar steeply dipping interval 10s m thick, 100s m in length and continuing beyond 1.5 km depth. There is an alteration zone of K-feldspar, biotite and muscovite schists with enrichment of K, Rb, Ba, Au, Ag, Sb, As, Hg, Mo and V. Retrogression has strongly affected the peak metamorphic mineral assemblages and produced muscovite and chlorite schists and a wide range of low temperature ore minerals: the latter are not in equilibrium with the peak metamorphic assemblages.

The medium sized Renco gold deposit is situated near the southern margin of the Zimbabwe Craton and adjacent to the Proterozoic Limpopo Orogen. It comprises several planar ore zones within high metamorphic grade felsic and mafic gneisses that include two pyroxene granulite assemblages distal to the ore zones. The ore zones are higher strain planar intervals of 1 – 2 m thickness that continue for 100s m. Quartzo-feldspathic veins are concentrated within the ore zones where they are inferred to be the result of local partial melting. Near faults and folds retrogression has modified the peak metamorphic mineral assemblages and led to some local redistribution of gold.

16.2.2 Goldfields in High Metamorphic Grade Domains of the Yilgarn Craton, Australia

The Archean Yilgarn Craton of Western Australia is bound on its north and east by the Proterozoic Capricorn Orogen and Albany Fraser Orogen, respectively (Fig. 16.3). Goldfields in the Yilgarn Craton but near these Proterozoic belts have amphibolite to granulite facies mineral assemblages, local partial melting and gneissic to schistose fabrics.

To the northwest of the Yilgarn Craton, the Plutonic goldfield (10 Moz) and Glenburgh goldfield (1 Moz) are close to the margin of the Archean Yilgarn Craton and Paleoproterozoic Capricorn Orogen. At Plutonic, high pressures of regional metamorphism (600°C, 8kb or 80 Mpa) are atypical of the Yilgarn low pressure

metamorphism but characteristic of the Proterozoic metamorphism. Plutonic is within an Archean greenstone inlier of hornblende—plagioclase amphibolite surrounded by Proterozoic rocks; Glenburgh is hosted in garnet – biotite— quartz—cordierite gneiss of suggested metasedimentary origin and late Archean to Paleoproterozoic age.

On the eastern margin of the Yilgarn Craton the Tropicana goldfield of 8 Moz is in a greenstone belt adjacent to the Proterozoic Albany— Fraser Orogen. Metamorphic conditions of 600-770°C and 5kb (50 Mpa) are inferred reflecting the Proterozoic metamorphic effects.

The Big Bell deposit is in high grade rocks of a narrow greenstone belt in the northwest of the Yilgarn Craton. The sequence is predominantly felsic, mafic, and ultramafic gneiss and schist, and includes a strongly foliated, 20 – 50 m thick

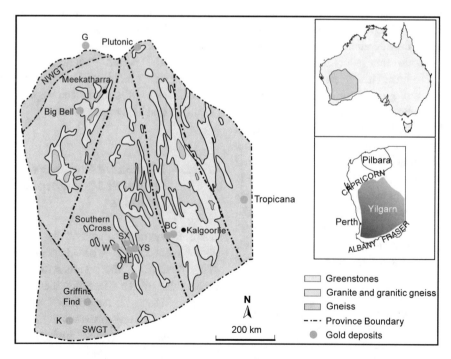

Fig. 16.3 Maps of the Yilgarn Craton showing the increase of metamorphic grade: green is lower grade to red colour being higher grade towards the south and west. The Southwest gneiss terrane, and Northwest Gneiss Terrane—SWGT and NWGT, respectively are along the western craton margin. Gold deposits in higher metamorphic grade domains are shown including B=Bounty,

BC=Bullabulling and Coolgardie, G=Glenburgh, K=Katanning, ML=Marvel Loch, SX=Southern Cross, W=Westonia, YS=Yilgarn Star. To its north, east and south, the Yilgarn Craton is adjacent to the higher pressure Proterozoic orogens (Capricorn and Albany – Fraser Orogens).

mineralised envelope (mine sequence) that is up to 500 m long and known to 1.5 km depth. Metamorphic conditions of 650 – 700°C and 4 – 5 kbars have been inferred. The mine sequence is quite heterogeneous with an orebody of biotite, muscovite, and K-feldspar schists (lode schists) that include folded and dismembered auriferous quartz veins and planar pegmatite dykes that have been boudinaged. Variants of the lode schist include the dominant K-feldspar-muscovite schist with quartz and pyrite, and biotite schist with cordierite, quartz, pyrite and pyrrhotite. Immobile trace elements such as Al, Ti, Zr and V in the biotite, muscovite and K-feldspar lode schists indicate that they were mafic rocks originally. The peak metamorphic conditions of upper amphibolite facies are indicated by the full assemblage of sillimanite – K-feldspar – biotite – cordierite – garnet – plagioclase – tourmaline – quartz, and the well-preserved high metamorphic grade mineral assemblages indicate that the mafic rocks had been altered by the time of peak metamorphism, i.e. the alteration was not part of retrogression despite the latter being moderately widespread to form chlorite and muscovite. The lode schists are enriched in K, Rb, S, Au, As, Sb, W, Ag and B. Carbonate minerals were lacking due to the high metamorphic grade except where they were introduced during retrogression.

In the Central Yilgarn area there are several medium sized goldfields of upper amphibolite to granulite facies grade. Marvel Loch comprises nine steeply plunging ore pipes in mafic amphibolite along the 80 km long Marvel Loch Fault (a complex shear zone system) and mined via declines off a 3 km long open pit. The Marvel Loch Fault in the mine area comprises gneiss and schist with cordierite, sillimanite, biotite, garnet, pyrrhotite and pyrite. Whole rock geochemistry utilising immobile trace elements suggest that the precursor to the schist and gneiss of the Marvel Loch Fault was a mafic composition altered prior to peak metamorphism.

The Southern Cross district includes several discrete deposits of 0.5 Moz (Golden Pig and Edwards Find), and widespread cummingtonite in hornblende amphibolite. Other nearby deposits such as Frasers have restricted mineralogical variation with mainly hornblende and plagioclase.

Within 100 km of Southern Cross town are Westonia, Yilgarn Star and Bounty (all ~1 Moz). To the east of Southern Cross and closer to Kalgoorlie, there are Coolgardie and Bullabulling goldfields in amphibolite facies domains (the latter with cordierite and sillimanite). The broader Central Yilgarn area has an endowment of ~20 Moz in deposits of amphibolite to granulite facies metamorphic grade, and there is a similar endowment in Yilgarn goldfields adjacent to Proterozoic orogens.

In the southwest of the Yilgarn Craton are mostly narrow granulite facies greenstone belts surrounded by granitic gneiss and migmatite. Griffin's Find deposit (0.05 Moz by 1990 but little since) has received considerable research attention with the definitive paper being Tomkins and Grundy (2009), and Katanning is an emerging resource of 1 Moz (30 t Au).

The Challenger deposit is 1000 km east of the Yilgarn Craton in the late-Archean Gawler Craton of South Australia. It is in granulite facies rocks and is especially interesting because of the documented partial melting of silicate assemblages, melting of sulfides, and remobilisation of gold in the pegmatite melts (Tomkins and Mavrogenes 2002).

16.2.3   Distinctive Features of Deposits in Higher Metamorphic Grade Domains

There are several features of these goldfields in higher metamorphic grade domains that distinguish them from the common deposits in the greenschist facies. Despite these differences though, four characteristics of gold-only goldfields remain—provinciality, enrichment, segregation, and epigenetic timing; the ore fluid type is not well characterised.

16.2.4   Textures

Hornfels and decussate textures are distinctive features around the gold deposits in amphibolite facies domains, and these are absent in the lower to middle greenschist facies deposits (Fig. 16.4; Fig. 16.5).

The hornfels texture arises from the amphibolite facies metamorphism and renders the rocks massive, hard and commonly with a distinctive ring when hammered. What is important in the formation of hornfels is that the higher temperature is not accompanied by on-going deformation, and so fine-grained minerals develop without alignment (schist and gneiss develops if deformation accompanies the heating). Rocks with a hornfels texture are widespread in the Southern Cross – Marvel Loch district even on the surface. A consequence of the hornfels texture at Big Bell is that some harder ores require extra crushing.

An example of a decussate texture is where a platy mineral like biotite grows without accompanying deformation and forms a randomly oriented mass with quartz and plagioclase. Instead of a polygonal aggregate as found in pure quartz rocks, it is impingement boundaries with biotite crystal faces that dominate (Vernon, 2004). In gold deposits such as Marvel Loch, the texture in biotite schist is one of interlayered biotite with quartz and plagioclase. Although the biotite is parallel to the planar fabric, it occurs as isolated islands on the planar surface quite unlike the interwoven texture of muscovite found in active deformation zones of the greenschist facies. The interpretation is that mineralisation

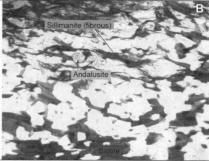

Fig. 16.4 Photomicrographs of rocks within the Marvel Loch F, Marvel Loch, WA: (**a**) cordierite grain several millimetres across with abundant inclusions of biotite and quartz; (**b**) biotite schist with quartz and both andalusite and sillimanite. The andalusite is interpreted as being metastable within the sillimanite stability field. Base of photomicrographs 2 mm and 3 mm, respectively.

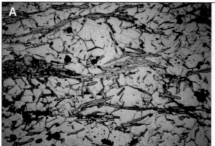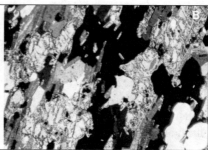

Fig. 16.5 Ores from Big Bell in the amphibolite facies: (**a**) Kfeldspar – muscovite – quartz – pyrite schist which is the main ore type; (**b**) cordierite – muscovite – biotite – quartz – pyrite schist which is also ore but typically lower grade. Cordierite comprises 30% of the field of view in **b** as the large porphyroblasts with common inclusions. The rocks in **a** and **b** were mafic rocks prior to gold-related alteration that pre-dated the amphibolite facies metamorphic peak. Base of photomicrographs ~4 mm.

has been metamorphosed (hornfelsed) after a pre-existing schistose fabric of aligned interwoven mica (probably muscovite with chlorite). Elevated temperature then converts muscovite-bearing assemblages to those with biotite, the planar orientation of the mica is retained but the interlocked muscovite of an active deforming zone is replaced by isolated aligned biotite grains defining the planar schistose, or transitional gneissic, fabric. The texture indicates where there has been high strain and active deformation in the past that has been overprinted by higher grade metamorphism (akin to contact metamorphism but more on a semi-regional scale).

The term 'skarn' has been applied to some of these calc-silicate assemblages which may contain pyroxene or garnet though not without considerable uncertainty as to what this name might be suggesting to different audiences. Given the confusion over skarn terminology, a modern metamorphic petrological approach is preferred that refers to specific minerals and assemblages (Appendix D).

16.2.5 Gold Grains

Unlike for deposits in the greenschist facies, many higher metamorphic grade deposits have gold grains physically trapped within silicate gangue minerals. Gold has been reported as blebs within orthopyroxene, diopside, hornblende, cordierite, garnet, feldspars, and tourmaline for deposits including Big Bell, Yilgarn Star, Tropicana and Challenger. A consequence of this texture is that fine grinding may be necessary to liberate the included gold, but these ores do not require roasting before the gold can be recovered with cyanide. Separately, in high-grade deposits including Hemlo, Renco, Tropicana and Big Bell, there are other gold grains growing in or across a retrograde schistosity indicating some remobilisation of gold on the scale of centimetres or possibly metres. The textural studies have led to considerable controversy regarding the timing of gold at the higher metamorphic grade deposits; in general, the late textural setting of some gold

grains does not imply a late timing of the deposit itself.

16.2.6 Mineralised Faults (Shear Zone)

The mineralised fault zones in the higher metamorphic grade deposits are planar, several metres thick and extend for 100s m. They typically have a proximal interval of abundant micas or K-feldspar that coincides with higher gold grade and a concentration of B, S, Ag, As, Sb and W. Major deposits usually have planar mineralised zones of more than one orientation and where these intersect it is common to have rich ore shoots. In all these respects, the mineralised fault zones in higher metamorphic grade domains resemble those in greenschist facies grade deposits although the former tend be to less visually conspicuous and lack strongly bleached proximal parts.

The internal composition of the mineralised fault zones in higher metamorphic grade deposits differ from their lower grade equivalents in texture and mineralogy. Some are schistose or gneissic with planar fabrics defined by biotite and amphiboles. Other textures are hornfelsic with partially randomly oriented biotite grains with decussate texture indicating mineral growth and textures outlived any active deformation. The difference in mineralogy reflects the higher metamorphic temperatures and a very different appearance is conveyed by biotite and diopside compared to the bleaching of the greenschist facies resulting from muscovite and carbonate alteration.

16.2.7 Migmatites, First Melts and Pegmatites

A feature of gold deposits in higher metamorphic grade domains is the prevalence of migmatite, partial melts, quartzo-feldspathic veins and pegmatite dykes depending upon the specific goldfields. At the highest metamorphic grades such as the granulite facies of the Southwest gneiss terrane of the Yilgarn Craton, migmatite

Fig. 16.6 Photomicrographs from the ore zone at Big Bell. The lobate textures at the margins of quartz and feldspar are characteristic of first melting and contrast with the polygonal grain boundaries between stable phases. The birefringent tabular grains in (**a**) are biotite. Base of photomicrographs 0.5 mm and 0.3 mm, respectively.

is widespread regionally including around Griffin's Find deposit. This migmatite is the product of partial melting of local and deeper rocks and likely to have involved ten percent or more of the sequence beginning to melt and migrate.

The first recognition that a gold orebody had begun to melt was Big Bell where microscopic aggregates of feldspar and quartz showed lobate margins indicative of initial melting (Fig. 16.6). Recognition of partial melting at several further gold orebodies has followed especially led by Andy Tomkins.

In amphibolite facies domains pegmatite dykes are common in open pits and underground where they have exploited some of the gold-mineralised structures for their ascent (Fig. 16.7). This overlap of pegmatite that has intruded the mineralised structures causes difficulties in exploration when deep drill holes intersect the position of mineralisation only to find a thick barren pegmatite in that position. The lack of pegmatite dykes in most of the deposits in the greenschist facies suggests that these magmas do not rise far in the crust. They are likely to be water rich and formed around 700°C rather than 800°C and above.

Where timing criteria are available these melts appear to post-date gold mineralisation. This applies at Renco in Zimbabwe, Hemlo in Canada, Borborema in Brazil, Marvel Loch, and Big Bell in Australia. Quartzo-feldspathic veins and pegmatite dykes are boudinaged whereas adjacent auriferous quartz veins are strongly folded and appear earlier. The preferential melting in and adjacent to mineralisation (Renco, Big Bell) follows from the gold-related alteration adding K, and this K significantly reducing the temperature required to melt these host rocks.

Fig. 16.7 (**a**) Pegmatite dyke (white) dipping at a moderate angle to the north (right) in the Marvel Loch open pit; (**b**) pegmatite dyke in underground workings at Marvel Loch mine; the coarse feldspar grains are approximately 5 cm in size.

16.2.8 Low Variance Assemblages and Many Coexisting Mineral Phases: Isochemically Closed Systems

Variance is a concept used in metamorphic petrology that helps to relate the number of phases (minerals) to the number of components (major elements or element oxides); it is not to be confused with variance as used in statistics.

The number of minerals and elements in a metamorphic system derives from the Gibbs phase rule (Mineral phases = Component elements - Degrees of freedom + 2) where degrees of freedom are usually taken as temperature and pressure in a closed isochemical metamorphic system. As an example, a one component system (e.g. H_2O) can have either one (water) or two (water and ice) or three (water, ice and steam) coexisting phases of H_2O in equilibrium. With the one phase there is freedom to vary both T and P (two degrees of freedom), with two phases it is possible to vary either of T or P but once one is specified the other is set (one degree of freedom); and with the three coexisting phases the system must be at the triple point of H_2O with no degrees of freedom where in this case T and P will be fixed at $0°C$ and 0.6 kPa or about 0.006 atmospheres. The system with two degrees of freedom might be described as a high variance system; the triple point with no degrees of freedom would be a low variance system but with multiple coexisting phases.

The number of minerals in equilibrium in a metamorphic rock will increase as the number of component elements increases. For example, a clean quartzite (virtually all SiO_2) is likely to be a mono-mineralic system of quartz, but a metamorphosed mafic rock (with essential SiO_2, TiO_2, Al_2O_3, FeO, MgO, CaO and Na_2O as components) will typically comprise several minerals in equilibrium and be more useful than the quartzite in constraining metamorphic conditions of T and P (Fig. 16.8).

In the above discussion it is implicit that the system is closed or isochemical as might be approximated by middle to high grade regional metamorphism. This is a low fluid-to-rock ratio environment of buffering where the internal mineral assemblages effectively dictate (buffer) the fluid composition. This is a situation of low variance and thus potentially many coexisting minerals depending upon the rock type.

16.2.9 Variance and Hydrothermal Open Systems

Hydrothermal ore deposit processes are different from isochemical regional metamorphism. A hydrothermal ore forming process is one of infiltration (not buffering), high fluid-to-rock ratio, high variance, and alteration involving several mobile components controlled by parameters external to the system. The earlier expression of the Gibbs phase rule needs to be refined to consider only the immobile components dictated by the host rock, and not to count the mobile elements being added and subtracted during alteration by the ore fluid. Qualitatively, this leads to fewer effective (immobile) component elements and hence there will be few mineral phases in a major hydrothermal event. In the extreme, hydrothermal alteration can produce many unusual mono-mineralic rocks with quartz, mica, feldspar, or amphibole.

For gold deposits in amphibolite to granulite facies domains (Big Bell / Hemlo type), the phase rule and mineral variance arguments provide a potential approach to determine whether these deposits formed in a major hydrothermal event at the metamorphic peak, or under greenschist facies conditions (~350°C) and were then metamorphosed.

This mineral assemblage variance approach was used in studies of the Big Bell gold deposit in the northwestern part of the Yilgarn Craton where some complex mineral assemblages provided strong support for the formation of the deposit around 350°C followed by the overprint of higher-grade metamorphic conditions. One example is the equilibrium co-existence of sillimanite – K-feldspar – cordierite – biotite – garnet – plagioclase – quartz – tourmaline, and noting this rock also contains metastable andalusite.

Fig. 16.8 Marvel Loch F, Marvel Loch gold mine: (**a**) 2 m thick auriferous quartz vein on left with laminations of altered country rock within the vein; altered mafic rock on the right side of **a** containing diopside and biotite with hornfels texture; (**b**) altered mafic rock with a former schistose fabric that has been overprinted by a hornfels texture and decussate biotite. Sillimanite (prominent white flecks) has overgrown the former schistose fabric during the higher temperature peak metamorphism. The interpretation of **b** is that gold-related alteration modified the bulk rock composition of the mafic rock, and subsequent high-grade metamorphism stabilised sillimanite in that modified bulk rock composition.

The importance of using *equilibrium* assemblages is highlighted by the presence of sillimanite and andalusite in samples such as from Marvel Loch and Big Bell goldfields. Textural studies suggest that the rocks with these two alumino-silicate minerals have passed through the andalusite field and then progressed well into the sillimanite field during prograde metamorphism; the andalusite did not disappear immediately into the sillimanite stability field because of the kinetics of this solid – solid transformation. The presence of both minerals does not indicate peak conditions were on the andalusite – sillimanite boundary, and a higher temperature is compatible with other silicate mineral assemblages and presence of partial melts. A different reflection of low variance and multiple phases at Big Bell would be the seven Sb minerals although it has not been established that they are co-existing in equilibrium. It is self-evident that application of variance arguments requires careful petrography to recognise equilibrium textures and some less-common minerals.

Even more care is required using the mineral assemblage variance approach for more complex minerals. This is because their compositions can have a 'non-stoichiometric' composition that can impart a relatively high degree of chemical flexibility and wide P-T stability range. These are known as berthollide minerals with classic examples being pyrrhotite and hornblende and both are important in higher metamorphic grade domains. Pyrrhotite has a formula of $Fe_{1-x}S$ where x varies along a sliding scale between 0 to 0.2. Hornblende is a chemically complex amphibole that incorporates many component elements. A very simple amphibole is cummingtonite ($Mg_7Si_8O_{22}(OH)_2$), but continuous substitutions of elements for one another can take place when they have similar ionic size, such as Mg-Fe, Si-Al, Mg-Ca and Na can yield $NaCa_2(Mg, Fe)_5AlSi_7O_{22}(OH)_2$. Even more complex hornblendes can contain Fe^{3+}, K, F or even a vacancy at a structural site. This multi-component chemical flexibility and resulting complexity of hornblende mean that an assemblage of hornblende – plagioclase – ilmenite – quartz is a common mafic assemblage with many mafic component elements and only few mineral phases. Therefore, in amphibolite facies domains such as around Southern Cross in the Yilgarn Craton, hornblende – plagioclase assemblages alone cannot be used *a priori* as indicative of low variance assemblages, high fluid-to-rock ratio, with infiltration in a peak metamorphic hydrothermal event, because of the chemical flexibility and wide P-T stability of hornblende.

16.3 Processes Leading to Gold Deposits in Higher Metamorphic Grade Domains

Several different processes have been suggested to explain the formation of the gold-only deposits in high metamorphic grade domains. Four of these have been considered in the past and are discussed here:

- Syngenetic formation on the seafloor akin to VMS deposits
- Metamorphosed deposits formed through metamorphic devolatilisation during greenschist facies conditions (as per Chapter 15) and then followed by peak amphibolite – granulite facies of metamorphism. These are referred to as Big Bell / Hemlo type deposits (Phillips and de Nooy 1988; Tomkins and Grundy 2009; Powell et al. 2017)
- At the peak of metamorphism formed by deep crustal to mantle derived fluids – referred to as the continuum model (Groves, 1993; Goldfarb and Groves 2015)
- During retrogression as the sequence cooled from its metamorphic peak (Wilkins, 1993)

The syngenetic model for these deposits is rarely advocated but was in vogue during the 1980s.

16.3.1 Big Bell/Hemlo-Type Deposits: Metamorphic Devolatilisation Followed by Higher Grade Metamorphism

The Big Bell / Hemlo-type gold deposits are inferred to have formed as gold-only deposits in the greenschist facies as described in Chapter 15. This occurred around 350°C and 5 – 10 km depth during active deformation involving brittle – ductile structures of many forms and orientations (Fig. 16.9a). Auriferous quartz veins and alteration haloes of carbonate minerals, muscovite and pyrite were accompanied by elevated concentrations of Au, Ag, As, Sb, S, W, B, Se, Te and Hg.

Following deposit formation, increase in the temperature led to the metamorphism of country rock, host rocks and ore to reflect the high metamorphic grade (Fig. 16.9b). Carbonate minerals became less stable, pyrrhotite more abundant at the expense of pyrite, and muscovite is replaced by biotite and K-feldspar. The process essentially involved loss of H_2O, CO_2 and some S. If these deposits approached 700°C, partial melting may occur with additional loss of a granite component. The alteration halo once metamorphosed does not bear any one-to-one correspondence to the original alteration halo—meaning that each original zone formed at 350°C does not correspond to a new zone at the higher temperatures, these are metamorphosed alteration assemblages (Fig. 16.10). Apart from the expulsion of minor volatiles and any effects of partial melting, this metamorphism of the existing deposit involves minor chemical changes, and is rock buffered, meaning low variance mineral assemblages. The typical lack of on-going deformation during the higher temperatures means that hornfels textures are common, and high-grade mineral assemblages are variably preserved.

For many deposits, the process of establishing a pre-peak metamorphic origin is not straightforward as illustrated by the example of Bronzewing (Chapter 9; Fig. 9.2) where sophisticated petrology was required to demonstrate that the temperatures of formation of the deposit (330 – 375°C) was less than the peak of metamorphism (440°C).

16.3.2 Overprinting Metamorphism Compared to the Gold-Formation (Devolatilisation) Process

Taking a gold deposit that has just formed at 350°C and heating it to 550°C is a form of devolatilisation but with significant differences from the process invoked to generate the auriferous fluid (Chapter 15). Both are rock-dominated processes of low fluid-to-rock ratio involving reactants of hydrous minerals, carbonates, sulfides and gold; however, the major difference

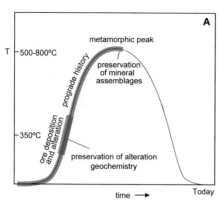

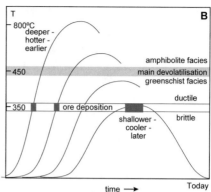

Fig. 16.9 (**a**) Temperature – time path of a Big Bell / Hemlo-type gold deposit. Upon burial a gold deposit is formed around 350°C accompanied by geochemical alteration of the country rocks and generation of distal and proximal alteration mineral assemblages. Further prograde heating to 500 – 800°C is essentially isochemical but the greenschist facies minerals are converted to assemblages reflecting the amphibolite and granulite facies conditions; (**b**) temperature – time paths for four different levels within the crust. In this modelling, deep parts of the crust reach their peak of metamorphic temperature before shallower levels reach their peak. Consequently, devolatilisation can form auriferous fluids at depth that then rise and form gold deposits higher in the crust before that higher part has reached its peak. This scenario is applicable to heating from below by the mantle upon crustal thinning beneath a greenstone belt.

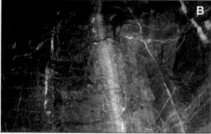

Fig. 16.10 Mineralised fault zones in two mafic host rock: (**a**) the bleached interval reflects intense carbonate alteration; the inner part of the bleached interval is the proximal high strain zone with auriferous quartz veins and intense K alteration generating muscovite. Location: minor mineralised interval on 4 level, Lake View mine 412 EXC, Kalgoorlie; (**b**) major mineralised fault zone in the Marvel Loch goldfield. There is a white central quartz vein surrounded by altered mafic rock with a hornfels texture. The pink colour is from decussate-textured biotite. The carbonate minerals have broken down during the amphibolite facies overprint, but sulfides are still widespread. Although the cm-scale alteration in **a** is visually more impressive, it is only the metres-scale alteration and vein in **b** that is commercially significant.

is the starting mix of components (Table 16.2). Column A is for the metamorphism of altered mafic rocks to generate a gold-only fluid (Chapter 15) with the starting concentration of H_2O, CO_2, S and Au in the source. Column B is for the metamorphism of a pre-existing gold deposit (Chapter 16) illustrating that A and B are dealing with completely different rocks undergoing metamorphism. The difference is exemplified by the expression of ratios of Au to the major volatile components.

The heating of a pre-existing gold deposit (B) is essentially the breakdown of ore including its high proportion of Au. The ratio of the reacting components means that the H_2O and CO_2 will be exhausted early, and the loss of gold may be

Table 16.2 Metamorphism of an existing gold deposit.

	A	B
	Background altered mafic rock as in the devolatilisation process	Metamorphism of an existing gold deposit
Starting material prior to metamorphism		
H_2O (ppm)	50,000	50,000
CO_2 (ppm)	20,000	50,000
H_2S (ppm)	2000	20,000
Au (ppm)	0.002	10
Ratio of main components to Au		
H_2O:Au	25,000,000	5000
CO_2:Au	10,000,000	5000
S:Au	1,000,000	2000

minimal. In Chapter 15 it was suggested that most of the Au might be extracted into the gold-only fluid from its background of 2 ppb, and this appears feasible; however, it is a different proposition completely if gold ore is a reactant and there are very different ratios of volatile elements to gold.

In practice it is not possible to determine how much gold has been lost during overprinting metamorphism of a gold deposit, but qualitative calculations indicate that it may be minor. To remove all the gold during metamorphism of a gold deposit (B) seems unlikely given the relatively large amount of Au in B compared to H_2O available to form fluid. To remove all the Au in B would require a gold concentration in solution of 5000 times that required to remove all the Au in A (25,000,000 / 5,000) which seems unlikely.

If instead of the gold deposit being heated to 550°C it is heated to 700°C, as some deposits have been, the host rocks will develop appropriate mineral assemblages for upper amphibolite to granulite facies conditions, and gneissic textures. Any remaining carbonates and sulfides may be minor, and partial melting of certain country rocks (e.g. clastic metasedimentary rocks) will consume H_2O in hydrous minerals; the ore zone with its elevated K is also a candidate for early onset of melting. Separately from silicate melting, the melting of sulfide assemblages to form a sulfide melt can also occur when an ore deposit is metamorphosed, and this can lead to gold redistribution (Tomkins et al. 2004).

16.3.3 Evaluation of the Continuum Model: Formation of Deposits up to 850 °C

The continuum model advocates a source of auriferous fluid deep in the granulite facies or mantle, with the fluid then depositing gold through the crust as it ascends 20 – 25 km to amphibolite and greenschist facies conditions. The proposition that many gold deposits formed at 600, 700 and 800°C at their respective peak metamorphic temperatures was widely promulgated as the continuum model. Some of its basic geology observations and inferred geological processes have been seriously challenged by Tomkins and Grundy (2009).

The interpretation that any of these gold-only deposits formed at the highest metamorphic temperatures such as 800°C leads to untenable conclusions and is not considered a sustainable concept. The major problem with the continuum model becomes evident above 700°C as the rock sequences undergoing partial melting cannot sustain a separate aqueous fluid necessary to form a gold deposit; instead, the fluid reacts to form further melt. A problematic aspect of the continuum model centres on the unreasonable calculated fluid compositions once the model is assumed.

The continuum model and its deep crust and mantle origin for aqueous auriferous fluids has recently been incorporated into the orogenic model for gold-only deposits. This is one reason the orogenic terminology has been avoided.

16.3.3.1 Partial Melting

The science of aqueous fluids and partial melts has been used in gold geology since the early 1980s with implications that a significant aqueous fluid event (as required to form a major gold deposit) does not simply pervade a metamorphic sequence at $700 - 800^{\circ}C$ and deposit gold, rather if it was there it would combine with feldspars, micas, and quartz to generate further partial melt.

A mix of quartz, feldspars and an aqueous fluid can partially melt just below $700^{\circ}C$; the limitation on this early melting is the role of an aqueous fluid that may be absent (no melting by this pathway), or it may be minimal and hence soon consumed.

A more widespread partial melting process in the crust can begin in appropriate rock types around $700^{\circ}C$ in the absence of an aqueous fluid phase. This fluid-absent melting derives H_2O from hydrous minerals—initially muscovite, and then biotite and hornblende at higher temperatures. Clastic metasedimentary rocks generally begin to melt early, and unaltered mafic rocks only above $800^{\circ}C$. However, alteration of mafic rocks to form muscovite or biotite as noted for Big Bell can significantly lower their first melting temperature as they gain a source of H_2O and K during their alteration.

In the environment of a crustal sequence that has begun to partially melt the introduction of any aqueous fluid from an external source leads to further partial melting until that fluid is consumed. The aqueous fluid cannot exist in equilibrium with a silicate melt, as would be required in the continuum model. This means that auriferous aqueous fluids cannot form, migrate, and generate gold deposits in the domain of partial melting, hence making the continuum model untenable.

16.3.3.2 Unreasonable Calculated Fluid Compositions

One approach to test the continuum model has been to assume it is correct and then follow through the implications of that assumption showing that the implications are untenable, and the model can be considered falsified.

The deposits used as type examples on which the continuum model was erected have peak metamorphic sulfide assemblages from which an equilibrium fluid composition has been calculated. The approach to these calculations is not assuming that there is a large amount of aqueous fluid; it is determining what composition of fluid would be in equilibrium with the mineral assemblage at the peak temperature.

For the Griffin's Find deposit, the type-example upon which the high temperature end of the continuum model is based, the sulfide assemblage at peak metamorphic temperature is in equilibrium with an aqueous fluid of 0.6 to 30 molal H_2S (Mikucki 1998 using the temperature of $700^{\circ}C$ assumed by advocates of the continuum model). This is an extraordinary result requiring some discussion because even an aqueous fluid of 10 m H_2S equates to 300,000 ppm S. For a more likely peak temperature of $800^{\circ}C$ the result would be more extreme and well outside the range of any reasonable crustal ore-forming fluid.

The fluid compositions implied by the continuum model would suggest that the auriferous fluid at granulite facies (with 300,000 ppm S) loses 99% of its S whilst migrating upwards to the greenschist facies by which stage it has 250 ppm S. With gold being complexed to S, these figures would suggest that virtually all the gold would be deposited in the granulite and amphibolite facies with none remaining to the greenschist facies. The distribution of gold in these terranes suggests almost the opposite. The above calculations are not being challenged here; the initial assumption of the continuum model appears to be the source of the error.

The continuum model was hailed as a panacea in the late 1980s. In the process, deposits such as Big Bell that had be shown to have formed before the peak of metamorphism were labelled as enigmatic and anomalous, and therefore excluded from further consideration. The continuum model for gold deposits continues to be widely promoted even after a decade of serious challenges to the model in the public domain. Despite the wealth of publications and citations the continuum model and mantle version of the orogenic model are not supported by many metamorphic petrologists.

16.3.4   Evaluation of the Retrograde Origin for Deposits in High Metamorphic Grade Domains

These deposits in high metamorphic grade domains pre-date retrogression because:

- the gold-related alteration zones, as defined by the K-rich ore intervals within mafic rocks, record the peak metamorphic conditions.
- the boron-bearing silicate mineral, tourmaline, and hence gold-related B addition, is texturally part of the peak metamorphic assemblage.
- barren quartzo-feldspathic and pegmatite dykes formed from partial melting at or near peak metamorphism cross-cut mineralisation and appear less deformed than auriferous quartz veins.
- gold grains occur as inclusions in several peak metamorphic minerals including pyroxenes, garnet, cordierite, feldspars and hornblende.
- there are high temperature sulfide assemblages such as loellingite despite there also being common thermal resetting of earlier sulfide minerals.
- there has been sulfide melting in some deposits.
- there is no causative correlation between degree of retrogression and gold concentration. However, this needs to be monitored carefully because detectable retrogression does correlate with rock type, and rock type does correlate with gold concentration.

Methods used to suggest that these Big Bell / Hemlo-type deposits are retrograde in their timing are generally conducted at inappropriate scales such as the observation of gold grains overprinting retrograde fabrics. *This method indicates the timing of the gold grain and not the gold orebody.*

Snapshot

- Metamorphism of existing gold deposits can profoundly modify their appearance.
- Existing gold deposits can be metamorphosed to higher temperatures

---

in contact aureoles, adjacent to younger orogens, or as a metamorphic extension of their main process of formation.

- Textural and mineralogical changes occur in response to the higher temperature.
- Partial melting of country rock and ore zones has occurred above 700 °C.
- The co-existence of partial melting and hydrothermal fluids is unlikely above 670 °C.
- The continuum model that advocates gold deposit formation between 500 °C and 800 °C is not supported by principles of metamorphic petrology.

Bibliography

Doyle MG, Catto B, Gibbs D, Kent M, Savage J (2017) Tropicana gold deposit. In: Phillips GN (ed) Australian ore deposits. The Australasian Institute of Mining and Metallurgy, Carlton, pp 299–306

Froese E (1971) Graphical representation of sulfide-silicate phase equilibria. Econ Geol 66:335–341

Gazley MF, Duclaux G, Pearce MA, Fisher LA, du Plessis E, Murray S, Hough RM (2017) Plutonic goldfield. In: Phillips GN (ed) Australian ore deposits. The Australasian Institute of Mining and Metallurgy, Carlton, pp 307–312

Goldfarb RJ, Groves DI (2015) Orogenic gold: common or evolving fluid and metal sources through time. Lithos 233:2–36

Greenwood HJ (1975) Buffering of pore fluids by metamorphic reactions. Am J Sci 275:573–593

Groves DI (1993) The crustal continuum model for late-Archaean lode-gold deposits of the Yilgarn block, Western Australia. Mineral Deposita 28:366–374

Groves DI, Goldfarb RJ, Gebre-Mariam M, Hagemann SG, Robert F (1998) Orogenic gold deposits: a proposed classification in the context of their crustal distribution and relationship to other gold deposit types. Ore Geol Rev 13:7–27

Holland TJB, Powell R (2001) Calculation of phase relations involving haplogranitic melts using an internally-consistent thermodynamic dataset. J Petrol 42:673–683

Hughes MJ, Phillips GN, Gregory LM (1997) Mineralogical domains in the Victorian Gold Province, Maldon, and Carlin-style potential. Australasian Institute Mining Metallurgy, Annual Conference; pp. 215–227

Kuhns RJ, Sawkins FJ, Ito E (1994) Magmatism, metamorphism and deformation at Hemlo, Ontario, and the timing of Au-Mo mineralization in the Golden Giant mine. Econ Geol 89:720–756

Mikucki EJ (1998) Hydrothermal transport and deposition processes in Archean lode-gold systems: a review. Ore Geol Rev 13:307–321

Phillips GN, de Nooy D (1988) High-grade metamorphic processes which influence Archaean gold deposits, with particular reference to Big Bell, Australia. J Metamorph Geol 6:95–114

Phillips GN, Powell R (2009) Formation of gold deposits: review and evaluation of the continuum model. Earth-Sci Rev 94:1–21

Powell R, Holland TJB, Worley B (1998) Calculating phase diagrams involving solid solutions via non-linear equations, with examples using THERMOCALC. J Metamorph Geol 16:577–588

Powell R, Phillips GN, De Nooy D (2017) Yilgarn gold deposits in higher metamorphic grade domains. In: Phillips GN (ed) Australian ore deposits. The Australasian Institute of Mining and Metallurgy, Carlton, pp 291–298

Sawyer EW (1999) Criteria for the recognition of partial melts. Phys Chem Earth (A) 24:269–279

Stüwe K (1998) Tectonic constraints on the timing relationships of metamorphism, fluid production and gold-bearing quartz vein emplacement. Ore Geol Rev 13:219–228

Tomkins AG, Grundy C (2009) Upper temperature limits of orogenic gold deposit formation: constraints from the granulite-hosted Griffin's Find deposit, Yilgarn craton. Econ Geol 104:669–685

Tomkins AG, Mavrogenes JA (2002) Mobilization of gold as a polymetallic melt during pelite anatexis at the Challenger deposit, South Australia: a metamorphosed Archean gold deposit. Econ Geol 97:1249–1271

Tomkins AG, Pattison DRM, Zaleski E (2004) The Hemlo gold deposit, Ontario: an example of melting and mobilization of a precious metal-sulfosalt assemblage during amphibolite facies metamorphism and deformation. Econ Geol 99:1063–1084

Vernon RH (2004) A practical guide to rock microstructures. Cambridge University Press, Cambridge, p 594

Wilkins C (1993) A post-deformational, post-peak metamorphic timing for mineralization at the Archaean Big Bell gold deposit, Western Australia. Ore Geol Rev 7:439–483

Formation and Modification of Deposits at Lower Temperatures

17

Abstract

Once gold deposits are formed, they can be modified at low temperature through retrogression, weathering, and erosion. An extension of the erosion of gold mineralisation is its transport, sorting, and the development of gold placers. All these modifying processes need to be accounted for when determining the origin of the gold deposit itself. Modification can form or destroy deposits.

Keywords

Weathering · Clay minerals · Regolith · Retrogression · Erosion · Placers

A principle that has been followed for most ore deposits is to understand the most recent modifying effects and to then work backwards in time. This means to understand and account for any events that post-date ore formation before tackling the main (primary) ore-forming process itself. Typically, the approach involves understanding and removing the effects of weathering, retrogression, and metamorphism (Fig. 17.1). Most of the time in geology, this approach is second nature; for example, in a study of primary processes, those rocks that are weathered might be avoided during sampling. Following this might be the removal of any effects of mineral assemblage retrogression or allowing for it and accepting and understanding effects of overprinting high-grade metamorphism.

Techniques such as immobile element geochemistry may help to look through modifying events. A special issue when dealing with any ores containing sulfides is that these minerals are highly susceptible to modification by any later conditions. Note that the chapters of this book have been arranged in the order a deposit forms and then might be modified; however, a research project or industry evaluation of geology would work in the opposite direction.

17.1 Gold-Related Alteration, Retrogression, and Chemical Weathering

Alteration, retrogression and weathering are not identical, but there are aspects of each that overlap, and they are studied with similar scientific approaches.

Gold-related alteration, as it is being used here, refers to the loss and gain of elements during the primary gold deposit forming event. Subsequent higher-grade metamorphism does not usually change this geochemical signature except with respect to volatile components like H_2O, CO_2 and S, or via any partial melting.

Retrogression is the resetting of relatively high temperature mineral assemblages to different minerals stable at lower temperatures. It occurs when the high temperature mineral assemblage from an igneous or metamorphic rock experiences an event involving deformation and

N. Phillips, *Formation of Gold Deposits*, Modern Approaches in Solid Earth Sciences 21,
https://doi.org/10.1007/978-981-16-3081-1_17

Fig. 17.1 Modification
clock illustrating the
sequence of geological
processes that might
overprint a gold deposit
after its formation. Small
numbers refer to relevant
chapters.

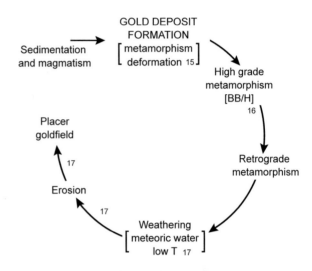

ingress of aqueous fluids at a lower temperature.
The transformation to retrograde minerals may be
complete but very commonly it is only partial,
producing a mix of high temperature and lower
temperature minerals that are out of equilibrium
with one another.

Weathering is usually considered separate
from retrogression and reserved for similar
resetting processes for near-surface temperatures
and conditions with involvement of meteoric
water. This usually means 25°C or thereabouts
and in equilibrium with the oxidising conditions
of the atmosphere, but weathering is gradational
with depth to processes at elevated temperatures.

17.2   Retrogression
and the Resetting of Mineral
Assemblages

If an igneous rock or prograde metamorphic mineral assemblage cools to surface conditions with
no external influences, there is a reasonable
chance that the high temperature minerals will
be preserved. If, however there is ongoing deformation and ingress of aqueous fluid to increase
the kinetics of the chemical reactions, the high
temperature mineral assemblage is likely to be
reset to reflect lower temperature conditions
such as those of the greenschist facies. This can

mean late biotite, chlorite and muscovite growth
and potentially schistose fabrics.

For many gold provinces of greenschist facies
grade, retrogression to sub-greenschist grade is
minor. For goldfields in amphibolite and granulite
facies domains, retrogression is more common
and can be heterogeneous with well-preserved
prograde assemblages juxtaposed with some
totally retrogressed rocks (Fig. 17.2).

Sulfide minerals are particularly susceptible to
retrograde resetting due to their metallic bonding
and fast reaction kinetics so their use in recording
mineralising conditions is difficult. Telluride
minerals are similarly prone to resetting and a
full sulfide and telluride list from a single deposit
can include early and late minerals that are not
part of a co-existing assemblage.

There are at least three consequences when
retrogression overprints a pre-existing gold
deposit. One is the likelihood of a mix of minerals
formed at different temperatures and out of equilibrium; and these may make little sense if there is
an attempt to interpret them as an initially
co-existing assemblage. A second consequence
is that radiometric dating to directly date
mineralisation needs to be within a well-
constrained framework that identifies what event
or process is being dated. The third consequence
is the prevalence of small-scale remobilisation of
gold resulting in grains overgrowing later structural fabrics leading to interpretations that the

Fig. 17.2 Partial retrogression of a biotite schist from Big Bell ore zone: (**a**) brown biotite has been partially retrogressed to pale chlorite that retains the tabular biotite form; (**b**) large cordierite grain with fine-grained chlorite and white mica along cracks. Base of the photomicrograph is 2 mm.

deposit is late: the more correct interpretation being that it was only the specific gold grains that recrystallised late.

As the temperature of retrogression approaches near-surface temperatures it blends into a process with many aspects in common with weathering.

A different form of modification of an existing gold deposit (with Cu) has been highlighted by Smith (2005) when describing the history of the Ordovician Ridgeway porphyry system in eastern Australia. Alteration related to the Cu-Au porphyry deposit had been mapped, but Smith points out that regional burial metamorphism followed, and therefore modified, the hydrothermal alteration assemblages (which are technically metamorphosed alteration assemblages). This regional metamorphism is likely to be a much wider occurrence than Ridgeway alone, significantly understudied, and important in older porphyry systems.

17.3   Weathering Processes at and Near the Surface

Weathering is, amongst other things, a chemical process involving the readjustment of rocks and minerals to near-surface conditions. Agents in this process can include water, biota, and the atmosphere (particularly oxygen but also CO_2). Weathering is enhanced by more reactive rock types and structural complexity such as faults and dykes, which can act as pathways for

meteoric waters to depths of hundreds of metres. It is a process during which igneous and metamorphic assemblages that were stable at elevated temperature and pressure are converted to reflect the surface temperature (about 25°C depending upon location and time of year) and atmospheric pressure. Differing degrees of weathering on the surface at the scale of metres laterally is the norm with some samples virtually unweathered and others extremely so. Weathering may also have a supplementary physical aspect that disaggregates existing minerals. Erosion can add to the near-surface transformation of rocks by removing weathered and unweathered rock material.

17.3.1   Regolith: The Product of Weathering

For many common igneous and metamorphic rocks, the changes during weathering result in clay minerals, quartz and Fe oxides, plus the dissolution or precipitation of carbonate minerals, any sulfides and some silica. The degrees of weathering seen on the surface are also reflected vertically with these mineralogical and physical transformations becoming less intense with depth approaching fresh rock. This whole near-surface system can be referred to as the regolith, i.e. "The entire unconsolidated or secondarily recemented cover that overlies coherent bedrock, that has been formed by weathering, erosion, transport and/or deposition of older material. The regolith thus includes fractured and weathered basement rocks, saprolites, soils, organic accumulations, volcanic material, glacial deposits, colluvium, alluvium, evaporitic sediments, concretions, aeolian deposits and groundwater." (Eggleton 2001). This comprehensive definition of the regolith can be summarised as being everything between fresh rock and fresh air (see Appendix B).

Few places offer the contemporary knowledge and extensive literature of a deeply weathered, intensely studied and highly auriferous region to match the Yilgarn Craton (though few places are as extensively weathered so the importance of this information varies globally depending on whether it is being applied in tropical, arid, humid, or glaciated terrains). In the Yilgarn there have been extensive studies over the last half century of weathering in and around gold deposits. Major compilations and reviews include regolith landscape evolution (Anand and de Broekert, 2005), variations with climate and rock type (Butt, 1989), regolith expression of ore systems (Butt et al., 2005), regolith synthesis of a gold province (Anand, 2000; Smith et al., 2000), regolith applications for gold exploration in the Yandal Gold Province (Anand, 2003), ferruginous material in the regolith, regolith of the Yilgarn Craton (Anand and Paine, 2002), and groundwaters and gold mineralisation (Gray, 2000).

The water table plays a major role in development of the regolith as oxidising conditions prevail above and reducing conditions below this level. The water table rises and falls over time with short term rainfall variation and longer-term climatic changes. In a simple regolith profile, elements above the water table may be oxidised (Fe, Mn) and those below may be reduced (Fe, Mn and C) with the boundary being a redox front that is important in regolith mapping and drill logging when it is termed the *base of complete oxidation* (BOCO). The profile becomes more complex if the water table has moved up and down over time, and this process moves Fe and potentially Au. Another effect of the regolith is the dissolution of quartz by downward migrating waters resulting in destruction of auriferous quartz veins above some but not all gold deposits.

The result of regolith studies at many gold deposits is an idealised profile (Fig. 17.3) that shows many possible situations but is more complex than normally found in a single pit or drill hole. An important feature of this column is that a large part of the profile may be below the dramatic visual change that is BOCO but still in the zone of chemical weathering referred to as saprolite.

Two aspects of the regolith that have been critical in many greenfield exploration programs are the landscape and sample media. The regolith surface can be conveniently divided into relict areas where the regolith surface is intact (R), erosional areas where some stripping away of the regolith profile has occurred (E), and depositional areas where there is younger cover over a regolith profile (D), giving what is known as the RED scheme (Fig. 17.4). The scheme is useful for considering gold deposits in the regolith and how to interpret various materials. The Fe-bearing particles on the surface tend to concentrate some of the rarer metals and so are useful as sampling media in geochemical exploration programs.

17.3.2 Weathering: How Deep?

The depth of weathering is influenced by many factors including time; four factors that will be focused on here for gold deposits are structures, reactive rock types, water supply and topographic relief. Ideal factors for deep weathering might include several km of topographic relief, high rainfall, highly reactive carbonate-bearing rocks and structural complexity, but measuring the extremes of deep weathering can be difficult.

One place where there is demonstrated weathering to 2 km depth is in the Caucasus Mountains of Georgia where weathering processes have dissolved and removed carbonate rocks for 2200 m vertically beneath the surface. The result is the very deep Veryovkina and Krubera (Voronja) cave systems which appear to have their natural base level near the Black Sea. Earlier tracing of the flow patterns of the waters with dye has been complemented by human exploration to measure and map the full depth.

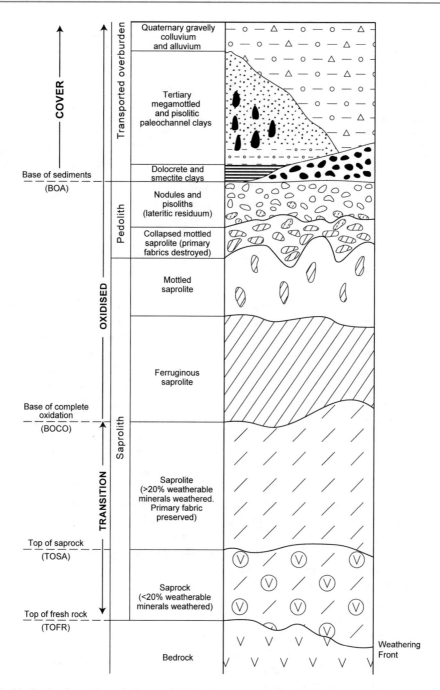

Fig. 17.3 Idealised column through the regolith based upon numerous studies in the Yilgarn Craton summarised by Anand and Paine (2002); the scheme is transferable to many other terrains. The scale is highly variable as the cover can be 0 – 100s m thickness, the oxidised zone above BOCO can be 10 – 100s m thick also, and the transition zone beneath BOCO can be 100s m thick, or even more in some parts of the world.

Fig. 17.4 A schematic cross-section showing some of the expressions of the regolith in a weathered district. A simplified subdivision can recognise relict, erosional and depositional environments and then guide exploration and the interpretation of gold mineralisation.

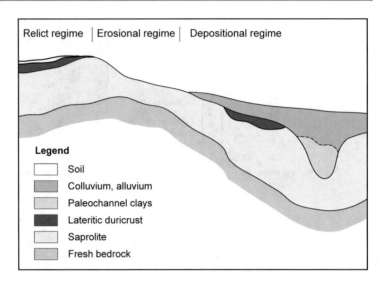

Penetration of meteoric water to 5 km depth is postulated in the genesis of some hematite-bearing iron ores, and here it has been enhanced by crustal extension (McLellan et al., 2003).

An example of meteoric waters percolating to great depth in and around gold deposits has been recognised in the Canadian cordillera where depths of 6 to 8 km have been suggested. Geochemical studies of the waters indicate that they are both temporally and chemically separate from the gold-forming event (Goldfarb et al., 1997; Kyser and Kerrich, 1991) and these are not epithermal deposits as earlier assumed from the isotopic measurements on the deep penetrating, but late, ground waters (Nesbitt et al. 1986).

Gold provinces in areas of high rainfall with topographic relief might be expected to have much deeper weathering than provinces of low relief and an arid environment even given long time periods. However, account needs to be taken of paleoclimate and even paleo-topography. The central to eastern parts of the Yilgarn Craton is an example of low rainfall and subdued topography today in which weathering that commenced millions of years ago in more humid climatic conditions exceeds 100 m depth.

Within many gold provinces, the deposits themselves are susceptible to deeper weathering than their surrounds. Gold-only deposits typically have a strong structural control and are sites of fracture network complexity. This complexity

within goldfields enhances ground water access from the surface during the weathering process in the same way that the structural complexity enhanced the introduction of gold-bearing fluids from below to form goldfields during the mineralisation event (Fig. 17.5). Gold deposits in greenschist facies domains have common pyrite and carbonate alteration haloes that facilitate chemical weathering. The pyrite is very susceptible to break down when in contact with oxidising ground waters, and the reaction generates acidic waters. Carbonate minerals are very susceptible to break down through reaction with these acidic waters. The roles of sulfide and carbonate minerals can be represented by simplified equations, viz:

$$FeS_2 \text{ (pyrite)} + 3.75O_2 + 3.5H_2O$$
$$= Fe(OH)_3 + 2SO_4{}^{2-} + 4H^+ \qquad (17.1)$$

$$CaCO_3 \text{ (calcite)} + 2H^+$$
$$= CO_2 + H_2O + Ca^{2+} \qquad (17.2)$$

17.3.3 Gold in the Regolith

Studies of ore zones in the regolith have demonstrated gold mobility and its redistribution, even into rock types younger than the gold

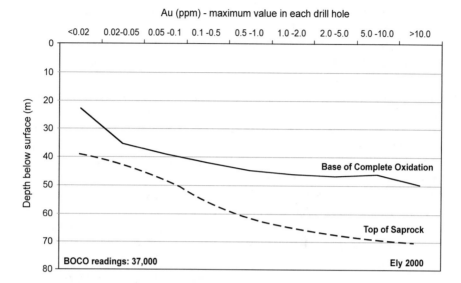

Fig. 17.5 Graph of the depth of weathering as recorded by BOCO and TOS plotted against maximum gold value in each drill hole. Higher gold grades coincide with deeper weathering. This plot is based on 37,000 drill holes in the Yandal Gold Province of Western Australia (Ely 2000).

mineralisation event (Fig. 17.6). To overlook such migration can distort the interpretation of age relationships, genesis, and tectonic setting. What can then follow are inappropriate exploration methods, flawed mine planning, and sub-optimal commercial decisions.

The behaviour of gold in the regolith is influenced by its oxidation state (Au^{3+}), climate and groundwater composition. In more oxidising conditions such as above BOCO, the $AuCl_4^-$ species is stabilised by higher salinity as found in southern parts of the Yilgarn Craton. This situation leads to secondary gold nuggets becoming silver poor as soluble Ag is removed leaving high fineness gold (Mann, 1984; Webster and Mann, 1984). Redox changes across the water table can lead to gold solution, migration, and precipitation, and in some situations to substantial

Fig. 17.6 Cross-section of the regolith showing different landscape situations and three mineralised systems. Gold is mobilised from the primary mineralised systems to give pockets of enrichment, intervals of depletion in some clay zones and broader dispersion haloes. BOA and BOCO are as defined in Fig. 17.3 and Appendix B.

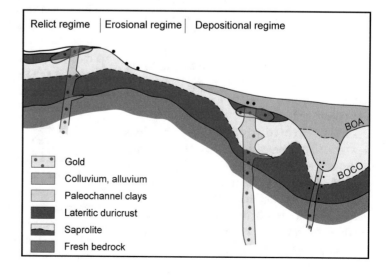

enrichment and nugget formation (Mann, 1998). A different environmental setting of high rainfall in the rugged highlands of Papua New Guinea does not allow any build-up of salinity or acidity and both Au and Ag are soluble as thiosulfate complexes, i.e. $Au(S_2O_3)^{3-}_2$. Secondary gold in these environments is more silver rich than its primary precursor. Regardless of the climate, any solution of gold in the regolith environment can lead to gold dispersion, such that gold mineralisation is spread over a larger area even if it may be at lower grade, and the mineralisation then bears a less-specific relationship to its primary geological controls (be they veins, other structures, or special rock types). However, the same dispersion may lead to a wider exploration target.

17.4   Distinguishing Gold-Related Alteration from Weathering

There is considerable overlap between the criteria being used to identify types of gold deposits (e.g. epithermal) and the characteristics of weathering and the regolith. For example, the products of weathering around many goldfields include several clay minerals, silica, and Fe oxides and oxy-hydroxides. The same features are mentioned as important in epithermal deposits. Interpreting the gold geology literature requires knowing how researchers and industry geologists are using the non-diagnostic criteria. For some professional communities, understanding supergene gold migration and the regolith is part of the basic training whereas in other communities the possibility of gold migration during weathering (oxidation) may not have been considered.

There is also overlap between the descriptions of weathering geometry and that of epithermal systems. Weathering around gold deposits can generate a V-shaped interface such as BOCO being deeper at the mineralisation site (see Fig. 17.5). This is not dis-similar to the shape described for several epithermal systems including those in which mineralisation is described as 'telescoped'.

17.4.1   Clay Products of Weathering

Clay minerals show a characteristic trend during the weathering of goldfields from more complex sheet silicates (chlorite and micas) and clays (illite, smectites) towards kaolinite reflecting the removal of mobile cations especially K. This weathering progression in metasedimentary precursors is commonly represented by the change from chlorite-bearing assemblages to illite and kaolinite. Similar examples of progression towards kaolinite occur around Archean goldfields.

The structure and composition of kaolinite varies through regolith profiles, but changes are not directly correlated with the regolith changes of BOA and BOCO (Cudahy 1997; Anand and de Broekert 2005). In the unconformably overlying weathered sediments above BOA, there is Fe-poor disordered kaolin with an abrupt shift to more ordered kaolinite in weathered basement. However, this trend could be reversed in different groundwater regimes. Also noted by Cudahy was the variation in kaolinite crystallinity with poorly crystalline kaolinite being found both towards the base of the regolith in saprock (i.e. slightly weathered material below the zone of oxidation or BOCO, see Fig. 17.3; Fig. 17.4) and near the top of the regolith within cover sequences. Between these intervals, kaolinite is well crystallised in the saprolite, but less so in the mottled zone and duricrust which may also have multiple generations of kaolinite.

The same variants of clay structure and composition are products of inferred hydrothermal alteration that is associated with some lower temperature ore deposits, and Cudahy attempted to provide a guide to distinguishing the contrasting settings. He found that less-ordered kaolinite and clays with higher incorporated Fe were usually associated with weathering, but the converse of more ordered kaolinite might be related to either hydrothermal alteration or weathering. Overall, it does not appear that any property of kaolinite easily relates to a unique origin for that mineral, and no simple mineralogical parameter could universally distinguish hydrothermal kaolinite from weathering kaolinite (Cudahy, 1997).

Stable isotopic systems using H and especially O isotopes provide a viable way to recognise meteoric waters, but the difficulty remains of demonstrating whether those waters were part of the primary ore formation or some later process. Some gold genetic models invoke meteoric waters (paleo-ground waters) in the actual deposit formation, and these may have a meteoric signature that may be indistinguishable from modern or paleo-weathering waters. Thus, the alternatives of primary low temperature mineralisation and a subsequent overprint by weathering, may both have a stable isotope signature of meteoric water. The difficulty of convincingly recognising hypogene kaolinite is widespread and there are many examples in economic geology where unambiguous hypogene kaolinite is yet to be demonstrated, such as for many inferred epithermal systems.

Although the meteoric waters start at surface temperatures, there is ample evidence they are heated after burial including emerging at the surface as hot springs. Some leeway has been used here to continue using the term weathering and referring to these meteoric waters even though they may be considerably evolved and above 25°C.

17.5   Erosion of Mineralisation, Transport, Sorting, and Gold Placers

Placer gold deposits have been a very important deposit type historically, were easily explored for, and regularly mined by individuals or small groups. They were the source of much of the world's gold up to the 17th century and mined in many countries. Gold placers continued to be important well into the 20th century until they faced mounting challenges from alternative land uses such as farming and urbanisation and came under environmental scrutiny. Gold placers are still an important source of gold in eastern Siberia (Goryachev and Pirajno 2014). The largest gold placers are concentrated around the Pacific margins (Fig. 17.7) including southern New Zealand, Victoria, Far East Russia, Alaska and NW Canada, California, Colombia, eastern Bolivia and the far south of Chile.

Gold placers form from the erosion of gold mineralisation with release of discrete gold grains and gold-bearing particles that move downhill and then down river systems. Gold placers depend upon the separation of particles with gold from voluminous sand, silt, clays and rock fragments and this can be achieved by settling the gold whilst transporting much of the waste material farther downstream. Relatively high energy streams can remove the waste material, but the settling and retention of gold relies on two physical factors. The first factor is the high density of gold of 19.3 $g \cdot cm^{-3}$ which is more than 5 times that of most of the stream detritus. The second factor is a high particle mass (which helps particles to settle) compared to particle surface area (which helps particles to continue migrating in the flowing water). The volume (related to the cube of the particle diameter) to surface area (related to the square of the particle diameter) increases as particles with gold become larger. Gold particles in placers are commonly 0.5 – 5mm in diameter and this size is partly due to primary grain shape and size and partly due to supergene effects of solution and grain remobilisation (McLachlan et al., 2018). Consequently, gold placers require a source of particulate gold which mostly means from eroded mineralisation within 0-20 km. Furthermore, background levels and nanoparticles of gold do not contribute to placer deposits because they do not meet the physical requirements that allow sorting and deposition. All the largest gold placers are in non-marine settings, and the largest of those in a marine setting at Nome Alaska is an order of magnitude smaller but not insignificant.

The gold placers of South Island, New Zealand have been studied in detail exposing a complex role for regional uplift, supergene weathering, stream diversion and capture, and multiple river re-orientations. The coarser gold is over 3 mm in diameter and 0.5 g but closer to 2 mm in distal regions (Craw et al. 2015). The source appears to be auriferous quartz veins in low to medium grade metamorphic metasedimentary rocks of Mesozoic age. The only major primary gold deposit in the

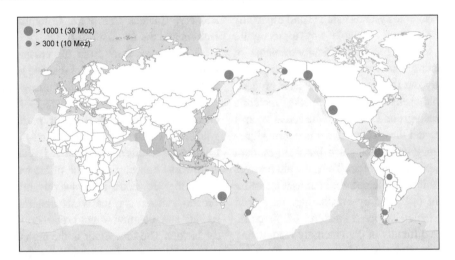

Fig. 17.7 Map of the large Phanerozoic placer goldfields illustrating the arrangement around the Pacific Ocean. Auriferous quartz veins in metasedimentary rocks appear to be the source of most of the particulate gold forming the placers during the late Mesozoic and Cretaceous periods. All settings are non-marine except for the offshore part of the Nome goldfield in Western Alaska.

region is Macraes of 10 Moz (300 t Au) and primary gold from here has been significantly modified in the supergene zone by both chemical and physical processes. These changes have resulted in gold particles of millimetric size entering the placer system indicating that the growth and enlargement of gold particles in the weathered zone is important in the placer process (Craw et al., 2006; Craw and McKenzie 2016).

The placers of eastern Siberia remain an important source of world gold production and are being mined in linear steep-sided valleys for 10s km. An example of such mining is the placer field downstream from the major primary goldfield of Natalka of 60 Moz (1800 t Au; Fig. 17.8 a-d). As for Otago placers in New Zealand, the source appears to be auriferous quartz veins in metasedimentary rocks of Mesozoic age.

Placers of the Victorian Gold Province yielded 50 Moz (1500 t Au) from 1851 and show very clear relationships with the further 30 Moz (1000 t Au) of primary gold production from auriferous quartz veins in Paleozoic metasedimentary rocks. Many placers were mined successfully in their own right, and others were followed upstream to

discover primary auriferous quartz vein deposits. The western portion of the Lachlan Orogen in Victoria contains the largest primary goldfields (Bendigo, Ballarat, Fosterville, Stawell) and has produced most of the placer gold (Fig. 17.9). In addition to this regional correspondence of primary deposit and placer, many primary deposits were found by following alluvial gold directly upstream to auriferous quartz veins. Travel distances of the alluvial gold was usually 0-20 km though 50 km has been suggested for some systems. This correspondence between placer gold and its primary source, and the general travel distance of 0 – 20 km, has given confidence to the process of adding back the placer gold to obtain a more realistic endowment of primary goldfields. Over the last 40 myr, there have been multiple periods of erosion of gold from the Paleozoic quartz veins, concentration in streams and then cover by younger sediments and basalt flows resulting in buried placer gold deposits known as deep leads.

Gold placers in marine sediments are rare globally and one of the largest is at Nome Alaska as part of a 10 Moz (300 t Au) goldfield straddling the hinterland, beach, and offshore environments

Fig. 17.8 Russian Far East, Yana – Kolyma Gold Province near Natalka gold deposit: (**a**) auriferous quartz veins in black shale; (**b**) Natalka gold mine; (**c**) valley with placer workings that continue for 20 km downstream; (**d**) modern placer workings in the valley.

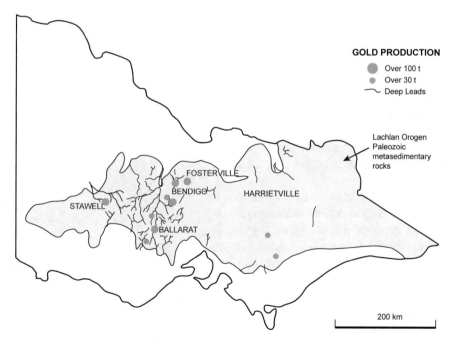

Fig. 17.9 Map of the Victorian goldfields, SE Australia showing the area of outcropping Paleozoic metasedimentary rocks in light blue, location of major goldfields based on the inferred endowment prior to erosion, and the location of the major placer systems. The close correspondence of primary deposits and gold placers on this regional scale is complemented by correspondence on a district scale, and the global scale. Deep leads are river valleys with alluvial gold near their base that are subsequently covered by younger alluvium or basalt flows. Modified from Hughes et al. 2004.

Fig. 17.10 Sketch of the geological settings around Nome that account for much of its 10 Moz Au production. From the hills to Nome beach is approximately 5 km, and the auriferous gravels extend a similar distance offshore.

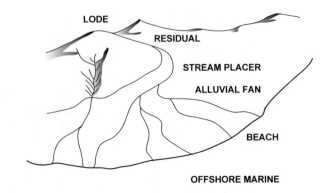

Fig. 17.11 Nome goldfield, Alaska: (**a**) low hills contain auriferous quartz veins that have shed detrital gold into creeks; (**b**) assessing gold concentrations by panning; (**c**) coastline with beach (bottom right corner), thawing ocean and offshore dredge.

(Fig. 17.10; Fig. 17.11). Total production from Nome includes settings such as auriferous quartz veins in carbonaceous schist and marble several km inland, and along small creeks leading from these primary occurrences. There is still active mining immediately behind the beach, and dredges have been used to mine gold for 5 km offshore.

None of the world's large placer goldfields can be traced back to a province of volcanogenic massive sulfide deposits, nor to copper – gold deposits. Given that the largest Cu – Au deposits are well over 1000 t Au at grades around 0.5 – 1 g/t Au it is unlikely the absence of large placer fields downstream from these is due to lack of gold. The more likely reason is the size of gold particles—these can be millimetres in diameter for auriferous gold-only quartz veins but are typically much smaller and in aggregates with sulfide and even telluride minerals in Cu-Au deposits, e.g. 0.1 mm in Cu-Au porphyry deposits.

Snapshot

- It is important to remove the effects of any later modifying events before trying to determine the origin of a primary gold deposit.

- Weathering of ores may yield two fluid signatures (primary ore forming and meteoric), but they may be of different times and not a sign of mixing.

- Features used to characterise deposits as being disseminated, such as carbonate destruction and clay assemblages, are a common product of weathering.

- In many goldfields, it is difficult to convincingly demonstrate that kaolinite is hypogene.

- Most gold placers are linear, in non-marine settings and formed from eroded mineralisation.

(continued)

- None of the world's large placer goldfields can be traced back to a province of volcanogenic massive sulfide deposits, nor to copper—gold deposits. All arise from gold-only goldfields.

Bibliography

Anand RR (2000) Regolith and geochemical synthesis of the Yandal Greenstone Belt. In: Anand RR, Phillips GN (eds) Yandal Greenstone Belt, vol 32. Australian Institute of Geoscientists Bulletin, Perth, pp 79–111

Anand RR (2003) Importance of regolith for gold exploration in the Yandal Gold Province, Yilgarn Craton, Western Australia. In: Ely KS, Phillips GN (eds) Yandal Gold Province, vol 1. CSIRO Explores, Canberra, pp 27–52

Anand RR, de Broekert P (eds) (2005) Regolith landscape evolution across Australia. CRC LEME, Perth, p 354

Anand RR, Paine M (2002) Regolith geology of the Yilgarn craton. Aust J Earth Sci 49:3–162

Butt CRM (1989) Genesis of supergene gold deposits in the lateritic regolith of the Yilgarn Block, Western Australia. Econ Geol Monograph 6:460–470

Butt CRM, Robertson IDM, Scott KM, Cornelius M (eds) (2005) Regolith expressions of Australian ore systems. CRC LEME, Perth, p 431

Craw D, McKenzie D (2016) Macraes orogenic gold deposit (New Zealand): Origin and development of a world class mine. Springer, Cham, p 127

Craw D, Youngson JH, Leckie DA (2006) Transport and concentration of detrital gold in foreland basins. Ore Geol Rev 28:417–430

Craw D, McKenzie D, Grieve P (2015) Supergene gold mobility in orogenic gold deposits, Otago schist, New Zealand. N Z J Geol Geophys 58:123–136

Cudahy TJ (1997) PIMA-II spectral characteristics of natural kaolins, CSIRO exploration and mining report 420R for AMIRA, Perth, WA, p 62

Eggleton RA (ed) (2001) The regolith glossary. Co-operative Research Centre for Landscape Evolution and Mineral Exploration, Perth, p 144

Ely K (2000) Bedrock lithologies, mineralisation and deformation and their relationship to weathering in the Yandal Belt. In: Anand RR, Phillips GN (eds) Yandal Greenstone Belt, vol 32. Australian Institute of Geoscientists Bulletin, Perth, pp 115–123

Goldfarb RJ, Newberry RJ, Pickthorn WJ, Gent CA (1991) Oxygen, hydrogen, and sulfur isotope studies in the Juneau gold belt, southeastern Alaska:

constraints on the origin of hydrothermal fluids. Econ Geol 86:66–80

Goldfarb RJ, Phillips GN, Nokleberg WJ (1997) Tectonic setting of synorogenic gold deposits of the Pacific rim. Ore Geol Rev 13:185–218

Goryachev NA, Pirajno F (2014) Gold deposits and gold metallogeny of Far East Russia. Ore Geol Rev 59:123–151

Gray DJ (2000) Chemistry of groundwaters in the Yandal Greenstone Belt. In: Anand RR, Phillips GN (eds) Yandal Greenstone Belt, vol 32. Australian Institute of Geoscientists Bulletin, Perth, pp 145–155

Henley RW, Adams J (1979) On the evolution of giant gold placers. Trans Inst Min Metall B 88:41–50

Hughes MJ, Kotsonis A, Carey SP (1998) Cainozoic weathering and its significance in Victoria. Aust Inst Geosci Bull 24:135–148

Hughes MJ, Phillips GN, Carey SP (2004) Giant placers of the Victorian Gold Province. Econ Geol Newslett 56(1):11–18

Ilchik RP, Barton MD (1997) An amagmatic origin for Carlin-type gold deposits. Econ Geol 92:269–287

Kyser TK, Kerrich R (1991) Retrograde exchange of hydrogen isotopes between hydrous minerals and water at low temperatures, in Stable isotope geochemistry: a tribute to Samuel Epstein. Geochem Soc Spec Publ 3:409–422

Mann AW (1984) The mobility of gold and silver in lateritic weathering profiles: some observations from Western Australia. Econ Geol 79:38–49

Mann AW (1998) Oxidised gold deposits: relationship between oxidation and relative position of the water table. J Geol Soc Aust 45:97–108

McLachlan C, Negrini M, Craw D (2018) Gold and associated minerals in the Waikaia placer gold mine, northern Southland, New Zealand. N Z J Geol Geophys 61:164–179. https://doi.org/10.1080/00288306.2018.1454482

McLellan JC, Oliver NHS, Ord A, Zhang Y, Schaubs PM (2003) A numerical modelling approach to fluid flow in extensional environments: implications for genesis of large microplaty hematite ores. J Geochem Explor 78-79:675–679

Mitchell CJ, Evans EJ, Styles MT (1997) A review of gold-particle-size and recovery methods: overseas geology series, Technical Report WC/97/14, British Geological Survey; pp 1–32

Nesbitt BE, Murowchick JB, Muehlenbachs K (1986) Dual origins of lode gold deposits in the Canadian cordillera. Geology 14:506–509

Rotherham J (2000) Gold mineralisation in the oxide-transition zone-Mt Joel prospect, Yandal belt. In: Anand RR, Phillips GN (eds) Yandal Greenstone Belt, vol 32. Australian Institute of Geoscientists Bulletin, Perth, pp 273–281

Sillitoe RH (2005) Supergene oxidized and enriched porphyry copper and related deposits. Economic Geology 100th anniversary volume, pp 723–768

Smith RE (2005) The ridgeway gold-copper deposit: a high grade alkalic porphyry deposit in the Lachlan Fold Belt, New South Wales, Australia—a discussion. Econ Geol 100:175–178

Smith RE, Anand RR, Alley NF (2000) Use and implications of paleoweathering surfaces in mineral exploration in Australia. Ore Geol Rev 16:185–204

Webster JG, Mann AW (1984) The influence of climate, geomorphology and primary geology on the supergene migration of gold and silver. J Geochem Explor 22:21–42

Part IV

Examples of Gold Deposits

Carlin, Witwatersrand, and Some Other Gold-only Examples

18

Abstract

Goldfields in greenstone belts and metasedimentary rocks have been used to build models for the origin of gold-only deposits, but these two types are only a sub-section of gold-only deposits globally. It is now possible to place the important Carlin goldfields of Northern Nevada USA and the gold porphyry and low-sulfidation deposits into a similar context of thermal events, deformation, and hydrothermal fluids. In addition, some initial observations about the Witwatersrand goldfields of South Africa illustrate their obvious differences but also similarities. In superficial appearance each grouping of deposits is quite different, but all share most of the five characteristics of gold-only deposits already described.

Keywords

Deep weathering · Carlin Gold Province · Epithermal deposits · Witwatersrand gold

Six informal groups of primary deposits account for virtually all gold-only production:

- Archean and Paleoproterozoic greenstone belts
- Clastic metasedimentary sequences (slate belt, turbidite hosted)
- Carlin Gold Province of USA (and possibly some similar occurrences)
- Auriferous veins in igneous host rocks (gold porphyries included)
- Low sulfidation epithermal deposits
- Witwatersrand goldfields of South Africa.

Introducing these well-known types of gold has been deliberately left until late in the story. If we had started our study by dividing all gold deposits into ten main types, we would likely have ended up with ten different types and ten ways in which they formed—somewhat like a self-fulfilling prophecy made even more probable if different research groups were studying each of the ten types.

For the question "do these deposits have economic base metals?", the answer is in the negative for all six groups. The choice of these six is neither scientifically rigorous nor ideal but it does address groupings in common usage. Many other names such as orogenic and mesothermal have been applied to sub-sets of gold-only deposits and these names may have their uses, but their lack of diagnostic defining features (just as for some of the above six) means classification can be controversial and overlapping. It is both interesting and useful to know the tectonic environment in which each group of deposits formed, but as this requires knowing how and especially when each formed, tectonic setting is not the first classification to be made.

The uncertainty in some of these deposit groups is illustrated by new depictions of Carlin

N. Phillips, *Formation of Gold Deposits*, Modern Approaches in Solid Earth Sciences 21,
https://doi.org/10.1007/978-981-16-3081-1_18

deposits involving continuums with porphyry, distal disseminated, and epithermal systems. These may encapsulate useful descriptive aspects but do not reflect fundamental differences in origin. Suggested origins for these deposit types have included metamorphic, magmatic and meteoric fluids but it is unlikely that all of these are fundamental to deposit formation.

18.1 Carlin Gold Province, Nevada USA

The Carlin Gold Province refers to many goldfields in Northern Nevada that share some geography, geology and age characteristics. Within this area, there are over 100 medium to large goldfields hosted mostly in low to medium grade metasedimentary rocks of Paleozoic age. The province covers an area of 300 km by 300 km with many major goldfields arranged in linear patterns that are up to 80 km long and 5-10 km wide (Fig. 18.1). This is a *gold-only* province in that its goldfields are gold-rich with subordinate silver and common As, Te, Sb, W and Hg. Apart from the obvious provinciality of the deposits in Northern Nevada, these deposits also share the other characteristics of gold-only deposits of extreme gold enrichment, segregation of gold from base metals, epigenetic timing and inferred low salinity ore fluids.

The Carlin Gold Province is used here to refer to gold-only deposits in Northern Nevada USA. The various goldfields of the province share geographic location and several geological characteristics; Carlin Gold Province is used as a term in the same way as Abitibi Gold Province and Victorian Gold Province.

18.1.1 Distribution and Discovery of Goldfields

The eponymous Carlin deposit was discovered in 1961, started producing in 1965 and was followed by further discoveries and mines. For the first three decades of production, the economics of mining was greatly assisted by the substantial depth of oxidised ores. The province has become the main source of gold from the USA and very important on a global scale. The endowment of the province is 250 Moz (7900 t Au) and annual production peaked at 8.9 Moz (277 t Au) in 1998 but has declined since then (Table 18.1; Muntean 2018). The Goldstrike operation which consists of Betze-Post, Rodeo and Meikle mines on the Carlin Trend, is one of the world's three largest gold mines producing over 2 Moz (60 t Au) in 2018, and 42 Moz (1300 t Au) since commencing in 1986. The combined operations of the eponymous Carlin mine produce almost 1 Moz pa.

The first mining of these deposits was by open pit with extraction of gold through heap leach methods, but this approach did not proceed deeper than the change from bleached oxidised ores to the porous black ores that are refractory. In their early years, the deposits were considered as epithermal and unlikely to have great depth extent.

A shift in the mid-1980s from the shallow (epithermal) thinking coincided with major discoveries at depth. Around the Goldstrike property these included Screamer, Betze, Deep Star, Deep Post and Meikle (Bettles 2002). Developing the refractory ores at depth has required pressure oxidation and roasting to extract the gold, and over time large mines needed to go underground.

18.1.2 Goldfield Geology

The main host rocks for these goldfields are Ordovician to Devonian metasedimentary rocks that include shale, black shale, siltstone, silty limestone, and limestone of 455 – 325 Ma age. The tectonic setting of Northern Nevada during this period was that of a Proterozoic to Devonian passive margin along western North America overlying Proterozoic continental crust. In this setting was deposited a succession of shallower water carbonate sequences and deeper water clastic sedimentary rocks including turbidites with minor basalt. These rocks have been multiply deformed with shear zones and faults, regional folds controlling outcrop patterns, and low to medium grade metamorphism.

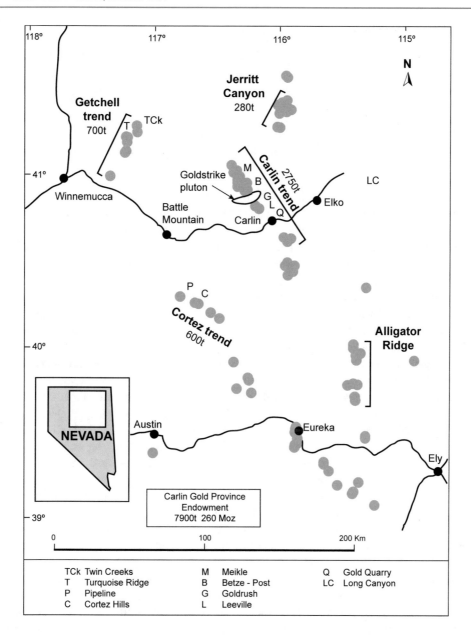

Fig. 18.1 Map of Northern Nevada showing major goldfields aligned along several linear trends of 10s km length.

Gold occurs throughout this Paleozoic succession, and important Carlin-type gold districts are in the shallow water carbonates, others in the deeper water turbidites, and some districts are in both. There is also a spread of deposits that spans host rocks of Cambrian to Devonian age though single districts tend to be concentrated in a particular setting (either deeper water basin, slope, platform, or shelf) and time period. In specific districts, most gold mineralisation may be within one host unit, but in a different district on the same trend mineralisation may be restricted to other units. For example, along the Carlin Trend, the hosting units become systematically younger over 80 km heading south: in the northwest they are 415-370 Ma, and in the southeast,

Table 18.1 Production and endowment, Carlin Gold Province.

	Production through 2016		Endowment	
	t	Moz	t	Moz
Carlin Trend	2700	88	3700	120
Cortez	625	20	1600	50
Getchell	700	23	1400	45
Jerritt Canyon	280	9	400	13
Carlin Gold Province	4600	150	7900	250

Muntean 2018; figures through 2016, rounded

they are 360-345 Ma (Berger and Theodore 2005). A less common lithological setting is that of gold ore in Paleozoic basalt at Twin Creeks mine in the Getchell area. Several of the goldfields display a spatial association with various younger igneous rocks ranging in age from 160 to 40 Ma, including lamprophyre dykes and granitic intrusions (157-160 Ma), and felsic dykes intruded around 40 Ma. Minor gold is reported from Late Paleozoic metasedimentary rocks and from the margins of later igneous rocks. Gold has been mined from the Jurassic Goldstrike Pluton with 0.1 t Au produced from shallow oxidised ore near the margins of this intrusion (Fig. 18.2).

Several periods of compressional tectonism characterise the Phanerozoic history and include the Antler Orogeny (Late Devonian), Humboldt Orogeny (Carboniferous), Sonoma (Triassic) and Sevier (Cretaceous) Orogenies, and these have given rise to the gross geometry of much of the Carlin Gold Province. Cenozoic deformation dictates the current geomorphology of Nevada which is a result of extension around 40 Ma and the Basin and Range Orogeny (10-20 Ma).

The Antler Orogeny involved east – west compression that displaced deeper water basin rocks along the major low angle Roberts Mountain Thrust towards the east and over shelf limestone. A foreland basin developed farther east of the Roberts Mountain Thrust with highlands to the west shedding sediment into that basin. Older parts of the Paleozoic sequence have been juxtaposed above younger parts with many of the goldfields immediately below, in or just above, the Roberts Mountain Thrust.

Gold mineralisation is structurally hosted, despite the local importance of stratigraphy. There is a wide range of structures including faults, shear zones, breccias and bedding related features, and many have an origin during the Paleozoic (Antler) deformation with a complex later history of repeated reactivation. There is a significant variability in ore shapes from being discordant, steep and fault-controlled, to more-conformable and sheet-like. Brittle-ductile low angle thrusts, high angle reverse thrusts with preserved brittle-ductile kinematic indicators, and high angle normal faults are all important. The brittle-ductile structures are pre-Jurassic in original age, but likely to have been reactivated; the normal faults are mostly post-Jurassic (Peters, 2004). The complex history of these structures is well illustrated by the Post Fault within the Goldstrike mine with over one km of displacement: this fault has had multiple periods of movements that are pre-Jurassic, syn-Tertiary and probably Recent. It has also been the site for intrusion of dykes during the ~30 – 40 Ma extension, the locus of major mineralisation in its footwall, and site of current meteoric water activity.

The metamorphic grade is reflected in muscovite-chlorite-quartz mineral assemblages suggesting greenschist facies approaching 300°C or above. This is consistent with the phyllitic fabric development in pelitic rocks, and the brittle-ductile deformation seen in shear zones in open pits. Local structural features, such as pencil cleavage after amphibole indicates that the metamorphic grades may have extended to upper greenschist facies in parts of the province. The very low metamorphic grade reported for some Carlin goldfields is not based on fresh rock but instead on clay assemblages that likely reflects the deep penetrating ground waters (weathering) at temperatures well above 25°C.

Fig. 18.2 Column showing geological events in Nevada from 500 Ma to the present. Carlin-type gold is abundant throughout different facies and ages of the Paleozoic metasedimentary rocks. Virtually all the gold is mined from Paleozoic rocks, but the small dot around 150 Ma represents gold produced from the oxidised interval of the Jurassic Goldstrike pluton. The Antler Orogeny deformed the whole Paleozoic succession and established many of the structures that were reactivated during later orogenic events. Meteoric waters of 40 Ma and younger pervaded the metasedimentary sequence coinciding with extension. The uplift of the Sierra Nevada mountains west of the Carlin Gold Province led to a more arid climate in Nevada, less stream runoff, and filling of valleys with alluvium.

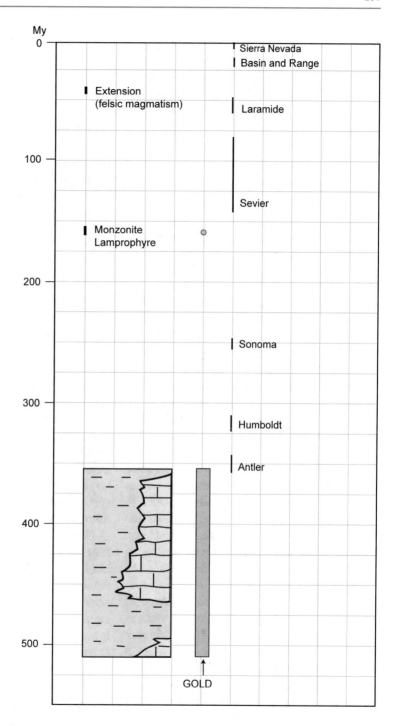

There is a distinctive vertical zonation in Carlin goldfields and mineralisation is highly variable in a vertical sense despite there being remarkable uniformity between goldfields over much of the province. In addition to any younger cover, the near-surface part of goldfields is an interval of mostly bleached material dominated by clays and Fe-oxides with free-milling (non-refractory, and thus easily leached) gold ores but only minor quartz veining. Beneath the bleached material is an interval of porous grey to black material that includes dark ores that are refractory (non-cyanide leachable) with a paucity of quartz veining. A third distinct ore type is an interval of quartz veins and carbonate alteration exemplified by the Screamer section of the Goldstrike deposit.

18.1.3 Bleached Interval with Free-Milling Oxide Ores

The upper interval of virtually all goldfields in the Carlin Gold Province is bleached, contains oxidised ore and gangue material dominated by clays and iron oxides. This bleached interval is present in several geomorphic settings including hills well above the basin floor (Jerritt Canyon) to beneath Tertiary cover in basins (Gold Quarry, Sleeper), and is a topography-wrapping feature.

The depth of oxidation is up to several hundred metres in open pits, and similar material is recorded from drilling to 800 m (Table 18.2). These intervals of deep oxidation are described from the literature (Pipeline, South Pipeline, Carlin, Twin Creeks), and recorded during drill core inspection (Rain), mine maps, and several mine visits (Fig. 18.3; Fig. 18.4), and there are unconfirmed reports of similar bleached material to 1500 m in drill core. In several locations the bleaching is deeper adjacent to ore and structures. In all these cases, this oxidation is equated to the upper component of the regolith profile. The shape of these zones, including narrowing into structures with depth, reflects the influence of the surface during downward penetration of meteoric waters. The bleached material from drill core from 800 m depth is remarkably similar to bleached material from near surface in open pits, i.e. lacking sulfides, carbonate minerals and quartz veining, being highly porous and friable, and with clays and iron oxides. Some variation in clay crystallinity and type is expected if the regolith profiles globally are a guide (see Ch17.4.1). The term 'supergene oxidation' is used in the Carlin Gold Province to approximate the bleached interval described here.

18.1.4 Bleached to Porous Black Interface: Base of Complete Oxidation (BOCO)

The base of complete oxidation is easily recognised in many of the open pits of the Carlin Gold Province by the dramatic colour change (Fig. 18.3; Fig. 18.5). Shallow bleached material gives way to dark grey to black material with depth. This boundary is broadly horizontal, but in detail is controlled by rock types, and especially by structures. A consequence of this control is that wedges of unoxidised material can lie above bleached material giving reverse BOCO successions. The interface coincides with the transition from free-milling ores to refractory ores.

18.1.5 Porous Black Interval with Refractory Ores

The base of complete oxidation (BOCO) marks a redox front and the change from abundant iron oxides above, to a new assemblage of carbonaceous material, orpiment (As_2S_3), realgar (AsS), arsenian pyrite and high overall As concentration. Dating of galkhaite [(Cs Tl)(Hg Cu Zn)$_6$(As Sb)$_4$S$_{12}$], a paragenetically-late mercury sulfosalt mineral, yields ages around 39 – 40 Ma from the porous black interval of refractory ores.

Not all minerals change across BOCO and features common to the bleached and porous black intervals include the clay minerals, a continuation of the meteoric water isotopic signature, and a continuation of the destruction of carbonate minerals (decalcification, decarbonation). The lower boundary of the porous black ore is poorly known.

18.1.6 Calcite Interval (Screamer Section of Goldstrike)

The Screamer section is deep and towards the west of the Goldstrike complex with some significant differences compared to most of the Carlin ores. Auriferous quartz veins are common with

Table 18.2 Depth of oxidation at various goldfields in the Carlin Gold Province.

	Depth of oxidation (m)	Comment
Pinson	600	
Getchell	600	
Twin Creeks	120	500 m near sulfides
Meikle	500	
Goldstrike	200	
Carlin	250	
Mike	500	
Gold Quarry	400	Deeper on structures
Rain	800	Possibly 1500 m in core
Pipeline	245	
Pipeline South	550	

Fig. 18.3 Oxidised (bleached) material in the upper parts of open pits: (**a**) Goldstrike open pit with porous black material beneath the capping of oxidised bleached rocks near the surface; (**b**) Goldstrike open pit with a complex interleaving of bleached rock and porous black rocks delineating a complex BOCO determined by depth, bedding and structures; (**c**) Gold Quarry open pit with wedges of porous black material surrounded by bleached rock illustrating a control on BOCO by bedding and structures; (**d**) Twin Creeks open pit, which is 100 km from Goldstrike, at the stage of transitioning from oxidised bleached material to the deeper porous black material. Different ore processing methods are required for the oxidised and non-oxidised ores and the boundary between the two could be mapped as BOCO. Photo 18.3a by Juan La Riva Sánchez reproduced with permission.

calcite, pyrite and arsenopyrite in the interval at the expense of the decarbonation. The change to an ore type with abundant carbonate minerals has significant implications for the Goldstrike mining operation when it needed to develop longer term plans to treat ores with 10% carbonates.

18.1.7   Explanation of Weathering from the Bleached Interval

Several lines of evidence suggest that the oxidation recorded in the bleached interval is supergene-related and not part of any hypogene

Fig. 18.4 Drill core from deposits of the Carlin Gold Province: (**a**) relatively intact core of metasedimentary rocks from 500 m depth as expected in most drilling deeper than ~100 m; (**b-d**) core from near Rain mine 100 km south of Goldstrike. Highly broken, friable and incoherent material from core drilling from 600 m, 700 m, and 800 m depth, respectively.

Fig. 18.5 Various rock units from in and around deposits of the Carlin Gold Province: (**a**) Gold Quarry open pit with bleached rocks of clays and Fe oxides overlying dark material containing arsenian pyrite and carbonaceous material. There is a wedge of Cenozoic sediment in the top centre; (**b**) pit highwall with upper oxidised bleached rock and mostly porous black material from Betze Post pit; (**c**) ore below the level of oxidation (BOCO) with realgar, orpiment, arsenian pyrite and carbonaceous material from Betze Post pit; (**d**) unweathered and unaltered sample of the Goldstrike Pluton, a massive granitic rock with horn-blende and plagioclase.

mineralisation event (Bakken and Einaudi, 1986, p. 396):

- oxidation boundaries cut young faults
- oxidation boundaries are very sharp and post-date most veins
- pyrite is preserved where encapsulated in quartz, so oxidation of pyrite postdates the formation of quartz
- bleached zones broadly wrap topography in the province.

Establishing the bleached oxidised interval as supergene-related (weathering) and hence part of the regolith was a turning point in Carlin gold geology marking the departure from epithermal ideas. What is interesting is that the epithermal ideas that dominated Carlin gold geology for two decades were built on classical observations of quartz and carbonate vein textures that were being used globally to recognise epithermal gold systems. The lesson from the Carlin Province is that these textures may not be diagnostic of epithermal systems but can form more widely through weathering processes such as in Archean greenstone and slate-belt gold deposits.

There are several mineralogical changes that have been recorded in the Carlin Gold Province approaching bleached oxidised material:

- distal fresh silty limestone becomes weak to moderate decalcification (dolomite halo), then strong decalcification, and finally decarbonation; these all reflect breakdown of carbonate-bearing assemblages
- distal smectite-kaolinite-illite becomes weak kaolinite, then intense kaolinite
- distal quartz veins give way to proximal jasperoid bodies with nearby gold mineralisation.

Aspects of this same progression are found globally as products of increasing intensity of weathering around gold deposits; this weathering interpretation is invoked here too.

The weathering history of the gold deposits of Nevada is strongly influenced by climate which in turn is influenced by the topography. Most of Nevada including the Carlin Gold Province is part of the Basin and Range terrain of southwest USA—essentially many north-south mountain ranges with intervening valleys. The elevation of much of Northern Nevada is constrained between 1000 m above sea level (or less) for valley floors filled with young sediment and lakes (cover regime of regolith) and 3400 m of the hill tops (erosional regime of regolith). Farther west near the Nevadan border but mostly in California are the Sierra Nevada mountain range rising to 4300m. The climatic effect of the Sierra Nevada range is to cause orographic rainfall in California as moisture from the Pacific Ocean is lifted to height immediately west of these mountains, in turn placing much of Nevada in a rain shadow today. The Carlin Gold Province receives around 250 mm (10 inches) of annual rainfall. However, the Sierra Nevada range was probably less than 1 km above sea level until it was elevated 1.5 to 2 km during the last five million years. Without the mountain range along the western border of Nevada moisture from the ocean could migrate inland where annual rainfall was likely much higher (Fig. 18.6; Fig. 18.7). One consequence of the higher rainfall was more scope for deep weathering within the Carlin Gold Province and Northern Nevada which began prior to 30 Ma and prior to valley filling. As the Sierra Nevada mountains rose and the climate became arid, river flows decreased and whole valley areas silted up with deep alluvial cover.

The great depth of the oxidised interval of the weathering profile in the Carlin goldfields, as revealed by 100s m of bleached rock above BOCO (Table 18.2) is a consequence of this earlier paleo-rainfall coupled with soluble calcite-bearing rocks, structural complexity around gold mineralisation and substantial relief across Nevada. Onset of deep weathering is inferred to be in the Eocene (30 – 40 Ma) coinciding with the regional extension that followed 300 myr of dominantly compressional events. Most of these factors are the same characteristics that have led to deeper weathering in goldfields globally and typically enhanced by gold-related sulfide minerals which reacted during weathering to generate acid ground waters (Chapter 17).

Fig. 18.6 Cross-section of California and Nevada. At 40 My ago (top) there was limited topography along the California – Nevada border so moderate to high rainfall in Nevada and potential for large scale groundwater penetration. Today (bottom), the Sierra Nevada mountain range has risen along the State border and confined Nevada to a rain shadow and isolated much of that State from ocean derived moisture.

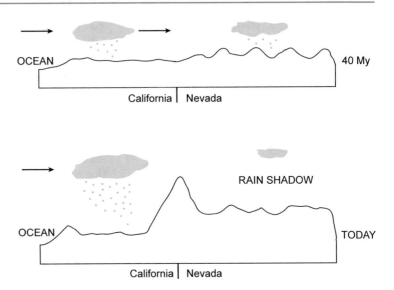

Fig. 18.7 Schematic diagram depicting the deep weathering environment in Nevada from 40 Ma. This was a period of higher rainfall than today and the first major crustal extension after 100s myr of compressional events. Rainfall on mountains entered fractures in the rocks to percolate downwards, react with any pyrite, and dissolve carbonate. This whole process is greatly enhanced around gold mineralisation and led to very deep weathering in parts of Northern Nevada. A modern analogue to this environment might be local parts of the Caucasus mountains along the border of Georgia and Russia which rise to 5600 m above sea level and have extremely deep weathering expressed by cave systems over 2 km deep that bottom with a base level close to the Black Sea. Artwork by Alice Coates, reproduced with permission.

18.1.8   What Geological Features Characterise the Carlin Deposits?

Outside the Carlin Gold Province of Nevada, there has been less success defining what is a Carlin-type deposit globally as many of the features being used in Nevada are non-diagnostic.

In a review of the Carlin deposits of Northern Nevada, Teal and Jackson (1997) identified four characteristics that distinguished these deposits (sulfidation of Fe-rich rocks, argillic alteration, silicification, carbonate dissolution). The first of these, sulfidation of Fe-rich rocks had already been noted in other gold provinces including those in Archean greenstone belts. As such, sulfidation is not diagnostic of Carlin goldfields, but is a common feature in many gold-only deposits. All three of the other distinguishing features used in Nevada are features of the weathering environment globally, and hence they are not diagnostic of Carlin goldfields either. Two decades later carefully compiled lists of features found in deposits of the Carlin Gold Province are more detailed but still utilising non-diagnostic features found in many other goldfields globally, and it remains a challenge to specify what is distinctive about a Carlin deposit (Muntean and Cline, 2018; Muntean, 2018):

- Tectonic setting
- Carbonate host rocks
- Replacement mineralisation, with structural and stratigraphic ore controls and a lack of veins
- Hydrothermal alteration characterised by dissolution and silicification of carbonate and argillisation of silicates
- Ore paragenesis characterised by auriferous, arsenian pyrite formed by sulfidation during replacement, where the majority of gold is invisible, in the form of Au^{+1} in the pyrite, followed by late orpiment, realgar, and stibnite
- Au-Tl-As-Hg-Sb-(Te) geochemical signature in both the ore and ore-stage pyrite that is low in Ag (Ag/Au $<$ 1) and base metals
- Temperatures and depths of formation ($\sim$180°–240°C; $<\sim$2–3 km)
- Lack of clear relationship with upper crustal intrusions, as exemplified by lack of mineralogical or elemental zoning at scales of $<$5 to 10 km laterally and $<$2 km vertically.

This is an important and comprehensive summary of what the leading researchers observe in the deposits of the Carlin Gold Province. Rather than being a criticism of their summaries, the practical difficulties using these non-diagnostic criteria globally questions whether there is any theoretical basis for considering the deposits of Northern Nevada as a significantly different group of gold deposits at all. Long term difficulties establishing the genesis of Carlin-type deposits re-enforces this opinion.

18.1.9   Consequences of Deep Weathering in the Carlin Gold Province

It is common, globally, for the regolith profile to include an interval beneath BOCO of rock that has interacted with groundwaters to form clays but without the oxidising component to convert ferrous minerals to Fe oxides. Sulfide minerals may be partly stable below BOCO in a transition from primary ore assemblages to a bleach interval lacking sulfides completely (Fig. 18.8). In the Carlin Gold Province, the porous black interval including the refractory ores is interpreted as the lower part of the regolith beneath BOCO but above fresh rock. As for the bleached interval, formation of the porous black interval is attributed to percolation of meteoric waters since $\sim$40 Ma. What is reflected as BOCO on pit walls today is a result of redox changes across a water table, but there has been a relatively complex vertical migration of that horizon over the last 40 myr. The stability of selected sulfide minerals in the porous black refractory ore is quite compatible with a reducing weathering origin.

A consequence of the porous black refractory interval being part of the regolith is that sampling from here for geochronology, stable and radiogenic isotope studies, geochemistry, and mineralogy is a study of weathering processes. Other studies such as that of fluid inclusions in quartz may be more robust through weathering.

Describing Carlin type mineralisation as *disseminated*, which is commonly done, might be correct because there is a lack of quartz veins, but it is emphasising a feature that has probably arisen from silica dissolution during weathering rather than during a primary gold mineralising event. In exploring for another Carlin Gold Province globally, the disseminated character found in Carlin regolith may not be important, and potentially misleading.

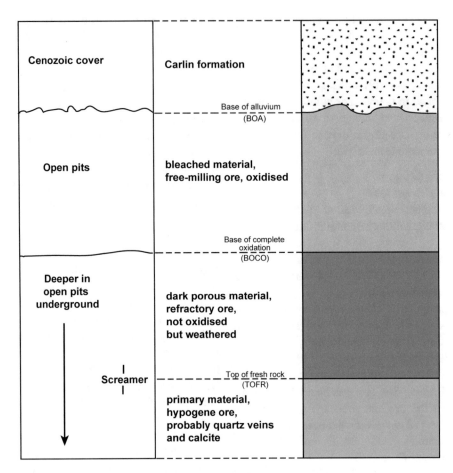

Fig. 18.8 Regolith column applicable to Northern Nevada. The left column shows the positions of open pits and deeper ores. The centre column shows major regolith mappable boundaries in drilling and pit mapping especially BOA and BOCO. Most parts of mining operations have not reached TOFR (after Phillips et al. 1999).

The carbonate content of ores in Carlin-type deposits is critical not just for processing ores, but also for determining the genesis of Carlin goldfields. Some ores at depth such as those of Screamer have large amounts of primary carbonate as part of the primary mineralisation. The loss of calcite and dolomite has been a later process related to weathering, and carbonate as part of primary mineralisation is compatible with H_2O-CO_2 fluid inclusions.

The protracted period of weathering in Nevada poses one difficulty interpreting the gold province, in that correlation of gold with many geological features may either originate from the time of gold introduction or may result from redistribution during Cenozoic weathering. For example, correlation of gold distribution with Tertiary faults, Tertiary igneous activity, or decalcification, may reflect the main gold introduction or may reflect modification of gold well after it was introduced. The latter is preferred here.

18.1.10 Role of the Goldstrike Pluton

The Goldstrike Pluton is one of several Jurassic age intrusions within the Carlin Gold Province. It is approximately 6 km by 2 km, elongate oblique to the Carlin Trend and dated at ~158 Ma (Arehart et al., 1993). The composition of the pluton is variable from granite to diorite with plagioclase, orthoclase, hornblende, quartz and biotite. It is generally medium to coarse grained with massive texture, unaltered throughout much of the pluton (Fig. 18.5d) but foliated around margins and cut by faults locally. The Goldstrike Pluton has contact metamorphosed rocks for a few hundred metres to its north and south to produce calc-silicate assemblages and hornfels

(Bettles, 2002). It has exploited part of a 70 km long pre-existing NNW-trending fault zone of the Carlin Trend but has not been significantly deformed by this structure (Fig. 18.9; Fig. 18.10). These features suggest that the Goldstrike Pluton is a *stitching pluton* (Chapter 7) intruded across the Carlin Trend after much of the strain (deformation) on that zone had taken place.

Within 5 km of the margins of the Goldstrike Pluton there is 1500 t Au in gold-only deposits in metasedimentary rocks; this is more than virtually

Fig. 18.9. Map of the Carlin Trend, some goldfields, structures and Goldstrike pluton. This pluton is massive granitic rock of ~158 Ma (figure adapted from Berger and Theodore, 2005).

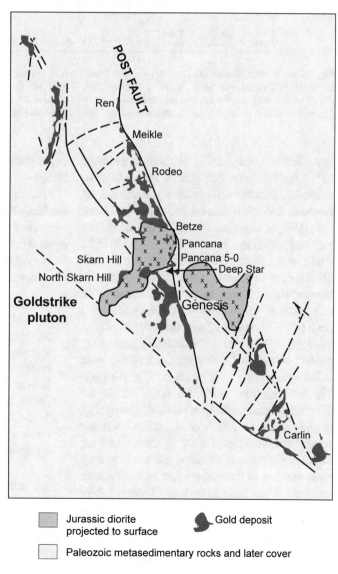

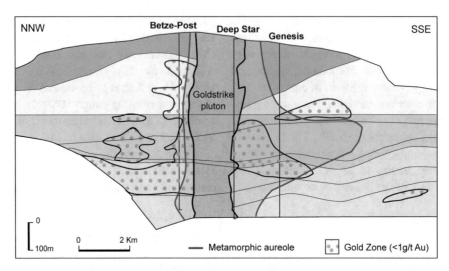

Fig. 18.10 Section through the Goldstrike pluton (pink) based on correspondence with Dr Steve Garwin who shows in this diagram the interval of contact metamorphism and distribution of gold at approximately 1 g/t Au. Although the pluton is not shown as mineralised here on this scale, it has been reported as weakly mineralised in its outer 15 m of weathered granitic rock (Teal and Jackson, 1997; Jory 2002).

any other intrusion globally. In contrast, a total of 0.09 t Au has been mined from the Goldstrike Pluton from some small open pits (Pancana, Lost Pancana and 5-0) where the ore has been described as supergene and deposited in the bleached oxidation interval just above BOCO (Heitt et al., 2003). The Betze deposit is well-developed immediately north of the pluton in metasedimentary rocks, but anomalous gold only extends 15 m into the outer (weathered) parts of the pluton itself and along faults (figure 5 of Bettles, 2002). This distribution of gold in the regolith around and above the Goldstrike Pluton is interpreted here as a function of weathering. The occurrence of 'undivided skarn' within the 6.8 g/t gold grade contour is especially intriguing (figures 4 and 5 of Heitt et al., 2003), as is the widely distributed gold mineralisation in all the adjacent clastic and chemical metasedimentary rocks but not in the pluton. Just as for Maldon and Stawell goldfields in Victoria (Section 7.2.6), the literature descriptions of mineralised skarn at Goldstrike appear to suggest gold mineralisation occurred before the hornfels event and is in equilibrium with and part of the hornfels mineral assemblages; this is compatible with the larger scale structural relationships of a stitching pluton.

One might expect the Goldstrike pluton to contain many millions of ounces of gold if it was already in place before one of the world's greatest gold mineralising events in the Carlin Gold Province. Factors that would make the pluton favourable for gold include:

- Being in the middle of the Carlin Gold Province, in the heart of the north Carlin Trend adjacent to major NNW-trending structures and ideally situated in the footwall to the Post Fault
- Having a favourable hornblende-plagioclase-quartz bulk rock composition to bring about deposition of gold
- Having a rheology that differs from the surrounding metasedimentary rocks.

The lack of primary gold in the pluton is best explained if the pluton was intruded and crystallised later than gold introduction. As a stitching pluton, some geochemical signatures of the pluton may be present in some of the gold ores and vice versa, but this does not mean that the source of the gold was the granitic rock.

18.1.11   Genetic Constraints

There is a contrast between the enviable track record of successful exploration for gold in the Carlin Gold Province versus the lack of coherence in models suggesting how it has formed. Unrivalled scientific resources have been available to unravel the Carlin Gold Province over the last half century and a vast amount of high-quality data have been generated. Despite this the gold is regularly described as difficult to explain or different from other global gold deposit types. Nearly four decades after the discovery of the first Carlin-type gold deposit, there was acknowledgement that "Carlin-type deposits do not fit neatly into any one of the models proposed for them." (Hofstra and Cline 2000 p. 164). Two decades later the uncertainty remains. In such a situation it may be a revision of basic assumptions that is needed rather than collection of further geological data. The aim would be to find a genetic model that explains *and predicts* many of the Carlin features as listed below (gold distribution, relationship to structures, significance of dating, relationship to contact metamorphism).

Virtually all the gold in the Carlin Gold Province (well over 90 %) has come from rocks of Paleozoic age. Of the minor amounts of gold in younger rocks some is in weathered rocks and appears to be from redistribution in the regolith. Some may be in younger igneous rocks that have intruded and incorporated existing mineralisation though this remains untested.

Gold mineralisation is related to structures of Antler age (350 Ma) including the Roberts Mountain Thrust with mineralised stratigraphy and structures correlated to the orientation of pre-Cenozoic, probably Paleozoic, events (Rhys et al. 2015; Griesel et al. 2020). This suggests that the main gold introduction was either Cenozoic and exploited reactivated Antler-age structures or the gold was introduced into actively deformed Antler-age structures that were later reactivated. The Carlin alteration and structures have been described as extensively reworked which is better explained if the main Carlin mineralisation was early and then overprinted by reworking including weathering.

Consistent ages around 30 – 40 Ma are generated from Carlin ores and appear reliable. Given these analyses are using weathered samples, the age is likely to reflect the period of weathering rather than primary Carlin gold introduction. Despite the large number of radiometric dates around 30 – 40 Ma, there has also been many much older dates such as 117 Ma and 93-100 Ma (Arehart et al., 1993). In other provinces, the older dates might be interpreted as closer to the mineralisation age, and the 30 – 40 Ma as resetting. The interpretation of geochemistry including stable isotopes and radiometric dating also needs to account for weathering; most sampling has used accessible bleached and black material from open pits and underground and as such is informing about the weathering event. Ore fluids trapped as fluid inclusions in quartz are potentially more robust during resetting; they are complex, but include low salinity, H_2O-CO_2-H_2S fluids like those found in many other gold-only deposits (Kuehn and Rose, 1995).

The relationship of gold mineralisation to Jurassic plutons is complicated by weathering, but underground maps of mineralisation in 'skarn' (Heitt et al., 2003) suggest that gold ore has been contact metamorphosed adjacent to the 158 Ma Goldstrike pluton.

18.1.12   Genetic Synthesis

There are many variants on how the Carlin gold deposits may have formed with different ages, different roles of weathering, different sources of ore-forming fluid, and different tectonic settings. These parameters are quite inter-related though, and lead to two main options for Carlin gold formation:

- Option A. Weathering is a near-surface feature, radiometric ages of 30 – 40 Ma reflect the main *Eocene* gold event, igneous activity including magmatic ore fluids are inferred to be critical, and the tectonic setting is one of extension. If these features are correct, then Carlin gold is syn-Eocene and very different from other global types.

- Option B. Weathering is exceptionally deep, radiometric ages of 30 – 158 Ma reflect resetting and post-date gold introduction, ore fluids are metamorphic in origin, and the tectonic setting is one of compression probably in the Paleozoic. Carlin gold is *pre-Eocene*, deeply weathered, but resembling many other global gold types.

The role of weathering in the Carlin Gold Province alone is enough to separate options A and B for Carlin gold genesis:

- If weathering is minor and confined to the upper sections of the oxidised bleached interval, then the main gold introduction event is likely to be *Eocene* around 30 – 40 Ma
- If weathering includes the deep oxidised and porous black intervals, then the main Carlin gold introduction is *pre-Eocene*, and likely to be Paleozoic.

18.1.12.1 Eocene Age for the Main Carlin Gold Introduction: Option A

There is extensive literature advocating a 30 – 40 Ma Eocene age for the introduction of Carlin gold mineralisation that has been well summarised (Hofstra and Cline, 2000; Muntean and Cline, 2018). The Eocene age is also an assumed starting point and basic assumption for much current research and literature. The Eocene age (option A) is not developed further here because, as alluded to throughout this sub-chapter, it does not appear to adequately explain the inferred deep weathering, gold distribution, importance of Paleozoic structures and common pre-Eocene radiometric ages. The case for linking the main Carlin gold mineralising event to Eocene igneous activity requires the black refractory zone to be unweathered (not demonstrated by the petrography and mineralogy of this zone). As discussed in Chapter 10, it has not been possible to identify any essential role for silicate magmas in the formation of gold-only deposits. Magmatic processes in general cannot adequately explain the gold enrichment, segregation of gold from base metals or the nature of the ore fluids.

18.1.12.2 Pre-Eocene Age for the Main Carlin Gold Introduction (Paleozoic): Option B

The origin for Carlin gold being advocated here is like many other global gold provinces except that it incorporates substantial deep weathering well after gold formation. The origin of the Carlin Gold Province is explained as numerous deposits formed in metasedimentary rocks and less common igneous rocks during Paleozoic deformation, and then overprinted by an Eocene weathering event during regional extension. The large-scale stitching geometry of the Goldstrike pluton on the Carlin Trend, its lack of significant mineralisation except in weathered rocks of this pluton, and gold in inferred contact metamorphic rocks support an age for Carlin gold introduction before the intrusion of the Goldstrike pluton (which was around 158 Ma). The preferred timing of gold introduction is the Antler Orogeny around 350 Ma when Paleozoic sequences of western Nevada were thrust over sequences to the east with movement along the Roberts Mountain Thrust. Gold mineralisation is concentrated today immediately above and below this thrust. The Paleozoic event introduced gold in a similar way to that described for many other global provinces in metasedimentary rocks (i.e. slate belt, shale – greywacke type; but noting the significant limestones in the Carlin Province). Dominant mineralisation involved auriferous quartz veins with carbonate and sulfide alteration assemblages.

The idea of the main Carlin gold introduction event being much older than Cenozoic (i.e. Paleozoic) is not new and had support from some radiometric ages and structural geology. Aspects of this model of pre-Eocene gold introduction that appeared insurmountable in the past included the role of the Goldstrike Pluton, but this is now re-interpreted with a greater understanding of weathering. This would explain the gold plume in the bleached zone above the Goldstrike Pluton and near the shallow Pancana deposit.

The appearance and features of the Carlin gold deposits today result from two factors after their inferred Paleozoic formation. One factor was multiple deformation of the deposits involving reactivation of the alteration and structures during Sonoma, Humboldt, Sevier and Laramide orogenies affecting western USA. The second factor was deep weathering starting around 40 Ma in a climate of greater rainfall than today. The conditions adjacent to the primary gold deposits of reactive sulfide and carbonate bearing rocks and multiple structures favoured particularly deep weathering which led to deep oxidation of gold ores. As descending meteoric waters increased in temperature, they became silica undersaturated and were able to dissolve quartz veins and redistribute gold especially above the water table.

Since 1990 the upper bleached zone has been generally accepted as the product of weathering and would be interpreted as forming in oxidising conditions above a paleowater table. This bleached material is continuous to nearly one kilometre depth with the implication of very deep weathering in and adjacent to ore zones. The black refractory zone continues for 100s m below the bleached zone and is inferred to be the part of the regolith formed below the paleowater table in reducing conditions. The calcite interval (Screamer) is at depth and incompletely delineated.

This explanation has three interesting aspects. One is the Antler foreland basin setting akin to the Witwatersrand foreland basin and goldfields of South Africa: the full significance of this parallel is not yet clear. The second aspect is the coincidence of the mid-Paleozoic age with the major goldfields of the Tethyan region of China through to Uzbekistan, the auriferous Variscan period of Portugal and western Europe, and Victorian goldfields of SE Australia. The third aspect is the multiple published Paleozoic radiometric ages that may support a Paleozoic gold event; in this scenario, the 30 – 40 Ma radiometric ages may be reliable data, but they reflect radiometric resetting by weathering during Cenozoic extension.

18.2   Low-S Epithermals, Au Porphyries, Breccia Pipes, Disseminated Sediment Hosted and Carbonate Replacement Deposits

Low sulfidation epithermal gold deposits occur in many areas of younger rock sequences and are associated with orogenic environments of active tectonism and volcanic activity. They occur in districts that typically contain a range of gold-only deposits referred to as gold porphyry-, breccia hosted-, disseminated- and carbonate replacement deposits. Although these varieties may appear quite distinctive in the field, they also share common characteristics such as their provinciality, enrichment, segregation from base metals, epigenetic timing, and low salinity ore fluid. Examples include the Northern Andes, North American Cordillera, Japan, Indonesia, Papua New Guinea, NE Queensland, Fiji, North Island New Zealand, Romania, Spain and the southwest Pacific (Fig. 18.11).

Host rocks for low sulfidation epithermal deposits include rhyolite, dacite and andesite, basalt, gabbro, granitic rocks, tuffs, diatremes, lavas and pyroclastic rocks; and they occur in stocks and domes. Most are associated with veins including stockworks. An example of the diversity in a single region is the NE Queensland Charters Towers area of Australia (Fig. 4.11) where the Charters Towers deposit is in granodiorite, Ravenswood in felsic to intermediate to mafic intrusive rocks and Mt Wright in rhyolite and granite breccia. Pajingo is within volcanic rocks and intermediate high-level intrusive rocks with mineralisation continuing to 500 m depth (Parks and Robertson 2005), and Mt Carlton in a structurally complex area with intermediate to felsic volcanic rocks and metasedimentary rocks. Additionally, Kidston and Mt Leyshon are two of the world's largest gold deposits in breccia pipes.

The epithermal deposits were originally defined as the lower temperature hydrothermal ore deposits formed at shallow depths of 100s m

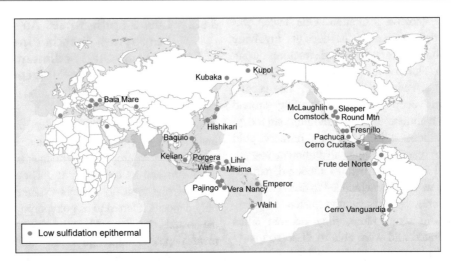

Fig. 18.11 Map of some larger gold-only deposits that have been described as low sulfidation epithermal deposits.

or less, but the term is now applied to a much wider range of deposits. They occur in districts where they have been interpreted by some as a continuum with distant disseminated deposits, carbonate replacement deposits and gold porphyry deposits—all gold-only types; and because of their juxtaposition in districts, there is common reference to gradations or transitions from the porphyry to epithermal environment.

Features used to recognise low sulfidation deposits include their low temperature of formation, their shallow depth of formation, various clay mineral assemblages and some vein textures including vugs and comb, colloform and crustiform banded textures. These textures can also occur adjacent to mesothermal textures in single samples. These mineralogical and textural identifying features of epithermal deposits are generally non-diagnostic in that they are also found in Carlin deposits, Archean deposits in greenstone belts and Paleozoic deposits in metasedimentary sequences. In each of Carlin, Archean and Paleozoic sequences, deposits have been interpreted as epithermal based upon textures only to be revised later with deeper exploration and mining when the epithermal interpretation became untenable, and a weathering interpretation preferred.

Unlike some other gold types, there is a different emphasis in the study of epithermal deposits

in that they can, and have been, classified relatively quickly based upon their surface features. In contrast, many exploration programs in Archean greenstone belts and slate belts would look past the near surface quartz vein textures and clay mineral assemblages in any classification exercise. This may explain why many early interpretations of an epithermal system in Archean sequences have been revised over time as different ore and gangue material becomes accessible for study.

18.2.1 Genesis

The study of these gold deposits in terms of local magmatic rocks and near-surface textures has assisted in exploration and ore delineation and the techniques are applied with success worldwide. However, these studies are not on the ideal scale to understand the formation of a whole province, and this may have led to the wide variety of origins and implicated magmas.

The tradition of linking epithermal gold deposits to proximal or distal magmatic bodies dates back a century but in the intervening period there is a much better understanding of the range of ore fluids available in the crust, and the way various elements are transported. Global syntheses indicate a wide range of magmatic rocks proximal to these gold deposits and no indication of a specialised magma accounting for epithermal deposits. Auriferous quartz veins indicate that the immediate host rock had crystallised before fluid introduction to emplace the veins and gold, and so it is common to advocate deeper igneous source rocks. However, once the immediate igneous host is eliminated as a potential source of fluid or gold, further options for the distant unexposed source are unconstrained and could include a variety of igneous or metamorphic rocks. Geochemistry including tracing with isotopes rarely resolves the source given the elements being used are not being transported with Au (see Chapter 12) or are not diagnostic of the gold source. For example, H and O isotopes may confirm a meteoric water activity but the isotopes do not inform whether this event is ore-related or later modification of ore. There has been little conclusive forward modelling of igneous processes to show how quartz – feldspar – mica rocks can melt and evolve to produce auriferous fluids on a large scale.

Once the focus is moved to a scale beyond the immediate host rock, the immobile elements and non-gold related isotope systems, quartz vein textures and clay assemblages provide few constraints on where a deeper fluid may come from. Such a fluid needs to be able to explain the provinciality of these deposits (Fig. 18.11), enrichment of gold, segregation of gold from base metals, epigenetic timing, and low salinity nature of the ore fluids (Fig. 18.12a and b). A metamorphic fluid can account for these features; but it is not clear that any magmatic processes is able to. Importantly, an early step will need to be the delineation of the boundaries of weathering, and extent of later meteoric fluid ingress.

The epithermal gold deposits were originally described as forming at low temperature and shallow depth. In the light of further observations over time, there has been definition broadening for these deposits; for example, deposits that formed at considerable depth are now included. The need for regular definition broadening of what constitutes an epithermal deposit likely indicates the lack of a *sound theoretical basis* for the classification. The need to introduce continuums between several gold deposit types indicates the lack of diagnostic and practical

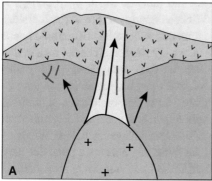
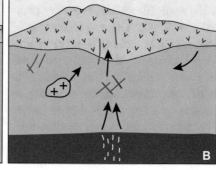

Fig. 18.12 Two possible models for the fluid and gold sources for some low sulfidation epithermal and related deposits. In both models, auriferous veins have been derived from greater depth and emplaced into mixed igneous rocks: (**a**) fluid and gold derived from unexposed magmas; (**b**) fluid and gold derived from metamorphic processes. The magmatism and metamorphism are responses to the same thermal event, and both produce aqueous fluids, but the fluids will not be easy to distinguish by timing or by geochemical analysis including isotopic methods. In both scenarios, fluid access is a response to the mechanical properties of the rocks including competency and fluid pressure; it is the chemical properties that influence gold deposition.

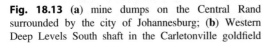

Fig. 18.13 (a) mine dumps on the Central Rand surrounded by the city of Johannesburg; (b) Western Deep Levels South shaft in the Carletonville goldfield

which reached 3777 m depth in 1977, now renamed Mponeng mine and it extends over 4 km below the ground surface.

criteria to recognise a low sulfidation epithermal deposit.

18.3   Witwatersrand Goldfields of South Africa

The Witwatersrand goldfields of South Africa have produced more gold than any other single continent and have dominated global production throughout the twentieth century. At their peak in 1970 they produced 30 Moz (or 1000 tonnes) of gold but that figure has progressively declined to around 3 Moz pa (100 tonnes) in 2020. Production has come from seven major goldfields based around Johannesburg (Fig. 18.13) and extending 300 km to the south (Brock and Pretorius 1964; Phillips and Law 2000). The great attraction of the Witwatersrand gold ores has been the combination of large tonnages, high gold grade, and exceptional lateral continuity. This continuity is a function of gold mineralisation occurring as planar orebodies of a few centimetres to metres thickness on laterally continuous unconformity surfaces of several 100 km^2.

The gold is mined from planar ore zones within reef packages that immediately overlie unconformity surfaces. The reef packages comprise conglomerate, clean quartzite, and shale, and are interpreted as marine transgressions. The major reefs and most of the gold production has come from the upper 3 km thick upper Witwatersrand (Fig. 18.14; Central Rand

Group). However, there is substantial mineralisation distributed throughout the whole Witwatersrand Supergroup and underlying basement in thick conglomerates, cross-cutting shear zones and breccias, and quartzite immediately below unconformities.

There are several important types of gold ore in the Witwatersrand goldfields (Fig. 18.15). The highest-grade ores comprise carbon on unconformity surfaces as carbon seam, flyspeck carbon or small veins, and with or without quartz pebbles and pyrite. These ores regularly have substantial uranium as uraninite (UO_2), brannerite (mixed oxide including uranium and titanium, $U_{0.5}Ca_{0.3}Ce_{0.2}Ti_{1.5}Fe_{0.5}O_6$) or coffinite (hydrated uranium silicate, $U(SiO_4)_{1-x}(OH)_{4x}$). The classic Witwatersrand gold ore of the literature is a conglomerate with well sorted, round quartz pebbles, some pyrite and a siliceous matrix that includes 0.01 to 0.05 mm sized gold grains (locally referred to as banket ore, or oligomict conglomerate). Less common are conglomerate ores in which pyrrhotite or arsenopyrite dominate the sulfide assemblage. A further ore type has abundant lithic fragments in a polymict conglomerate, and these lithic fragments are mostly BIF, shale, and granite clasts. Other important ore types comprise pyritiferous sandstone in some mines, and less commonly albitic quartzite may be the main host locally. Surrounding the ore intervals is an alteration halo of mineral assemblages with pyrophyllite, muscovite, chloritoid, tourmaline and pyrite. This halo is up to 300 km along the basin

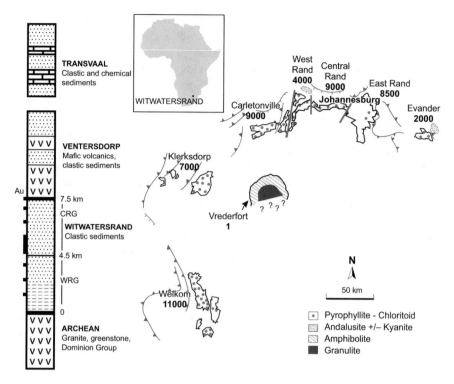

Fig. 18.14 Stratigraphic column showing the position of the 3000 – 2700 Ma, 7 km thick Witwatersrand Supergroup with most of the gold produced from the upper 3 km section. Map of the Witwatersrand Basin showing seven major goldfields with their production (bold, in tonnes Au). Also shown are major faults in the footwall of each goldfield, and the metamorphic grade which increases towards Vredefort.

Fig. 18.15   a) mineralised Ventersdorp Contact Reef conglomerate above unconformity, and overlying some mineralised quartzite, Elandsrand (Kusasalethu) gold mine; **b)** quartz pebble conglomerate hosting high grade gold, Main Reef, Durban Deep mine, Central Rand goldfield in Johannesburg.

margin linking all goldfields, 50 km wide into the basin, tens to hundreds of metres thick embracing the reef zones and focused on the upper Witwatersrand succession.

The Witwatersrand deposits match four of the main characteristics of gold-only deposits discussed in Chapters 4–8, namely provinciality, enrichment, segregation from base metals, and

H_2O-CO_2 bearing ore fluid types recorded in fluid inclusions in auriferous quartz veins. For the fifth characteristic of *timing*, there has been long standing disagreement on the timing of gold introduction. There are two diametrically opposed ideas as to how these deposits formed and it is not the plan to investigate the debate in this book. Instead, the intention is to summarise some of the Witwatersrand features so they can be compared with all other global gold deposits. The aim is that it ultimately becomes possible to use the knowledge of all gold deposits to gain insight into Witwatersrand gold so that improved approaches might lead to some exploration success. Both the placer model and hydrothermal model have been discussed since the earliest years of Witwatersrand gold mining. The numbers of their adherents have waxed and waned as new data has emerged, and as new personalities have arrived and departed.

18.3.1   Classic Placer Model of Pretorius

The description of the Witwatersrand goldfields that dominates modern textbooks, lectures, and the perceptions of non-experts is the placer model provided by Pretorius (1981); and this idea dominated from the early 1960s up to the Wits centenary in 1986 in what was described as the golden age of sedimentology. The classic Pretorius representation of the placer model shows the Witwatersrand basin, its seven goldfields and six 50 km wide overlapping alluvial fans feeding into each goldfield and interfingering with marine sediments (Brock and Pretorius 1964). The representation was published in 1964 and was used then to explain some of the gold distribution using the contemporary sedimentological model in an extensional tectonic setting. This 1964 model derives detrital gold from sources in Archean greenstone belts with the assumption that the Witwatersrand Basin was Proterozoic in age and received that gold-rich detritus. A major strength of this model was, and remains, its uniformitarian analogy with modern alluvial systems that can concentrate detrital gold eroded from older mineralisation. The well sorted, mature quartz pebble conglomerate of many Witwatersrand reefs is a vivid parallel for some modern streams with gold placers.

Despite them being the poster child of the placer model, the overlapping alluvial fans are unlikely to exist at all (Fig. 18.16) and the original authors never claimed that their margins had been walked out, mapped underground, or compiled from drilling. In fact, much of the Rand goldfield (West, Central, East goldfields) was mined out and inaccessible by 1964. The Brock and Pretorius alluvial fans were simply the authors' view of how a Witwatersrand Basin might look if it was assumed that the gold was introduced according to their contemporary sedimentological placer model. Obviously, the schematic of alluvial fans is compatible with that model because that is how and why it was created in the first place.

Today, the schematic is hailed as support for the placer model for the distribution of Witwatersrand gold which is a circular twist because the schematic was developed from the gold distribution.

Not surprisingly, there have been many changes in Witwatersrand geology since the Brock and Pretorius 1964 model:

- A foreland basin compressional setting is now inferred rather than extension.
- Rather than being Proterozoic, the Witwatersrand Supergroup is Archean (2700 – 3000 Ma) and pre-dates the major gold introduction into greenstone belts globally (2600 – 2700 Ma).
- Modern placers are linear (few 100 m wide by 20 km length) whereas Witwatersrand reefs are planar (50 km wide by more than 10 km long).
- Alluvial fans appear absent and a braided stream – delta setting is favoured (Els 1998).
- Migration of aqueous fluids through the Earth's crust is widely accepted today as is our understanding of the aqueous geochemistry of gold from experiments and thermodynamics.

Fig. 18.16 Widely used schematic of overlapping alluvial fans (labelled and shown in orange) feeding into the major Witwatersrand goldfields (Brock and Pretorius, 1964). This remains a highly influential schematic despite it only being developed to convey the authors' understanding of their placer model for Wits gold. It does not match new data since 1980; e.g. "No convincing evidence for alluvial-fan deposits (*sensu stricto*) in the Witwatersrand Supergroup has been found" (Els, 1998).

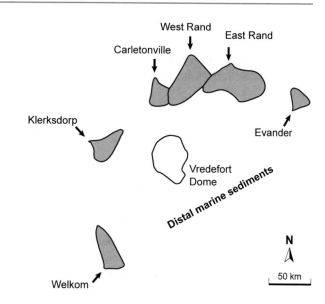

The generic placer model for Witwatersrand gold is still keenly supported today in part because it has been flexible enough to absorb all these changes through its ongoing modifications. However, in modifying and preserving the placer model, these modifications have not only increased the complexity of the model but have done so in a way that has not aided the discovery of any new goldfield. Furthermore, the modifications have moved the Wits placer model well away from any modern placer analogues – an irony for a model that was originally built upon a uniformitarian approach.

18.3.2   Hydrothermal Model

The alternative to the placer model, which is the introduction of gold into the reef packages in aqueous fluids, is analogous to gold introduction into Archean greenstone belts and its timing is inferred to be within the same 2600 – 2700 Ma time interval. The main fluid channelways are the large thrust faults in the footwall of each goldfield. Deposition of gold was around 300 – 400°C as determined from fluid inclusion studies and mineral assemblages, and the deposition occurred in response to sulfidation of Fe in the conglomerates (e.g. magnetite, ilmenite) and

reduction from interaction with carbon, e.g. Carbon Leader reef with its 1-2 cm thick seam of carbon. Evidence of the hydrothermal event resides in the 30,000 km^3 alteration envelope extending to all goldfields.

Resistance to the hydrothermal model arises from the sheer size of the Witwatersrand goldfields and the volumes of auriferous fluid required. Also, the concepts of large volumes of metamorphic fluid and its migration for kilometres through the crust is foreign to many.

In the 1950 – 1980 period several Witwatersrand gold mines were producing 1 Moz Au pa at grades of 10 – 20 g/t. This luxury meant that there were few challenges for mine geologists who needed to follow high-grade reefs and test major unconformities with drilling. Unwisely these successes and the huge profits were taken as proof that the Witwatersrand gold geology was completely understood, and alternatives were not worth considering. This confidence led to a Tertiary education system completely focused on one perceived correct answer and meant that graduates were not equipped to evaluate alternative versions of Witwatersrand gold geology. The warning signs were clear even as South African gold production peaked at 30 Moz (1000 t Au) in 1970 as there had been no new Witwatersrand goldfield discovered since 1951. The half century after 1970 was a continuation of non-discovery, and despite projections of the Witwatersrand producing 600 tpa until 2060, gold production has fallen to 100 tpa.

18.4   Summary of Different Groupings of Deposits

This chapter attempts to merge many gold-only occurrences by introducing some classifications of gold deposits that are less than ideal for discussions of their common genesis but remain very much in common usage. As an example, modern discussions about a continuum between low-sulfidation epithermal deposits and Carlin-type deposits suggest a sub-optimal classification in that it lacks the practical aspect of clear differentiation which is probably a consequence of it lacking a strong theoretical basis.

With different groups of researchers each focusing on one deposit type or province there is a tendency to make such deposits a group on their own. This can divide into those deposits the researchers are most familiar with and those they are not familiar with; hence for each research group their classification scheme appears adequate from an internal perspective but has difficulty in a global context because their schemes are not universally transferable or applicable. For example, once studies of orogenic gold deposits begin by excluding deposits of the Carlin Gold Province it is almost certain that the conclusion or lasting perception will be that orogenic deposits and Carlin deposits are different. This might be convenient for those studying orogenic deposits and for those studying Carlin deposits, but it means neither field of research advances as well as it might. Similarly, once orogenic models exclude epithermal deposits the obvious similarities of provinciality, enrichment, segregation, ore fluids and epigenetic timing are lost.

Another example of numerous types reflecting local differences are when a single stratovolcano is pictorially linked to porphyry, skarn, carbonate-replacement, and epithermal deposits. Some of these are useful descriptive terms and

valuable in exploration but this is effectively emphasising differences between deposits and not similarities that might inform of the fundamental formational processes.

For the Witwatersrand, a common approach in contemporary gold studies has been to consider the Wits as different or unique and therefore to mention, then immediately dismiss, it in discussions of global gold. Going forward, a different approach advocated here would be to draw upon the knowledge of global goldfields and develop further some of the common characteristics. Just as for other gold deposit types, there are benefits in extending the scale of study to look and think well beyond immediate host rocks (i.e. beyond the conglomerates) when considering genesis.

It is interesting to speculate why some of the discussion of genesis in this chapter does not mirror the common opinions amongst explorationists and mine geologists in districts such as Nevada, the Pacific Rim and South Africa. One reason is that the different personal starting points which, for this book and its author, has meant a global context starting with the characteristics of gold provinces and deposits then combined with a strong theoretical basis on how gold works as a chemical element. A second reason for difference is the way in which the plethora of information on every gold province and type is selected, prioritised and where necessary discarded. Each of these decisions by researchers is influenced by personal starting points, different skills bases, previous experiences, and the already established institutional positions adopted by organisations that employ them.

Snapshot

- Several decades of mining in the Carlin Gold Province has been influenced by the deep weathering to at least 800 m depth.
- Unoxidised ores in the Carlin Gold Province are not the same as unweathered ores.

(continued)

- Samples that purport to inform about primary Carlin mineralisation should be unweathered.
- Much Carlin research including sampling has used weathered material.
- Carlin deposits are neither epithermal nor genuinely disseminated though these terms were given to the deposits after initial discovery.
- Gold has been mobile during weathering and paleo-weathering, and this has disturbed the primary gold distribution pattern and some critical field relations in the Carlin Gold Province.
- Four parameters provide favourable conditions for deep weathering in the Carlin Gold Province:
  - Soluble and reactive rocks that had calcite or dolomite, and these are now evident as the widespread rocks around mineralisation that have lost either carbonate mineral.
  - Structures such as fault networks that provide access for meteoric water infiltration, and this is exemplified around mineralisation as structural complexity.
  - Elevation in the form of topographic relief of 2000–3000 m in Nevada.
  - High rainfall which was more prevalent in Nevada before the rise of the Sierra Nevada mountains at 5–10 Ma.
- Gold deposits of the Carlin Province, low sulfidation epithermal deposits and those of the Witwatersrand goldfields, despite their most obvious differences, share provinciality, extreme gold enrichment, segregation of gold from base metals, low salinity ore fluid type, and for the first two, their epigenetic timing. In all three deposit types the timing of gold introduction could be better constrained.
- Low sulfidation epithermal deposits will be better understood once deep

weathering effects and ore-related alteration are clearly differentiated.
- To elucidate processes of formation of low-S deposits, it is essential to look beyond the local host rocks.
- Witwatersrand goldfields have been the major source of all-time global gold production and dominated gold production during the twentieth century.
- The downturn of Witwatersrand gold production since 1970 correlates with non-discovery.
- The placer model for Wits gold is based on uniformitarian comparisons with modern placer processes.
- Modified placer models, in addressing newer observations, have lost this uniformitarian component.
- The hydrothermal model for Wits gold is analogous to the genesis of other gold-only provinces.

Bibliography

Carlin, Low Sulfidation Epithermal and Some Other Deposits

Arehart GB, Foland RA, Nauser CW, Kesler SE (1993) ^{40}Ar/^{39}Ar, K/Ar, and fission track geochronology of sediment-hosted disseminated gold deposits at post-Betze, Carlin Trend, northeastern Nevada. Econ Geol 88:622–646

Bakken BM, Einaudi MT (1986) Spatial and temporal relations between wall rock alteration and gold mineralization, Main pit, Carlin gold mine, Nevada, U.S.A. In: Macdonald AJ, ed. Proceedings of Gold '86, an international symposium on the geology of gold, Toronto, 1986; pp 388–403

Berger VI, Theodore TG 2005 Implications of stratabound Carlin-type gold deposits in Paleozoic rocks of North-Central Nevada. Geological Society of Nevada: Window to the World; pp 1–36

Bettles K (2002) Exploration and geology, 1962 to 2002, at the Goldstrike property, Carlin Trend, Nevada. Soc Econ Geol Spec Publ 9:275–298

Chauvet A (2019) Structural control of ore deposits: the role of pre-existing structures on the formation of mineralised vein systems. Minerals. 9:56. https://doi.org/10.3390/min9010056

Cline JS (2018) Nevada's Carlin-type gold deposits: what we've learned during the past 10 to 15 years. Rev Econ Geol 20:7–37

Cook HE (2015) The evolution and relationship of the western north American Paleozoic carbonate platform and basin depositional environments to Carlin-type gold deposits in the context of carbonate sequence stratigraphy. In: Pennell WM, Garside LJ, eds. New Concepts and Discoveries, Geological Society of Nevada Symposium Proceedings, pp 1–80

Corbett GJ, Leach TM (1998) Southwest Pacific rim gold-copper systems: structure, alteration, and mineralization. Soc Econ Geol Spec Publ 6:237

Garwin S (2003) Geologic overview of the gold deposits of the Carlin Trend, Nevada, USA. In: Dunphy J (ed). Global Gold deposits, Centre for Global Metallogeny International Short Course, pp 143–151

Griesel G, Valli F, Essman J, Rhys D, Hart E, Phillips A, Berg E (2020) Five years later: progressing the Rhys model of fold and fault geometry on the Carlin Trend by the integration of mapping, sequence stratigraphy and lithogeochemistry. In: Vision for Discovery: Geology and Ore Deposits of the Basin and Range, 2020 conference proceedings, Geological Society of Nevada; pp 137–155

Heitt DG, Dunbar WW, Thompson TB, Jackson RG (2003) Geology and geochemistry of the deep star gold deposit, Carlin Trend, Nevada. Econ Geol 98:1107–1135

Hofstra AH, Cline JS (2000) Characteristics and models for Carlin-type gold deposits. SEG Rev 13:163–220

Jory J (2002) Stratigraphy and host-rock controls of gold deposits of the Northern Carlin Trend. In: Thompson TB, Teal L, Meeuwing RO, eds. Deep deposits along the Carlin Trend: Nevada Bureau of Mines and Geology Bulletin 111:20–34

Kesler SE, Fortuna J, Ye Z, Alt JC, Core DP, Zohar P, Borhauer J, Chryssoulis SL (2003) Evaluation of the role of sulfidation in deposition of gold, screamer section of the Betze-post Carlin-type deposit, Nevada. Econ Geol 98:1137–1157

Kuehn CA, Rose AW (1995) Carlin gold deposits, Nevada: origin in a deep zone of mixing between normally pressured and overpressured fluids. Econ Geol 90:17–36

Kyser TK, Kerrich R (1991) Retrograde exchange of hydrogen isotopes between hydrous minerals and water at low temperatures, in stable isotope geochemistry: a tribute to Samuel Epstein. Geochem Soc Spec Publ 3:409–422

Muntean JL (2018) The Carlin gold system: application to exploration in Nevada and beyond. Rev Econ Geol 20:39–88

Muntean JL, Cline JS (2018) Nevada's Carlin-type gold deposits: what we've learned during the past 10 to 15 years. Rev Econ Geol 20:1–5

Parks J, Robertson IDM (2005) Pajingo epithermal gold deposits, NE Queensland. In: Butt CRM, Robertson IDM, Scott KM, Cornelius M (eds) Regolith expressions of Australian ore systems. CRC LEME, Perth, pp 298–301

Peters SG (2004) Syn-deformational features of Carlin-type au deposits. J Struct Geol 26:1007–1023

Peters SG, Ferdock GC, Woitsekhowskaya MB, Leonardson R, Rahn R (1998) Oreshoot zoning in the Carlin-type Betze orebody, Goldstrike mine, Eureka County, Nevada US Geological Survey Open-File Report, pp 98–620

Phillips GN (2005) Deep weathering around gold deposits of the Carlin Gold Province. Geological Society of Nevada: Window to the World, pp 93–111

Phillips GN, Thomson D, Kuehn CA (1999) Deep weathering of deposits in the Yilgarn and Carlin Gold Provinces. Regolith '98: New approaches to an old continent, CRC-LEME, Perth, pp 1–22

Rhys D, Valli F, Burgess R, Heitt D, Griesel G, Hart K (2015) Controls of fault and fold geometry on the distribution of gold mineralization on the Carlin Trend. In: Pennell WM, Garside LJ, eds. New Concepts and Discoveries, Geological Society of Nevada Symposium Proceedings. pp 333–389

Teal L, Jackson M (1997) Geological overview of the Carlin Trend gold deposits. Soc Econ Geol Newslett 31:1 and 13–25

Ye Z, Kesler S, Essene E, Zohar P, Borhauer J (2003) Relation of Carlin-type gold mineralization to lithology, structure and alteration: screamer zone, Betze-post deposit, Nevada. Mineral Deposita 38:22–38

Witwatersrand

Barnicoat AC, Henderson IHC, Knipe RJ, Yardley BWD, Napier RW, Fox NPC, Kenyon AK, Muntingh DJ, Strydom D, Winkler KS, Lawrence SR, Cornford C (1997) Hydrothermal gold mineralization in the Witwatersrand Basin. Nature 386:820–824

Brock BB, Pretorius DA (1964) Rand basin sedimentation and tectonics. In: Haughton SH (ed) The geology of some ore deposits in Southern Africa, v. I. Geological Society of South Africa, Johannesburg, pp 549–599

Burke K, Kidd WSF, Kusky T (1986) Archaean foreland basin tectonics in the Witwatersrand, South Africa. Tectonics 5:439–456

Cadle A, Bailey A, Law J, Phillips GN (1987) How valid is the fluvial model for the Witwatersrand Basin? Society of Economic Paleontologists and Mineralogists, Midyear meeting, Austin, Texas, abstract, pp 4, 11

Coward MP, Spencer RM, Spencer CE (1995) Development of the Witwatersrand Basin, South Africa. In: Coward MP, Ries AC (eds) Early precambrian processes, vol 95. Geological Society Special Publication, London, pp 243–269

Dankert BT, Hein KAA (2010) Evaluating the structural character and tectonic history of the Witwatersrand Basin. Precambrian Res 177:1–22

Els BG (1998) The question of alluvial fans in the auriferous Archaean and Proterozoic successions of South Africa. S Afr J Geol 101:17–26

Frimmel HE, Groves DI, Kirk J, Ruiz J, Chesley J, Minter WEL (2005) The formation and preservation of the Witwatersrand goldfields, the World's largest gold province. Economic Geology 100th Anniversary Volume, pp 769–797

Jolley SJ, Freeman SR, Barnicoat AC, Phillips GM, Knipe RJ, Pather A, Fox NPC, Strydom D, Birch MTG, Henderson IHC, Rowland TW (2004) Structural controls on Witwatersrand gold mineralisation. J Struct Geol 26:1067–1086

Parnell J (1999) Petrographic evidence for emplacement of carbon into Witwatersrand conglomerates under high fluid pressure. J Sediment Res 69:164–170

Phillips NG, Law JDM (1994) Metamorphism of the Witwatersrand goldfields: a review. Ore Geol Rev 9: 1–31

Phillips GN, Law JDM (2000) Witwatersrand goldfields: geology, genesis and exploration. SEG Rev 13:439–500

Phillips GN, Powell R (2015) Hydrothermal alteration in the Witwatersrand goldfields. Ore Geol Rev 65: 245–273. https://doi.org/10.1016/j.oregeorev.2014.09.031

Phillips GN, Myers RE, Palmer JA (1987) Problems with the placer model for Witwatersrand gold. Geology 15: 1027–1030

Pretorius DA (1981) Gold and uranium in quartz-pebble conglomerates. Economic Geology 75th Anniversary volume, pp 117–138

Young RB (1917) The Banket: a study of the auriferous conglomerates of the Witwatersrand and the associated rocks. Gurney and Jackson, London, p 125

Gold-plus Copper-gold Deposits

19

Abstract

An outcome of analysing gold-only deposits is the importance of scale, looking beyond immediate host rocks to understand how the deposits form, and linking metamorphic and igneous petrology into economic geology models. A similar approach is taken with gold-plus deposits, namely looking well beyond the mineralisation, and seeking common features that may be essential during deposit formation. Many of the differences within the group of gold-plus deposits are local effects at the depositional or district site and are not reflecting fundamentally different processes of formation. A starting point when considering a mountain chain of abundant volcanic and plutonic rocks is to remember that *this is also a metamorphic belt*.

Keywords

Cu–Au · Porphyry · Epithermal · IOCG · Gold-plus

Gold-plus deposits differ from the gold-only deposits in two important characteristics: gold-plus deposits are associated with saline ore fluids and lack the segregation of gold from base metals. Otherwise, there are some similarities such as the gold enrichment, epigenetic timing and provinciality though this last factor may be on a different scale.

The differences within the group of gold-plus deposits give rise to terms such as iron oxide copper gold (IOCG) and Cu-Au deposits in ironstone, Cu-Au porphyry deposits, Cu-Au in siltstone, and high sulfidation epithermal deposits. Some of these differences reflect variation in host rocks and controlling structures which then lead to differences in ores and alteration assemblages. These terms may be highlighting important distinctions for exploration, mining and ore processing but they do not necessarily inform on the fundamental processes forming these deposits. The terms are used in this chapter to make links between current community usage and the genetic processes already discussed; but some of these terms may be hindering both scientific and commercial progress. There is a marked contrast between how easy it is to classify deposits into gold-only or gold-plus, and how tenuous the subdivisions are between most of these gold-plus variants. Several Cu-Au deposits are described in the literature using more than one of these terms such as porphyry and epithermal.

Recognising the importance of scale and keeping a focus on the similarity of characteristics, the gold-plus deposits are mostly described here as a single broad group noting differences where they arise. Volcanogenic massive sulfide deposits, in which gold is only a by-product, are not being considered.

© The Author(s), under exclusive license to Springer Nature Singapore Pte Ltd. 2022
N. Phillips, *Formation of Gold Deposits*, Modern Approaches in Solid Earth Sciences 21,
https://doi.org/10.1007/978-981-16-3081-1_19

19.1 Global Distribution of Cu-Au Deposits

The gold-plus deposits are concentrated in parts of the Pacific Rim, the Proterozoic of inland Australia, and southern and eastern Europe. Beyond these concentrations, two important additions are the large high-sulfidation epithermal deposit of Pueblo Viejo in the Dominican Republic and large porphyry Cu-Au deposits in Uzbekistan including the Kalmakyr deposit in the Almalyk mineral field (Fig. 4.17). Copper-Au deposits appear to be rare in the Archean with the possible exceptions being Boddington in the southwest of the Yilgarn Craton of Western Australia and the Carajas region of Brazil.

Porphyry Cu-Au deposits have variable Cu, Au and Mo contents and based on their gold contents, some of the largest include Grasberg in Irian Jaya (90 Moz Au), Bingham Canyon in Utah USA (60 Moz Au), and several around 50 Moz Au (Pebble in Alaska USA, Cadia in eastern Australia, Oyo Tolgoi in Mongolia, and Kalmakyr in Uzbekistan; Cooke et al. 2005). Each of these examples also contains substantial Cu, but the very largest porphyry Cu deposits north and east of Santiago in Chile, such as El Teniente, Chuquicamata and Escondida have relatively low gold concentrations though their ore throughputs are high and include by-product Au (Fig. 19.1). In general, the porphyry copper deposits in the Andes and southwest USA have more by-product Mo, and those in the SW Pacific have more Au (Corbett and Leach 1998; Cooke et al. 2005).

High sulfidation epithermal gold deposits are spatially concentrated around the Pacific Rim from Santiago through Northern Chile and into Peru including the Yanacocha deposit in Peru, Pueblo Viejo in Dominican Republic, Western Nevada USA, Philippines, the southeast of Australia and southern and eastern parts of Europe (Fig. 19.2).

The iron oxide – copper – gold (IOCG) deposits are interpreted liberally in the literature. The focus here is on those with significant Au which are less than ten districts globally with the main ones being in the Archean Carajas region of Brazil, the Proterozoic of inland Australia (Fig. 19.3), and the Cenozoic Andes in northern Chile (Williams et al., 2005). Once the requirement of having significant gold is removed, the IOCG group is more widespread globally.

The Carajas region of Brazil has an endowment around 20 Moz Au with deposits including Salobo and Cristalino spread over 100 km of the Amazon Basin. The Proterozoic IOCG examples in Australia comprise multiple deposits at each of Cloncurry – Mt Isa (Ernest Henry and Starra) and Tennant Creek in the north, and the Stuart Shelf including Olympic Dam in the south. In each of these mineral fields the deposits are spread over 100 km or more. Telfer in Western Australia is also Cu-Au producing from siltstone and has less Fe than the other examples (Vearncombe and Hill 1993). In the Central Andes, there are IOCG deposits from Lima in Peru to Santiago in Chile with the main ones being Candelaria (3 Moz, 100 t Au) and Manto Verde near Copiapo, 600 km north of Santiago. Although close to the main porphyry copper belt of South America which is also from Lima to Santiago, the IOCG deposits are generally to the west of this copper belt.

Boddington is in an Archean greenstone belt in southwest Australia producing over 0.5 Moz pa with minor by-product copper (previously ~40,000 t Cu pa production). Grades are low with 0.5 g/t Au and only 0.1% Cu compared to most porphyry Cu deposits having 0.5% Cu or better. By the definition used of whether it produces copper, Boddington qualifies as a gold-plus deposits but is marginal in several ways as it also has characteristics of an Archean gold-only deposit.

19.2 Features of Gold-plus Deposits

The gold-plus deposits have generally lower concentrations of gold than the gold-only deposits, but they have a commercial advantage of a wider range of saleable commodities. This range of commodities provides some buffer for when the price of one commodity might be down

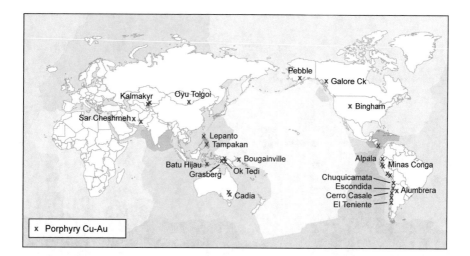

Fig. 19.1 Map of some larger porphyry Cu-Au deposits showing a concentration around sections of the Pacific Rim in the Andes, western North America, and SE Asia including Philippines, Indonesia, Papua New Guinea, and southeast Australia. The deposits labelled have over 10 Moz Au. Several major porphyry Cu deposits such as those in Arizona have minor gold and are not included.

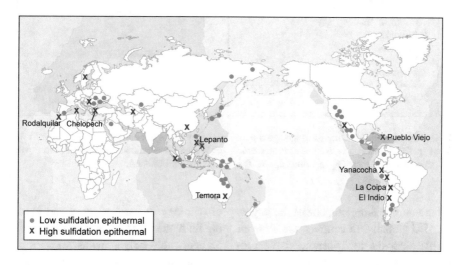

Fig. 19.2 Map of high sulfidation epithermal gold deposits especially in the Andes, SW Pacific and southern to eastern Europe (in red), with low-sulfidation deposits superimposed (in green). Although both occur in tectonically active belts, there is an imperfect antithetic spatial relationship between the high sulfidation and low sulfidation deposits. The high sulfidation deposits are spatially associated with the porphyry Cu-Au deposits and for some deposits both class names have been applied by different authors.

if there is decoupling from other commodity prices. Although many gold-plus deposits would not be economic based on their gold alone they can be highly attractive multi-commodity propositions of substantial tonnage. For example, Olympic Dam in South Australia has an endowment of 2500 – 3000 t Au in 10,000 million tonnes of ore under 0.3 g/t Au in both native form and in tellurides. However, revenue from Olympic Dam is 70% from Cu (1% grade) and

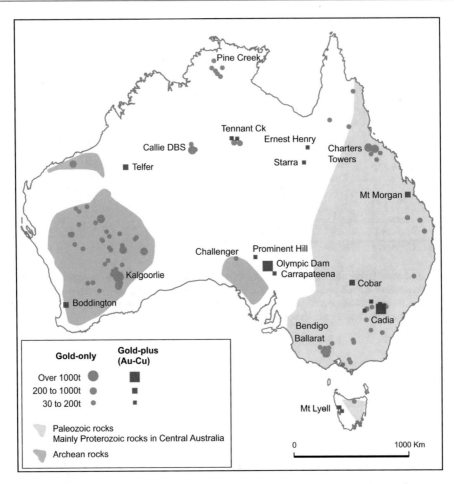

Fig. 19.3 Map of Australia showing the major gold-plus deposits, their spatial association with areas of Proterozoic rocks, and their antithetic spatial relationship to the concentrations of gold-only deposits in the Yilgarn Craton of Western Australia and Victorian Gold Province around Bendigo and Ballarat.

25% from uranium making this one of the largest producers of U globally. In contrast, the Tennant Creek Cu-Au deposits in ironstones have produced around 5 Moz Au in total, with significant Cu, Bi, Se and Ag but can be very high grade, e.g. Juno mine produced 0.8 Moz (25 t) at 56 g/t Au. The high-grade Mt Morgan in eastern Australia (Fig. 19.3) produced 8 Moz (260 t Au) whilst also being a major Cu producer.

As a group the Cu-Au deposits are epigenetic having formed as veins and breccias after host rocks have consolidated and lithified. The contrasting rheology of the host rocks and adjacent rocks has been important in localising veins and mineralisation. The structural geometry, role of rheology of various rocks and importance of ore fluid pressure have similarities to the controls on many gold-only deposits.

The ore fluids inferred from fluid inclusions from gold-plus deposits are saline to very saline with up to 40-60 wt% NaCl equivalent. This characteristic extends to IOCG, porphyry Cu-Au and high sulfidation epithermal deposits (Corbett and Leach 1998; Williams et al. 2017). These are oxidising fluids in equilibrium with hematite and with SO_2 (see Fig. 14.1) providing conditions that are highly favourable for gold transport as the oxidised Au^{3+} with the Cl ligand.

The relatively complex alteration geochemistry of gold-plus deposits contrasts with the small number of mobile elements in gold-only deposits. In Olympic Dam, for example, elements enriched in the deposit include transition metals, uranium, non-metals, rare earth elements and noble metals:

- Au Ag Sb Se Te W Mo As Bi
- F S C
- Sr Ba
- V Cr Fe Co Ni Cu Zn Cd
- P Nb Y U
- In Sn
- Rare earth elements.

Depletion of elements including K Rb Na Cs Al Hf Zr and Ti has been recorded at nearby Carrapateena (Williams et al. 2017) and at Olympic Dam (Ehrig et al. 2017).

Host rocks for gold-plus deposits are highly variable. In the Andes the largest Cu-Au porphyry deposits are in calc-alkalic rocks, but outside South America K-rich calc alkaline rocks are important hosts (Grasberg, Bingham, Kalmakyr) and alkalic rocks at Cadia. Overall, porphyry Cu-Au deposits are mainly in intermediate to felsic intrusions but also in mafic rocks; they comprise veins in porphyritic igneous rocks that have been emplaced at shallow level as relatively small plutons of 2 km or less diameter. Examples of porphyry Cu systems have been described that are up to 1.5 km deep and adjacent to related deposit types including skarn, carbonate replacement, sediment hosted deposits, cross-cutting and bedding-parallel veins, and high sulfidation deposits (Sillitoe 2010).

High sulfidation epithermal deposits are common in sub-aerial calc-alkaline sequences but are also found in a variety of igneous rocks including island-arc tholeiites in the case of the 40 Moz (1200 t Au) Pueblo Viejo deposit (Kesler et al. 2005). Iron oxide – Cu – Au deposits are in a variety of mostly sandstone, arkose, siltstone, shale, and chemical metasedimentary rocks including BIF.

The co-existence of gold-only and gold-plus deposits in some districts such as Tennant Creek and Central New South Wales (Fig. 19.3) is partly a function of the small scale of these maps and in detail the types are usually in different sequences.

In the Tennant Creek mineral field, there are Cu and Bi with Au in the deposits to the north, but gold-only in the deposits to the south. Their broad juxtaposition might be understood by learning from the variations in the mineralogical domains within the Victorian Gold Province (Fig. 4.10). In Victoria there are two domains in which there are elevated Ag and base metals in the ores of gold deposits, and these domains are sharply delineated from other domains by major faults, and presumably by contrasting fluid sources.

19.3   Genesis of Gold-plus Deposits

A feature common to all Cu-Au gold-plus deposits is the high salinity fluid recorded in fluid inclusions. This saline fluid distinguishes the gold-plus deposits from all gold-only deposits and helps to explain the contrasting element mobility patterns between the types.

The size of gold-plus deposits from less than 3 Moz to 90 Moz (2800 t Au) and the low background level of gold in crustal rocks indicate that the process of formation of these deposits is on a scale much larger than the immediate host rocks for the orebody. This is demonstrated by similar calculations as those used for gold-only deposits (Chapter 5); it requires much more than 200 km^3 of source rock at 2 ppb Au to provide 2800 tonnes Au. One implication of these calculations is that the source of the gold is well removed from its eventual mineralised host rock. There are published schematics designed to explain the origin of larger gold-plus deposits that implicate fluid activity through 15 km vertically of crust; however, over such distances, it is difficult to demonstrate that the source of these fluids was a magma.

Determining the origin for Cu-Au gold-plus deposits requires understanding the origin(s) of the highly saline fluids. The significance of the saline fluids is their potential to dissolve and transport gold given oxidising conditions that stabilise Au^{3+} as an Au-Cl complex, likely to be $AuCl_4^-$. Moderate, as opposed to highly acid, fluids are suggested by the lack of Pb and Zn in these deposit types. Three options are discussed for the enrichment of gold from near crustal values of 1-2 ppb to near-economic levels of 1 ppm or more (Fig. 19.4). These options address alternative sources of Au (silicate magma or the surrounding metamorphic rock sequence) and the alternate phases that might be fundamental to the partitioning of gold (silicate magma or saline aqueous fluid). Options considered are:

- Option A: Gold is partitioned strongly into a silicate magma
- Option B: Gold is partitioned into an aqueous fluid with both gold and fluid being derived from a silicate magma
- Option C: Gold is partitioned into an aqueous fluid with both gold and fluid being derived through metamorphism of halite-bearing metasedimentary rocks.

It is always possible that other options may exist. The derivation of saline fluids through partitioning into a non-evaporite metamorphic fluid is not considered as an option for gold-plus deposits because such a fluid would be of low salinity and would replicate neither the alteration assemblages nor the gold-plus base metal character.

Timing of release and infiltration of these saline fluids may be dictated by changes in tectonic activity regardless of whether magmatic or metamorphic sources are invoked. One example

is that of southwest Alaska where the change from convergent to dextral strike slip plate motion triggered numerous gold-only deposits by the release of metamorphic fluids at 55 – 56 Ma. Deposits were formed during this brief event over a strike length of 200 km (see Chapter 4).

19.3.1   Gold Is Partitioned Strongly into a Silicate Magma: Option A

Igneous rocks are widely linked to the formation of porphyry and other Cu-Au deposits (Sillitoe 2010). Initially the discussion here is restricted to *magmatic* processes being those in which the silicate magma played an essential role in deposit formation and in particular the partitioning of gold. Definition broadening of the term *magmatic* is avoided by disregarding shared thermal events or spatial associations where the magma is not involved in the partitioning of the gold and hence the magma is unlikely to be fundamental to deposit formation. One reason for this tight definition is the implications for genesis and exploration. If a specific magma is implicated in deposit formation, then there are quite different consequences than if it is only heat from a magma that is the critical factor.

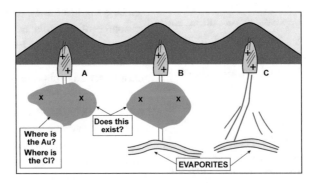

Fig. 19.4.   Three possible models to link the source of saline auriferous fluids with gold-plus deposits in igneous rocks such as porphyry Cu-Au deposits: (**a**) a saline auriferous fluid is evolved from a large magmatic mass and migrates to the small host pluton; the Au enrichment, source of Cl and mass balance remain problematic; (**b**) the large magmatic mass forms from melting sedimentary rocks that include evaporites thus providing a source of Cl

but the model still appears problematic with respect to Au and the type of magma produced; (**c**) metamorphic fluids derived from a sequence with evaporites that generates a saline auriferous fluid without requiring involvement of a magma in the ore formation process. In each example the role of the host pluton is that of a lithology with favourable rheology to fracture under high fluid pressures and favourable chemical composition for gold deposition.

Possible partitioning of gold into silicate magmas has been discussed in Chapter 10. Any proposed magmatic processes face the same challenges as for gold-only deposits which is how to achieve gold enrichment. The enrichment required to form gold-plus deposits is not quite as demanding as for gold-only deposits but is still 100 to 1,000 times background. This enrichment appears unlikely given that the gold concentrations of unaltered igneous rocks are routinely within one order of magnitude of crustal background and well below the enrichment needed to form any gold-plus deposits.

There are no large-scale examples from igneous petrology to show that the concentration of gold has increased by two to three orders of magnitude in a silicate magma. Just because copper occurs in the gold-plus deposits does not lessen this limitation. Attempts have been made to link magmatic rocks with porphyry Cu-Au deposits include the matching of stable isotope (O and H) systems from the ore environment, and these signatures are commonly compatible with a magmatic origin; however, more thorough testing is required to eliminate alternative isotopic matches that might be equally compelling such as the inevitable metamorphic fluids that must be in the environment.

19.3.2   Gold Is Partitioned into a Saline Aqueous Fluid Derived from a Specialised Silicate Magma: Option B

An alternative to the magmatic model for gold enrichment is to partition gold strongly into a hot saline aqueous fluid that has been derived from crystallisation of a silicate magma. This process does not involve the specific igneous rock that hosts the mineralisation because this rock has already crystallised before being fractured to allow auriferous vein ingress.

Highly saline and oxidising aqueous fluids at $300 - 500°C$ have the capacity to partition gold into solution as a chloride complex; concentrations of 10s of ppm Au are achievable and are equivalent to 10,000 times crustal background levels (Fig. 14.1). Furthermore, the same fluids will dissolve a wide range of elements that include Cu and thus fulfil requirements of being able to enrich both Au and Cu.

The origin of such a saline fluid is less clear given the scale of these deposits as the source will be well removed from the deposit and so relatively unconstrained. Options for a saline crustal fluid include specialised magmatic fluids and specialised metamorphic fluids; seawater and diagenetic waters are not considered further given the fluid temperatures of 500°C and above. The option of specialised metamorphic fluids is developed in Option C.

Generating Cu-Au deposits from saline fluids released by silicate magmas requires a major source of Cl in the magma (maybe 0.5 wt %), and hence available at the time of partial melting to enter a new magma when it forms. The sources of Cl are quite limited and, in the absence of evaporites, Cl may be bound chemically in biotite or hornblende, and upon melting these minerals can release the Cl into the magma (Finch and Tomkins 2017). Working backward in time, for biotite or hornblende to contain substantial Cl prior to partial melting requires a source of Cl at lower metamorphic grade for which options are scarce; micas and chlorite may incorporate some Cl in their structure, but this is not a normal feature of metamorphic terrains.

Invoking significant evaporites in the process of forming partial melts may provide Cl in the magma that later generates saline ore-forming fluids; however, this should also impose a distinctive character on the geochemistry and mineralogy of the magmas. Depending upon the minerals in the evaporite undergoing melting, sodic magmas with unusual major element compositions would be likely, easily recognised and an excellent guide for exploration; this does not appear to be the case. However once evaporites are invoked as the source for Cl, the direct metamorphism of the evaporites appears to be a more logical source for the saline fluids rather than necessarily involving a silicate melt phase. Forward modelling using thermodynamics would help to decide whether melts incorporating evaporites are a viable option and, if so, the extraction and concentration of gold from tens to hundreds of km^3 would need to be addressed.

The significant research effort devoted to characterising the igneous rocks hosting porphyry Cu-Au deposits and high- and low sulfidation epithermal deposits has suggested that calc-alkaline rocks are important hosts including some with high K, but so are some alkaline igneous rocks. However, these studies of the deposits may not be on the scale to address the source of any mineralising fluids or Au which are removed by kilometres from the hosting pluton(s). Instead, the types of igneous host rocks that contain these gold-plus deposits might be informing about rheological properties such as the propensity for various intrusions to hydraulically fracture and form auriferous veins under high fluid pressures. The learnings from Archean gold-only deposits show that small compositional variations between igneous rocks can lead to contrasting rheological properties and strong focusing of auriferous fluid into one specific rock type. The archetypal example is the Mt Charlotte deposit at Kalgoorlie where the majority of the 6 Moz Au is in the Unit 8 granophyre of the Golden Mile Dolerite because this granophyre hydraulically fractures under high fluid pressure due to its low tensile strength.

There are considerable challenges remaining to demonstrate that saline fluids derived from magmatic crystallisation are fundamental in the formation of giant Cu-Au porphyry deposits. There is a well-demonstrated spatial association between igneous rocks on a district and regional scale with these deposits, and element and isotopic signatures in and around ore and alteration are compatible with igneous systems. However, uncertainties arise in trying to link the partial melting, sources of Cl, mass balance of metals and ligands, and fluid flow regimes with an adequate source of gold. Establishing that a magmatic fluid is fundamental (essential) to forming these deposits also requires identifying and allowing for inevitable meteoric waters (this has been done) and metamorphic fluids (this has hardly been considered or attempted). One modern view in igneous petrology does not support the concept of supergiant bodies of magma (Clemens 2012) that has been depicted in some schematics of porphyry Cu-Au deposits in economic geology. Metamorphic petrologists have not embraced any mass balance calculations that support the large-scale generation of saline fluids through partial melting and silicate magma crystallisation processes.

19.3.3   Gold Is Partitioned into an Aqueous Fluid that Is Derived from Metamorphism: Option C

Copper-Au deposits occur in extensive metasedimentary sequences like the Proterozoic of Australia, and major volcano-sedimentary mountain belts such as the Pacific Rim. One possibility is that these two large groupings form from chemically similar saline fluids of quite different origins, being metasedimentary-derived and, as for option B, magmatic-derived, respectively. The alternative is that they all form from a similar process and the fluids migrate into quite different host rock successions.

The size of the larger Cu-Au deposits necessitates a scale of formation well beyond the host rocks and alteration zones. This means that in major mountain belts of volcanic rocks and intrusions it is important to consider not just the igneous province but the whole subsurface as a metamorphic environment. In this environment of elevated temperature, circulating fluids of various origins will influence deformation, element geochemistry, partial melting and mineral re-equilibration. This creates an option for the origin for high temperature saline fluids through metamorphism of evaporite sequences as has been described for Proterozoic sequences of Australia (Oliver et al., 1992; Morrisey and Tomkins 2020). Much of the research has been in the Cloncurry – Mt Isa region where a wide variety of ore deposits are in amphibolite facies terrains including several containing Cu and Au (Williams, 2005; Fig. 19.5). Two distinctive features of the region are an abundance of the mineral scapolite mapped over 100 km length and 5 – 10 km width, and the regional scale

Fig. 19.5 Gossan in the Pegmont area in the southeast of the Cloncurry – Mt Isa region of North Australia. The region includes Proterozoic gold-plus deposits of which some outcrop strongly like Pegmont, and others have been discovered under Cenozoic cover. The mineral scapolite is common in meta-evaporites in the region and implicated in the formation of saline fluids and multi-commodity ore deposits.

alteration involving addition of rare-earth elements, U, Th, Sr and Cr (Oliver et al., 1992; de Jong et al. 1998).

Scapolite is a silicate mineral found in contact metamorphic aureoles and higher metamorphic grade districts of calc-silicate rocks. Its composition is simplified to a calcic end member meionite $Ca_4[Al_6Si_6O_{24}]CO_3$ and a sodic end member marialite $Na_4[Al_3Si_9O_{24}]Cl$. The end members are conveniently thought of as the plagioclase composition with some added $CaCO_3$ and NaCl, respectively. The sodic marialite component is of interest here as one of the few minerals with essential Cl in metamorphic terrains, and in the example of the Cloncurry – Mt Isa region it is extensive. Although scapolite is widespread in this amphibolite facies terrain and present in other terranes, it appears rare globally once granulite facies conditions are reached.

The origin of highly saline fluids through the metamorphism of evaporite-bearing sequences is supported for the Cloncurry – Mt Isa region by the regional coincidence between the fluid and a spectrum of ore deposits formed from saline fluids. Additional support comes from the demonstrated scale of sequences with anomalous Cl, and the viability of a process to generate saline fluids. With 1000s km^3 of succession undergoing regional metamorphism the source of gold in saline fluids becomes plausible.

Whether evaporites are a viable option for some or all the Cenozoic Cu-Au deposits, IOCGs and high-sulfidation epithermal deposits of the Pacific Rim remains to be tested. One approach is to develop the metamorphic approach already applied in Proterozoic Cu-Au provinces by Oliver et al. (1992) and Morrisey and Tomkins (2020) combined with knowledge of modern and ancient evaporites (Warren 2010).

There are several variants amongst modern evaporites with different proportions of halite (NaCl), gypsum ($CaSO_4.2H_2O$), anhydrite ($CaSO_4$) and calcite ($CaCO_3$). By considering different starting mixes of these and other evaporite minerals it may be possible to explain some of the variation within the gold-plus deposits including within IOCGs. The subset of Cu-Au IOCG include some with significant Co (Cloncurry – Mt Isa), and Cordilleran porphyry Cu-Au deposits have common Mo and lesser Au; an extension may be to the Cu deposits of Central Africa with abundant Co but negligible Au.

Snapshot

- Gold-plus deposits are significant producers of Cu and Au especially from the Pacific Rim and Proterozoic of Australia.
- Cu-Au deposits include porphyry-, iron-oxide-copper-gold and high sulfidation types.
- There are many subjective subdivisions of Cu-Au deposits based on host rocks and textures.
- In many Cu-Au deposits, including porphyry Cu-Au, mineralisation is as veins and not immediately magmatic in origin.
- Cu-Au deposits are consistently associated with hot saline fluids.
- Saline fluids are a viable medium to concentrate and transport Cu and Au.
- A magmatic source for Au appears unlikely from mass balance calculations.
- Magmas are not the only source for Cl, and evaporites should be considered.
- Metamorphism of evaporites is implicated in the formation of Proterozoic Cu-Au deposits.

Bibliography

Clemens JD (2012) Granitic magmatism, from source to emplacement: a personal view. Appl Earth Sci 121: 107–136

Cooke DR, Hollings P, Walshe JL (2005) Giant porphyry deposits: characteristics, distribution, and tectonic controls. Econ Geol 100:801–818

Corbett GJ, Leach TM (1998) Southwest Pacific Rim gold-copper systems: structure, alteration, and mineralization. Soc Econ Geol Spec Publ 6:237

De Jong G, Rotherham J, Phillips GN, Williams PJ (1998) Mobility of rare-earth elements and copper during shear-zone related retrograde metamorphism. Geol Minjbouw 76:311–319

Ehrig K, Kamenetsky VS, McPhie J, Cook NJ, Ciobanu CL (2017) Olympic dam iron oxide Cu-U-Au-Ag deposit. In: Phillips GN (ed) Australian ore deposits,

vol 32. The Australasian Institute of Mining and Metallurgy, Monograph, Melbourne, pp 601–610

Finch EG, Tomkins AG (2017) Fluorine and chlorine behaviour during progressive dehydration melting: consequences for granite geochemistry and metallogeny. J Metamorph Geol 35:739–757

Kesler SE, Campbell IH, Smith CN, Hall CM, Allen CM (2005) Age of the Pueblo Viejo gold-silver deposit and its significance to models for high-sulfidation epithermal mineralization. Econ Geol 100:253–272

Morrisey LJ, Tomkins AG (2020) Evaporite-bearing orogenic belts produce ligand-rich and diverse metamorphic fluids. Geochim Cosmochim Acta 275: 163–187

Oliver NHS, Wall VJ, Cartwright I (1992) Internal controls on fluid compositions in amphibolitic-facies scapolitic calc-silicates, Mary Kathleen, Australia. Contrib Mineral Petrol 111:94–112

Phillips GN, Powell R (1993) Link between gold provinces. Econ Geol 88:1084–1098

Pohl WL (2020) Economic geology, principles and practice: metals, minerals, coal and hydrocarbons – an introduction to formation and sustainable exploitation of mineral deposits, 2nd edn. Schweizerbart Science Publishers, Stuttgart, p 755. [Chapter 4 is on salt deposits and evaporites]

Porter TM (2017) Cadia gold-copper deposits. In: Phillips GN (ed) Australian ore deposits, vol 32. The Australasian Institute of Mining and Metallurgy, Monograph, Melbourne, pp 755–758

Sawyer M, Whittaker B, de Little J (2017) Carrapateena iron oxide Cu-Au-Ag-U deposit. In: Phillips GN (ed) Australian ore deposits, vol 32. The Australasian Institute of Mining and Metallurgy, Monograph, Melbourne, pp 615–620

Sillitoe RH (2010) Porphyry copper systems. Econ Geol 105:3–41

Vearncombe JR, Hill AP (1993) Strain and displacement in the Middle Vale reef at Telfer, Western Australia. Ore Geol Rev 8:189–202

Warren JK (2010) Evaporites through time: tectonic, climatic, and eustatic controls in marine and nonmarine deposits. Earth-Sci Rev 98:217–268

Williams PJ (1998) Metalliferous economic geology of the Mt Isa eastern succession, Queensland. Aust J Earth Sci 45:329–341

Williams PJ, Barton MD, Johnson DA, Fontboté L, De Haller A, Mark G, Oliver NHS, Marschik R (2005) Iron oxide copper-gold deposits: geology, space-time distribution, and possible modes of origin. Economic Geology 100th anniversary volume, pp 371–405

Williams PJ, Freeman H, Anderson I, Holcombe R (2017) Prominent Hill copper-gold deposit. In: Phillips GN (ed) Australian ore deposits, vol 32. The Australasian Institute of Mining and Metallurgy, Monograph, Melbourne, pp 611–614

Part V

Discovery Case Histories and the Role of Science

Discoveries and the Role of Science in the Yilgarn Goldfields of Western Australia

20

Abstract

The gold boom of the 1980s was a period of substantial production increase in Australia and Nevada USA. In other parts of the world, gold production did not parallel these increases, or, in the case of South Africa, it declined significantly. The increased gold price, mechanised mining and processing technologies were helpful but cannot explain the contrasting performance of gold industries in different regions. In the Archean Yilgarn Craton of Western Australia, the 10-fold increase in annual production was built on exploration success discovering 200 Moz (6000 tonnes) between 1980 and 2005, and that discovery rate has continued to 2020. Much of the outperformance of discovery in the Yilgarn Craton can be explained in terms of scientific breakthroughs to assist gold exploration. Three breakthroughs in understanding the regolith and three more breakthroughs in primary gold geology were major contributors to greenfield and brownfield exploration successes, respectively.

Donald Thomson was the leader of exploration in the Duketon belt at the time of the activity at Famous Blue being described here; he and Julian Vearncombe are thanked for their recall of the day in 1996 and confirmation of this summary. In the larger scheme of Yilgarn and global exploration, the Famous Blue day is only a small event, but it nicely reveals some of the approaches that contribute to exploration success.

Keywords

Discovery · Exploration success · Yilgarn Craton · Duketon · Jundee

Knowing how goldfields form helps to satisfy curiosity but can also be of great commercial value. The Yilgarn goldfields of Western Australia have been an example of stand-out exploration success since 1980 attributable in large part to scientific breakthroughs and their application. The period of exceptional success in the Yilgarn Craton is not unique but ranks with that of the Carlin Gold Province of Nevada since 1980 (Bettles 2002) and several comparable but much earlier periods of discovery within the Witwatersrand goldfields of South Africa. Based on exploration success, the Yilgarn Craton is one of the leading gold-producing regions of the world and in terms of annual production has surpassed the Witwatersrand goldfields since 2013.

The Yilgarn Craton is a 1000 km by 1000 km area of southwest Australia comprised of deformed Archean greenstone belts and granitic rocks. These are dominantly of low to medium metamorphic grade with subordinate areas of high metamorphic grade. The craton has been a major source of nickel, gold and many other mineral commodities (Vearncombe and Elias 2017). The greenstone belts are dominantly ultramafic, mafic and felsic igneous rocks, clastic metasedimentary rocks, chert and BIF.

20.1 Yilgarn Gold Production

Gold was discovered in the Yilgarn Craton in the late 19th century (Kalgoorlie in 1893), production followed with a peak around 1903 and then a long decline to 1979. The 1980-90s gold boom in the Yilgarn Craton raised production to 5 Moz pa (150 tpa Au), and this production level has continued for thirty years (Fig. 20.1).

The 1980s gold boom affected different countries and provinces in different ways. A similar production pattern to the Yilgarn was followed in the USA based on the State of Nevada and its Carlin gold deposits, but otherwise it was not a pattern followed elsewhere. Other large countries with significant historical gold production either had lesser increases like Canada or remained steady like Brazil (Fig. 20.2). The Victorian Gold Province had been the world's leading gold producer in the 1850s and was Australia's leading all-time gold producer up to 1985, but then completely missed any benefits of the 1980s gold boom. The outstanding example of contrasting gold production is South Africa where gold production declined during the 1980s boom.

All gold producing regions have been influenced by variations in gold price, but the price fluctuations do not explain these production figures (Fig. 20.3). The gold price traded within a narrow range for the first part of the 19th century including being fixed by several governments until the 1970s. Two events in the 1970s dramatically changed that pattern of near-constant gold price. To be able to compare prices across decades, prices are expressed in constant US dollars reflecting the real price; this allows for the effects of inflation especially during the 1970 – 1980s. The 1st Oil shock in 1974 marked a large increase in the price of gold from which the real price never retreated. Then in 1979 the 2nd Oil shock built on the earlier gains and even though it retreated 50 percent from its peak remained much higher than pre-1970s.

The fluctuations in the price of gold are not matched by changes in Australian gold production. Production basically rose between 1980 and 1990 and then remained around that level. The gold price fell overall through the 1980s and then more sharply through the 1990s before rising again prior to 2010. The price increases in the 1970s led to greater interest in gold and investment in exploration globally, but many countries and regions gained no benefit.

South African production illustrates the inelastic nature of gold production (Fig. 20.2). Despite being the major gold producer throughout the 1970s, being leaders in many aspects of deep mine technology and having published Resources to support continued mining for decades, the Witwatersrand gold industry was unable to increase production and open new mines in response to the substantial rises of the gold price in 1974 and 1979. In real dollar terms (i.e. allowing for inflation) the gold price rose substantially during the 1970 – 1980 period and South African gold production fell from 1000 to 600 tonnes pa. South African gold production since the initial Witwatersrand discovery in 1886 has followed exploration success with a time lag of a decade for mine development but has not followed the gold price.

The changes in gold price do not parallel production in the Yilgarn goldfields, Witwatersrand of South Africa, Carlin Gold Province in Nevada, or the Victorian Gold Province. Furthermore, as the gold price is global it does not explain the production outperformance of the Yilgarn and Carlin Provinces, and so alternative reasons are sought.

Several factors have been beneficial for the growth of Yilgarn gold production including larger-scale mining equipment and carbon-in-pulp processing methods. However, access to new mining equipment and gold processing technologies do not explain the Yilgarn outperformance. The mining equipment used in the Yilgarn from 1980 was available globally and most had already been in use elsewhere. Processing technologies of carbon-in-pulp and carbon-in-leach were also available globally and had been in use well before 1980. There was also a substantial change before 2000 in the technologies available for exploration particularly airborne magnetic surveys, global positioning systems for location, and geographic information systems for

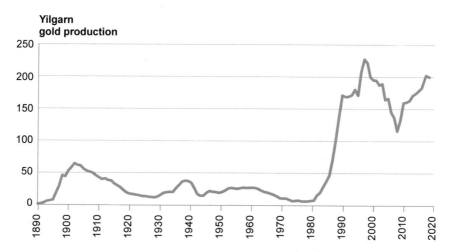

Fig. 20.1 Yilgarn production since 1890 with a peak in 1903 and decline until 1980. The production increase of the 1980s has been maintained through continued discovery. Data sources and calculation methods are in Phillips et al. 2019a.

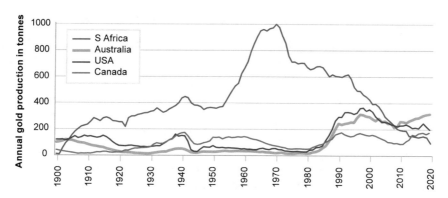

Fig. 20.2 Gold production of four major producing countries since 1900 using comparable data between countries and across years. The increases during the 1980s in USA and Australia contrast with steep decline in South Africa. The gold production data available for China are not comparable and so are not included here (see Appendix A).

data handling and representation. However, the gold price and new methods were available globally and so cannot adequately account for the exceptional discovery history of the Yilgarn goldfields since 1980 (Phillips et al. 2019a).

20.2   Yilgarn Gold Exploration Success

To explain the outperformance of Yilgarn gold production it is necessary to look beyond those global factors such as gold price and equipment. By compiling production and exploration successes (and non-successes) of various countries and their sub-regions there is a strong correlation between making gold discoveries and growing gold production (Table 20.1). Those countries that experienced large increases in their gold production did so because of significant exploration success. The converse also applied with South Africa having negligible exploration success and a long-term declining production profile. Data sources are in Phillips et al. (2019a and b).

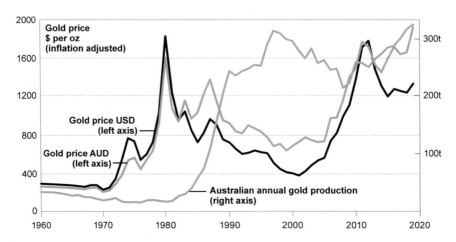

Fig. 20.3 The inflation adjusted gold price in US and Australian dollars is poorly correlated with Australian gold production. Fluctuations in gold price do not explain gold production trends in Australia, nor for South Africa, USA or Canada.

Table 20.1 Linkages between exploration success and gold production from the 1960s until the present.

	Higher gold price	Production pattern	Exploration success
Australia	Yes	Large increase	Major
USA	Yes	Large increase	Major
Canada	Yes	Increase	Modest
Brazil	Yes	Small increase	Minor
Philippines	Yes	Steady	Minor
South Africa	Yes	Long term decline	Negligible

This correlation between exploration success and production may appear trite but many explanations for the Yilgarn gold success simply attribute it to the gold price without any deeper thought: such a conclusion ignores that the Yilgarn experience of production increase was not universal and ignores that the gold price was global. By incorrectly attributing the Yilgarn production to gold price the next steps of trying to repeat the performance in Western Australia or another terrane will be mis-guided. A country attempting to raise its gold production would be better working towards exploration success than waiting for the next gold price rise; without any Resources to develop, the next price rise will come and go with no accrued benefit.

Gold production from the Yilgarn Craton has been built upon unrelenting discoveries over the last half century. As production increased from 1980, discoveries have been able to replace mined resources in most years and commonly add substantially to Economic Demonstrated Resources (EDR). As a result, even though the Yilgarn Craton produced ~6 Moz (200 t) in 2019, ongoing discoveries continue to replace that ore, and so maintain resources equivalent to a 30 – 40 years of production. This maintenance of resource life and production has spanned periods of high and low gold price with only 10 percent of years of negative discovery; these reversals are caused by revisions of EDR downwards due to price changes or Resource re-interpretations and have been easily countered in following years (Fig. 20.4).

The discoveries in the Yilgarn following 1980 were widespread with clusters around established mining districts like Kalgoorlie and Southern Cross (brownfield), but also new goldfields like the Yandal and Duketon greenstone belts in the

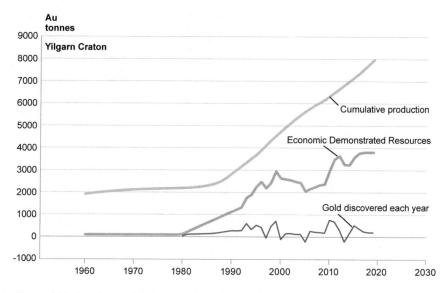

Fig. 20.4 The tenfold rise of annual Yilgarn gold production since 1980 of Fig. 20.1 has been built upon gold discoveries each year that have replaced production and increased the EDR.

NE, the Laverton goldfields south of the Duketon belt, and Kambalda – St Ives goldfield 60 km S of Kalgoorlie (greenfield; Fig. 20.5).

The discoveries contributing to the rise in EDR come from both greenfield and brownfield exploration (Vearncombe and Phillips 2019). Greenfield refers to exploration distant from known mineralisation, brownfield refers to extensions and additions in areas of known mineralisation. There is considerable overlap between the terms at the margins of near-mine exploration. In greenfield exploration there is usually little geological data to rely upon and weathered outcrop or Cenozoic cover. In brownfield exploration there are considerable geological data and most likely there is access to both fresh and weathered rocks.

Around 1980 there were approximately 2000 mines and old workings across the Yilgarn Craton which equates to one per 500 km², or an average spacing of several km between each. Greenfield exploration between those old workings relied extensively on the new regolith geoscience methods being developed through CSIRO since the 1970s. The understanding of gold dispersion in the regolith provided large geochemical targets, ferruginous lateritic material tended to concentrate gold and was an ideal sampling medium. Regolith landscape assessments meant

optimal exploration methods were tailored for different landscapes.

Discoveries using regolith geoscience soon led to production from new open pits, and once these could not be deepened any farther, exploration drilling at greater depth led to underground mines, much more primary gold geology and a source of detailed information for brownfield exploration nearby. Primary gold geology included the role of fluid – wallrock alteration, host rocks, and structural geology controls on mineralisation including quartz veins,

The production increase since 1980 and sustained levels around 150 tpa since 1990 have been achieved through exceptional exploration success. The exploration success has been built on a relatively small number of major greenfield discoveries and numerous examples of growing the known occurrences through near-mine discoveries and expansions (brownfield).

20.3   Six Scientific Breakthroughs Contributing to Discovery

The three scientific breakthroughs related to regolith (Table 20.2) were being developed by CSIRO in the 1970s and applied through the

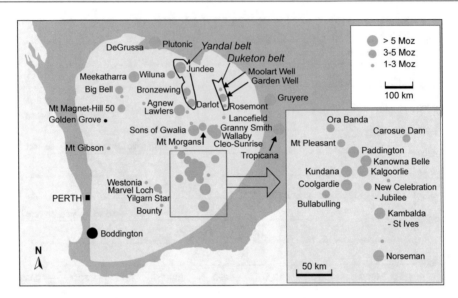

Fig. 20.5 Map of the Yilgarn Craton showing the major goldfields based on endowment. All are gold-only except for the VMS Golden Grove deposits, and noting that Boddington has produced minor Cu. The Yandal and Duketon greenstone belts, amongst several other centres, have been explored with success using the approaches outlined in this chapter.

1980s with considerable exploration success (Butt 1989; Smith et al. 2000; Anand and Paine 2002). The three scientific breakthroughs related to primary gold geology (Table 20.2) were developed at University of Western Australia from 1979 and were being applied from the early 1980s onwards (Phillips et al. 1983). The way in which these six scientific results were distributed was not uniform but was effective whether through industry collaborative projects (regolith) or open presentations and publications (primary gold geology). All the breakthroughs were well-embedded into Yilgarn gold exploration practices well before 1990 (Phillips 2011; Phillips et al. 2019b). Although the breakthroughs had significant global value, the willingness to consider and uptake each varied between countries.

Table 20.2 Scientific breakthroughs impacting upon Yilgarn gold exploration methods, and discovery (after Hogan 2004).

Regolith-related breakthroughs	
1. Landforms and exploration tactics	Different landforms require different exploration methods giving rise to a RED scheme (residual, erosional, depositional) applied to the differing landscapes
2. Gold dispersion	Gold is mobile in some regolith environments and as it migrates laterally can create a broad exploration target at low concentration levels
3. Sampling media	Selective collection of Fe-rich surface material and calcrete may enhance gold levels particularly compared to the use of kaolinitic clays
Primary gold deposits	
4. Structural control (epigenetic timing)	Yilgarn gold deposits were not formed syngenetically on the seafloor; structures guided auriferous fluids and influenced orebody geometry
5. Favorable host rocks (mechanical and chemical properties)	Tensile strength and the concentration of iron and carbon were important in making some host rocks more auriferous than others
6. Alteration haloes (carbonate alteration related to gold deposition)	Carbonate and sulfide alteration were part of the gold deposition event and hence could be used directly as exploration guides

Case history examples from the Yandal and Duketon greenstone belts in the northeast part of the Yilgarn Craton are described to highlight different aspects of science and its role in discovery. The selection of these two greenstone belts is made because the link of science and discovery was recorded at the time, but there would be many other parts of the craton where geoscience played an equally important role in discovery.

20.4 Applying the Six Scientific Breakthroughs in the Yandal Greenstone Belt

In 1985 the Yandal greenstone belt in the northeast of the Yilgarn Craton was a poorly outcropping area without significant gold production and rarely visited by geologists. Soon after, it was recognised as having high prospectivity for gold based on what geology was available in outcrop. By 1995 the discoveries of Jundee (8 Moz), Bronzewing, Darlot including its Centennial component, and then Thunderbox had completely revised the understanding of the region and put considerable focus on its potential (Phillips et al. 1998; Vearncombe 1998).

At the start of the 1980s gold boom Wiluna was one of the top ten all-time gold producers in the Yilgarn Craton (1 Moz) and only 50 km from the as yet to be discovered Jundee goldfield (Fig. 20.5). The Yandal greenstone belt was delineated by aeromagnetic surveys by the 1980s, but gold exploration was difficult because of extensive cover by Cenozoic alluvium and sands, and outcrops were variably weathered. Although it was classified as lacking high gold potential based on standard maps, field work established by 1987 that there were widely distributed anomalous gold, alteration characteristic of other Archean gold settings, some differentiated dolerite intrusions, and favourable large-scale structures and auriferous quartz veins (Fig. 20.6).

Bronzewing was discovered in 1992 after recognising its structural setting in a greenstone wedge between granitic rocks and drilling to fresh rock which revealed gold-bearing lateritic

residuum under 10s m of Cenozoic cover (Wright et al. 2000). One hundred kilometres to the north, Jundee was a gold-anomalous area of outcropping Archean rocks spread over several kms. Weathering was 10s m deep and mining operations were established to exploit oxidised ore through eight open pits.

Four parameters that had been derived from an understanding of gold genesis in Archean greenstone belts then highlighted the significance of some data that was collected during systematic early exploration prior to any mining. Those parameters were gold, alteration, lithology and structure:

- Gold was important in two ways. One was as outcropping areas where the 19th century prospectors had started small workings with limited success. These were not especially abundant, but this scarcity was attributed to the extensive Cenozoic cover over much of the belt. The second way was gold values at the anomalous but uneconomic levels of 0.01 g/t Au level from rock chip sampling of shallow drilling to bedrock
- Alteration was a sign of fluid interaction with country rocks and particular importance was given to sulfides, carbonates and further minerals reported adjacent to other greenstone gold deposits. One such mineral was chloritoid $((Fe,Mg)Al_2SiO_5(OH)_2)$, and although its role was not fully understood in gold formation, it was known from near Kalgoorlie and Leonora mines
- Lithology was important for two reasons, one being mechanical properties and rheological contrasts which could influence fluid ingress, the other was chemical and potentially could precipitate gold from solution
- Structure was important in many forms including folding, faulting, ductile fabrics and especially auriferous quartz veins and vein arrays.

By the mid-1990s and after the discovery of Bronzewing and Jundee goldfields, there were tens of thousands of drill holes in the region, and it was apparent that these holes were a resource that had not been fully exploited.

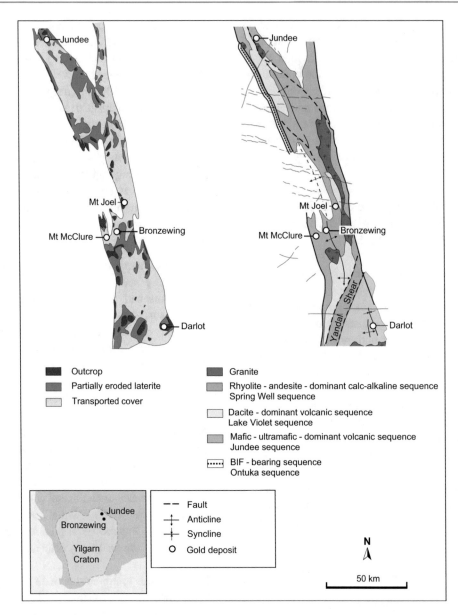

Fig. 20.6 Maps of the Yandal greenstone belt in the northeast of the Yilgarn Craton showing: (**a**) limited outcrop and extensive Cenozoic cover; (**b**) interpreted geology based on more than 50,000 regional drill holes to Archean basement beneath the cover. The four sedimentary – volcanic sequences are arranged in inferred time order and the youngest Spring Well sequence of calc-alkaline volcanic rocks formed at 2704-2690 Ma. Banded iron formation is restricted to the greenstone belt margin south and west of Jundee.

Logging sheets were developed to systematically record the critical information from the existing drill holes and to do so in a semi-quantitative way that could ensure consistency amongst more than 50 geologists. The scheme used a combination of gold, alteration, lithology and structure, and the results made it possible to prioritise various parts of the greenstone belt for further investigation and drilling.

One area prioritised during analysis of the relogged drilling was Jundee itself. Although already known as a large area of surface mineralisation, the logging and mapping of the regolith in shallow open pits identified differentiated dolerite sills thought to be akin to the Golden Mile Dolerite at Kalgoorlie (Travis et al. 1971), and further comparisons with Kalgoorlie revealed the unusual and favourable structural setting at Jundee that was known from Kalgoorlie and Timmins in Canada (Fig. 20.7). The mapping also confirmed that what was being mined in open pits at Jundee was a weathered gold deposit and not simply migrated supergene gold. A special study on Jundee began in May 1996 and was completed in July 1996. Drilling began immediately and the discovery of Barton Deeps was announced in August 1996 including an intersection of 8.5 m at 33.2 g/t Au from 321 m. Barton Deeps has become one of the most important gold lodes in the Yilgarn Craton outside Kalgoorlie and has proved a forerunner for other deep discoveries at Jundee.

The story of Barton Deeps highlights the considerable corporate commitment to support geoscience at Jundee and then to immediately test with drilling the ideas from the July report. Such Board action represented an important message about the company's commitment to geoscience and discovery. Behind the successes in the Yandal greenstone belt was a deep belief in good geology during exploration, and a simple corporate message. From the Board to the field office the aim was to make discoveries, and this simple unwavering message guided priorities and decision making.

The Yandal exploration story challenged another well-held belief in the industry that confidentiality of new ideas was critical. A degree of sharing of new ideas with other exploration companies and even tenement neighbours was encouraged and viewed as a potential positive externality. A flow of ideas could potentially help all, and a discovery by a neighbour would only help to enhance an area. As one example, a new idea was developed that gold deposits may lie in hidden greenstone belts beneath thin granitic sheets. As the company held extensive areas of granite in the Yilgarn Craton for its diamond search, here was an opportunity to apply a new idea on ground already held. The corporate

guidance was to apply the ideas of looking beneath granitic areas immediately and prepare for the idea to be shared at a major gold conference in North America in two months. Even if another company benefits from shared data nothing is lost by the originator in the process, and some pride can be taken from success elsewhere when the idea is tested by others (see Chapter 21).

Once discoveries are made and the discovery process is being recounted, the importance of ideas in changing the course of exploration are usually subordinated or forgotten. Instead, it is easier for those reporting to objectively report the type of drilling, soil survey or geophysics, and it is not uncommon for them to be unaware of the thought processes that prioritised the ground on which they have explored.

20.5   Scientific Re-Interpretation of the Duketon Greenstone Belt

The Duketon case history illustrates how a new geological interpretation can quickly revise perceptions of an area. In 1996, the Duketon greenstone belt appeared to be a low priority area near the NE margin of the Yilgarn Craton with, at best, moderate gold potential. A single day at the Famous Blue prospect (12th June 1996) with several geologists looking at drill core changed the perception of the region and was the harbinger of discoveries to come.

Gold had been known in the Duketon belt for a century and there were many small mines in this remote area; however, without any large deposits and with extensive regolith cover it received less attention than elsewhere in the Yilgarn Craton (Fig. 20.8).

A concerted effort to relog many drill holes from 1996 was supplemented by regional mapping of the belt. The mapping revealed what was initially described as an 80 km long unit like a diorite. The rock type description did not sound like a typical greenstone horizon, and its geometry hardly sounded like a granitic intrusion. However, this diorite was shown to be adjacent to established mineralisation at Christmas Well

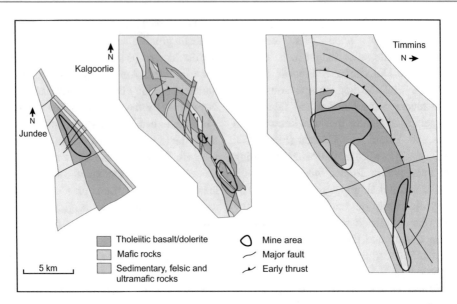

Fig. 20.7 Maps of Jundee and the largest Archean goldfields in Australia (Kalgoorlie) and Canada (Timmins including Hollinger, McIntyre and Dome) with a common orientation with respect to the far field stress field (Phillips et al. 1996). In each example, it is the contrast between competent and incompetent rock packages that has dictated the rock package response to deformation and focusing of auriferous fluids.

(Rosemont) and Famous Blue. At the latter, a resource of 300 Koz was known but the lack of a coherent understanding of the geology hampered drill programs and any detailed resource model.

Part of the reason for calling the unit a diorite was thin section petrography indicating it was an igneous rock with some quartz; this name was supported by an appropriate SiO_2 concentration from whole rock analysis. To resolve some uncertainties, a day was then devoted to looking through the Famous Blue core as a large group of senior and junior geologists. During this time the group's focus moved from the quartz to the Thomson-Widmanstätten texture of ilmenite and magnetite, i.e leucoxene. This texture was well-known in the main host rock for the Kalgoorlie goldfield, the Golden Mile Dolerite (Travis et al., 1971), so that information alone was enough to generate some interest.

In igneous rocks generally, the Thomson-Widmanstätten texture forms from magnetite dissolving several percent of Ti in its mineral structure at high temperatures. During cooling, less Ti can be dissolved in the magnetite structure, so a separate Ti-rich phase is formed that mimics the crystal structure of the enclosing magnetite. Temperatures over 1000°C that accompany basaltic (tholeiitic) magma enable the several percent of Ti to be dissolved in magnetite. The lower magmatic temperatures of felsic magmas including diorite only allow small amounts of Ti to be dissolved in magnetite and hence the Thomson-Widmanstätten textures are absent. Subsequent whole rock geochemistry and logging confirmed the Famous Blue host rock to be part of one or more differentiated dolerite sills and the 'diorite' terminology was abandoned. Relogging of drill core from Christmas Well confirmed the host there to be a differentiated dolerite sill, gave a new direction for targeting, and was followed by discovery of the deeper ores there. Renamed then as Rosemont, this deposit has an endowment of 1.6 Moz. The dolerite theme was extended to other prospects in the Duketon belt leading to discoveries in subsequent years of Moolart Well (1.6 Moz) and Garden Well (3 Moz; Balkau et al. 2017; Vearncombe and Elias 2017).

The case study illustrates several moments that can make the difference between discovery or failure.

Fig. 20.8 Map of the Duketon greenstone belt in the northeast of the Yilgarn Craton showing the main goldfields. Note that the Famous Blue prospect has been renamed as Gloster; Rosemont had a small pre-1996 production under the name of Christmas Well.

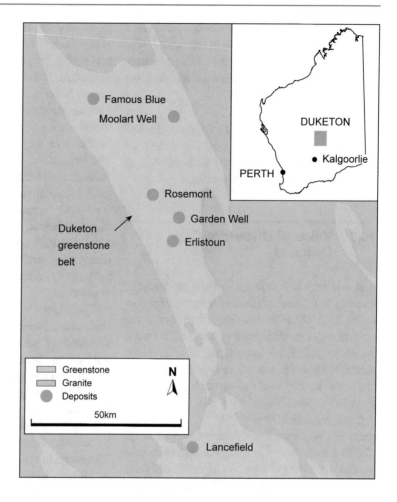

- Enunciating uncertainties and deciding to resolve them—in this case it was the geology of Famous Blue and the nature of the 'diorite' unit.
- Expecting success and being prepared to provide adequate time, personnel, and resources.
- Having a degree of scepticism and noticing little things that do not fit—in this case it was the 80 km-long diorite unit.
- Devoting time to brain storming—in this case staring at drill core, having unconstrained discussions, and not requiring every comment to be instantly justified. Many ideas would be wrong within a few minutes.
- Expecting high quality data so that small anomalies are treated as real and important to build on, rather than being rejected as outliers or dismissed as mistakes—in this case the mapping was trusted despite the initial interpretation being difficult to explain.
- Listening to others in the group—it can be the listener who creates the breakthroughs.
- Avoiding groupthink—if the initial question was so easy that the answer was obvious and could be imposed on the group, then it is hardly a breakthrough.
- Share credit—"who came up with the idea first?" is almost certainly the wrong question to be asking.
- Any of three sentences, if uttered at Famous Blue that day, may have changed the course of Duketon geology and delayed discovery:
  - "Lots of people have tried to sort out the geology and been unsuccessful, we should not waste any more time there"

- "I have drill rigs working today and I can't afford geologists talking in the core shed"
- "It cannot be a mafic rock because its chemical composition is plotted on this diagram as an intermediate rock".

In the quest for the leading-edge exploration tools, it can be forgotten that the difference between staying on a discovery path and failure can be some simple leadership actions.

20.6   Value of Discovery Case Histories

A case history of discovery provides the basis for learning and then replicating a successful formula. Such histories can also inform about the optimal ways to utilise geoscience and indeed a means to judge the importance of geoscience research. However not all case histories are of equal value, and when discovery histories are recorded several years after the event or by authors not directly involved in a discovery there will be distortions of the whole recording. The first such distortion of a case history written at arms' length is the focus on objective activities such as the type of drilling at the expense of the style of leadership, creation of a listening and questioning environment and small but critical decisions and breakthroughs. To some extent this distortion is only natural because the type of drilling, for example, can be easily verified; the role of various arrangements and ideas is more difficult to evaluate by those not involved, and ideas can be diminished in value well after the event. It is relatively uncommon to read about the important role of setting up the optimal corporate environment and team skill set, opportunities for brainstorming, chance for all to contribute, and refinement and promotion of new ideas.

A detailed analysis of 154 case studies of gold discovery highlights some of these difficulties of inaccurate reporting and especially the under-representation of intellectual activities (Brown and Vearncombe 2014). The two such activities that Brown and Vearncombe highlighted as being under-reported are structural geology and alteration which is curious because both are inextricably bound in genetic models on gold deposit formation. The absence of structure and alteration in case histories is here attributed to misreporting because structures in the form of quartz veins would always be noted and reported by geologists in the same way that pyrite would be reported. The importance of that pyrite is not because it is valuable, but because it might be a component of gold-related alteration. Hypothetically, a dangerous and expensive consequence of this misreporting may follow if the industry concludes that alteration does not contribute to gold discoveries with the next step of advising universities to replace alteration studies in their courses.

Unfortunately, the reporting of gold discoveries does not always adhere to the standards of normal peer review of a scientific publication. Invitation to document a discovery may be directed to the current property owner (not the owner or operator at the time of discovery) and corporate decisions are likely to dictate that authorship is restricted to current employees rather than those involved in the discovery. Therefore, in several but not all cases, academics and former employees are not included in discovery case histories, and for the consultant it is even less likely their contribution will be fully recorded. This authorship selection is quite different from best practice in academia, and the process distinctly under-represents input from researchers and consultants. Although accepted in parts of industry and by some conferences, the equivalent behaviour in academia is likely to be questioned.

Ideally there should be a direct connection between corporate strategy enunciated before a discovery, and the details of recording after a discovery. In a few circumstances when there is a complete mismatch between reporting of a discovery and earlier corporate reporting then either the company has not kept shareholders informed or the history of discovery has undergone a retrofit, i.e. parts have been added to the story after the event. The Yandal and Duketon examples are not chosen because they represent the most exceptional discoveries, but for the learnings available through the case histories. The discovery practices reported are consistent with the reporting of exploration strategy before any discovery was made and this can be verified in company reports and presentations.

The role and importance of geoscience including research is debated in industry and universities and therefore accurate case histories can provide the support to engage in applied research and to direct its focus. The linkage between geoscience and exploration discovery is not via the type of drilling or relative emphasis on geochemistry or geophysics but through nuanced decisions that are rarely recorded and usually lost in the course of time and corporate takeovers, restructuring, and staff changes.

- It is difficult to find accurate reports of the intellectual input including geoscience, thought processes, and nuanced decisions of the exploration team.
- Critical breakthroughs in interpretation are under-represented because they are rarely recorded and usually lost in the course of time and corporate takeovers, restructuring and staff changes.

Snapshot

- The 1980s gold boom led to a tenfold increase of gold production in the Yilgarn Craton.
- Carlin Gold Province in the USA followed a similar pattern, but many other regions of the world saw minimal production increase.
- Gold production over the last 40 years has not correlated with gold prices.
- On-going gold production correlates with sustained exploration success.
- Declining production correlates with lack of discovery, such as in South Africa.
- Scientific breakthroughs in regolith science and gold geoscience led to discoveries that irreversibly changed approaches to Yilgarn gold exploration after 1980.
- Understanding what is fundamental to gold deposit formation guides the development of exploration strategies, area selection, prioritisation of data collection and data organisation and interpretation.
- Most published discovery histories report on objective activity such as the type of drilling.

Bibliography

Anand RR, Paine M (2002) Regolith geology of the Yilgarn craton. Aust J Earth Sci 49:3–162

Balkau J, French T, Ridges T (2017) Moolart well, garden well and Rosemont gold deposits. In: Phillips GN (ed) Australian ore deposits. The Australasian Institute of Mining and Metallurgy, Melbourne, pp 261–266

Bettles K (2002) Exploration and geology, 1962 to 2002, at the Goldstrike property, Carlin Trend, Nevada. Soc Econ Geol Spec Publ 9:275–298

Brown L, Vearncombe JR (2014) Critical analysis of successful gold exploration methods. Appl Earth Sci (Trans Inst Min Miner Metall B) 123:18–24

Butt CRM (1989) Genesis of supergene gold deposits in the lateritic regolith of the Yilgarn Block, Western Australia. In: Keays RR, Ramsay WRH, Groves DI, eds. The geology of gold deposits: the perspective in 1988. Economic Geology Monograph; pp 460–470

Hogan L (2004) Research and development in exploration and mining: implications for Australia's gold industry. ABARE eReport 04(3):35

Phillips GN (2011) Gold exploration success: 33rd Sir Julius Wernher Memorial Lecture. Appl Earth Sci (Trans Inst Min Miner Metall B) 120:7–20

Phillips GN, Groves DI, Clark ME (1983) The importance of host-rock mineralogy in the location of Archaean epigenetic gold deposits. Geol Soc South Afr Spec Publ 7:79–86

Phillips GN, Groves DI, Kerrich R (1996) Factors in the formation of the giant Kalgoorlie gold deposit. Ore Geol Rev 10:295–317

Phillips GN, Vearncombe JR, Eshuys E (1998) Yandal Greenstone Belt, Western Australia: 12 Moz of gold in the 1990s. Mineral Deposita 33:310–316

Phillips GN, Vearncombe JR, Eshuys E (2019a) Gold production and the importance of exploration success: Yilgarn Craton, Western Australia. Ore Geol Rev 105: 137–150. https://doi.org/10.1016/j.oregeorev.2018. 12.011

Phillips GN, Vearncombe JR, Anand RR, Butt CRM, Eshuys E, Groves DI, Smith RE (2019b) The role of scientific breakthroughs in gold exploration success:

Yilgarn Craton, Western Australia. Ore Geol Rev 112: 103009. https://doi.org/10.1016/j.oregeorev.2019. 103009

Smith RE, Anand RR, Alley NF (2000) Use and implications of paleoweathering surfaces in mineral exploration in Australia. Ore Geol Rev 16:185–204

Travis GA, Woodall R, Bartram GD (1971) The geology of the Kalgoorlie goldfield. In: Glover JE (ed) Symposium on Archaean Rocks, Special Publication 3. Geological Society of Australia, Canberra, pp 175–190

Vearncombe JR (1998) Shear zones, fault networks, and Archean gold. Geology 26:855–858

Vearncombe JR, Elias M (2017) Yilgarn Craton ore deposits and metallogeny. In: Phillips GN (ed) Australian ore deposits. The Australasian Institute of Mining and Metallurgy, Melbourne, pp 95–106

Vearncombe JR, Phillips GN (2019) The importance of brownfields gold exploration. Mineral Deposita 55: 189–196. https://doi.org/10.1007/s00126-019-00897-1

Vearncombe JR, Zelic M (2015) Structural paradigms for gold: do they help us find and mine? Appl Earth Sci (Trans Inst Min Miner Metall B) 124:2–19

Wright J, Phillips GN, Herbison I (2000) History of exploration in the Yandal Greenstone Belt. Aust Inst Geosci Bull 32:187–195

Abstract

The path to a major discovery is far from uniform, and every discovery is different. Anecdotally it is about the seventh owner of a property who makes the big discovery, and to extend this tale, it is important not to be amongst the first six, nor waiting in line to be number eight. The siting of the so-called discovery hole is greatly over-rated in its importance, and the critical moments may have been much earlier and possibly protracted. The critical moment can even come well after drilling—there are multiple examples of a discovery having been drilled and not recognised until a new owner comes along. In this setting, the challenge of management is to engender conviction well before the big intersection is obvious and announced to the world, and this is best done using all scales of geology.

Fosterville Deeps, in the Victorian Gold Province, is a major gold-only discovery in the 21st century, and the input of three decades of external research played a significant role in understanding the opportunities at depth. Transfer of the research findings led to a change of direction of the company and a major discovery.

Keywords

Fosterville · Victoria · Discovery · Regolith · Uptake of ideas · Confidence · Long-term research

Fosterville, 20 km northeast of Bendigo in southeast Australia, was one of thousands of minor gold deposits in Victoria for a century following its discovery in 1894. From 2001, the discovery of its Phoenix Shoot marked an important conceptual shift with the confirmation of the opportunity termed here as Fosterville Deeps. Between 2015 and 2020, multiple drill results over 1000 g/t Au now mean that Fosterville was able to produce 0.6 Moz pa (2019) reflecting an especially rich ore zone. With operating costs near US$200/oz and gold prices around US$1500/oz that equated to a substantial profit. This is a strong platform on which to grow a major company if exploration continues to be successful.

A good question is 'why Fosterville' especially as there are 7000 or more gold mines, pits and workings in Victoria dating from 1851? This question has more than one correct answer. For example, the recognition and re-enforcement of the opportunity at Fosterville has been built on 30 years of research; and there have been multiple exploration campaigns by different owners (in keeping with the proverbial seventh-owner anecdote).

The documentation of major discoveries such as Fosterville Deeps is important because it can help inform future decisions such as whether to continue an exploration program, and ultimately how we direct any investment into research and training. Fosterville Deeps is attractive as a case history demonstrating the influence and value of research. Here is a three-decade history of

N. Phillips, *Formation of Gold Deposits*, Modern Approaches in Solid Earth Sciences 21,
https://doi.org/10.1007/978-981-16-3081-1_21

reporting research ideas in the public domain that illustrates how ideas evolve and how they inform corporate decisions and progress. For researchers, it is always rewarding when management acts on research ideas, and especially when there are positive outcomes as there have been at Fosterville.

21.1   Overview of Discovery at Fosterville

The State of Victoria in southeast Australia was the world's leading gold producer in the second half of the 19th century but it was very much in decline after 1900 and this gold province completely missed the 1980s global gold boom. From 1980, numerous negative industry and government views of any remaining potential in Victoria were re-enforced by expensive failures to re-start mines at Bendigo and Ballarat. The negativity regarding Victorian gold that existed in both Victoria and Western Australia was tempered by geoscience research that continued to highlight opportunities particularly within the international community. The stand-out of these opportunities in Victoria was the Fosterville goldfield which was identified from 1990-1 as a prime target based on its surface geology, global analogies, synthesis of the Victorian Gold Province and the application of regolith science and structural geology.

The discovery at Fosterville of Phoenix Shoot continuing to depth led to a significant conceptual shift of thinking (Fig. 21.1; a *paradigm* shift in the modern usage of the word). Rather than being a limited near-surface heap leach operation working an epithermal deposit, Fosterville emerged after 2001 as a potential major gold mine. On-going drilling added resources and in 2015 revealed the Eagle zone with consistent high gold grades, abundant visible gold, and extensive quartz veining. In 2017, the even-richer Swan zone was discovered near the Eagle zone. An Ore Reserve of 2 Moz at over 30 g/t was announced; and the Fosterville endowment is around 6 Moz (2020).

It might seem easy in hindsight to suggest that the Fosterville Deeps potential was obvious but much of the critical information was not available on the mine site or even within the community of the Victorian Gold Province, and instead required a global perspective (this explains why critical drill results were not followed up on site). It is informative to identify the positive signs leading up to the Phoenix, Eagle and Swan discoveries and understand why Fosterville was being elevated in the early 1990s above 7000 other Victorian gold prospects for its potential despite negligible all-time production to that time.

Harking back to the preface of this book, here is a recent example of the commercial value of understanding how gold deposits form. Fosterville Deeps provides an insight into how science is developed and transferred during exploration, and how the role of research may appear of less importance when looking backwards well after a discovery. The lead-up to exploration success begins around 1990 a decade before the Phoenix discovery. A major difference with the Fosterville history is that the research findings were published well-ahead of discoveries and were the impetus for the fund raising and renewed exploration at a time when the operation was planning for closure.

21.2   Geological Setting of Fosterville

Fosterville is within a sequence of Early Ordovician turbidites (490 Ma) of sandstone, greywacke, mudstone and black shale of the Bendigo Structural Zone (Boucher et al. 2015; Hitchman et al., 2017). This is a similar stratigraphic setting to the Bendigo and Ballarat goldfields which have all-time productions of 22 Moz (700 t) at 15g/t, and 14 Moz (410 t) at 12 g/t respectively. Regional deformation has generated tight upright fold axes at Fosterville trending NNW with a 300 – 500 m wavelength. The Fosterville Fault is an important locus of mineralisation and trends NNW, dips steeply west and is traceable for 30 km. This fault has been interpreted as being reverse with a sinistral component of movement; and there are similar parallel faults and some important cross-faults. The Ordovician

Fig. 21.1 Cumulative gold production from the Fosterville goldfields since 1894. The point of inflexion around 2005 coincides with the discovery of Phoenix Shoot and early mining of Fosterville Deeps referring to ores accessed by underground methods.

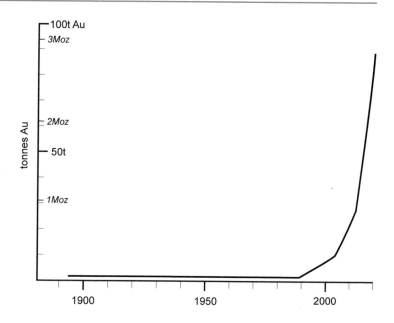

metasedimentary rocks include muscovite – albite – quartz – chlorite assemblages of greenschist facies, with increased siderite, ankerite and sulfide minerals approaching mineralisation.

The Fosterville goldfield extends 20 km along strike, is up to 3 km wide and is spatially associated with the Fosterville Fault and the parallel O'Dwyer Fault (Fig. 21.2). This is a gold-only deposit with pyrite, some stibnite and arsenopyrite but no economic base metals. Twenty-one open pits had been developed by 2001 to depths of a few 10s m to extract the oxidised ore, where sulfide minerals have broken down liberating Au and making it amenable for heap leaching. The regolith consists of variable alluvial cover of 10 m thickness or more above a weathered profile that extends for many tens of metres depth.

21.3   Small Open Pits, Free-Milling Ore, and a Heap Leaching Operation: 1991–2001

There was minor gold production at Fosterville during 1894 – 1909, in the 1930s and again in the 1980s (Table 21.1). During this latter period, Fosterville and several other prospects in Central Victoria were classified as epithermal gold deposits well into the 1990s (Ramsay et al., 1998), and this epithermal model played a part in guiding investment, exploration, and the interpretation of drilling.

From 1991 to 2001, mining from the open pits produced free-milling ore, i.e. gold that is not locked in gangue minerals and hence can be treated by heap leaching. This whole operation was relatively small both in grade and the amount of gold produced, and continuance of the operation depended upon finding new resources to offset the depletion. Throughout the heap leaching operation, the open pits were not deepened beyond the base of the oxidised material because the refractory ores beneath could not be treated by heap leaching methods.

A drill program started in 1994 to test for ore at depth with plans to follow-up any positive results. Drill hole SPD7 finished in mineralisation at 440 m depth with the last 55 m averaging 1.9g/ t Au. However, for five years the 1994 Drill Program remained without follow up and as the results were not considered material to the company's position they were hardly reported or discussed in company reports up to 2001. The drill hole became the basis of a detailed internal study to characterise the whole rock geochemistry and mineralogy of the metasedimentary sequence

Fig. 21.2 Map of the Fosterville goldfield with NNW-trending lines of the open pits from which oxidised ore was extracted for heap leaching. There were 21 pits when the mining of oxide ore finished in 2001 and they were generally 30 – 60 m in depth and spread over 10 km.

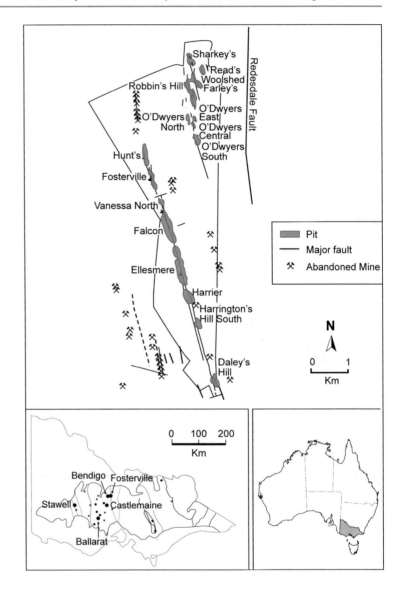

Table 21.1 Fosterville production and mining methods.

Period	Production (t)	Mining method
1894–1909	1.6	Underground
1930s	<1	Retreatment
1988	<1	Retreatment
1991–2001	7.5	21 open pits to 60 m, heap leach
2005 to the end of 2015	31	Underground, sulfide ore
2016	4	Underground, sulfide ore
2017	5	Underground, sulfide ore
2018	8	Underground, sulfide ore
2019	20	Underground, sulfide ore
2020	20	Underground, sulfide ore

(Arne et al., 1998). Meanwhile, the open pits were rapidly depleting, and the future of Fosterville was hardly positive.

21.4   Steps to a Major High Grade, Underground Mine

At the same time as the open pits were depleting, the Vicgold Research Project was progressing to understand the opportunities within the Victorian Gold Province and especially at Fosterville. The project began semi-formally in 1991 as a collaboration with Dr Martin Hughes with the research being on a global scale that involved merging various phases of research with the Fosterville mine-scale observations. By using a global approach, a positive interpretation was placed on the results of the 1994 Program especially hole SPD7, and this interpretation would underpin the bright future for the mine that could advance beyond the depleting open pits.

The research generated within several institutions contributed to the Vicgold Project and was combined to build an understanding of the potential of the Fosterville goldfield. These institutions included University of Ballarat, the Key Centre of Economic Geology in Townsville, Australian Research Council, Johnson's Well Mining NL, University of Western Australia / Great Central Mines *Giant Goldfields Project*, CSIRO *Gold Provinces Project*, and multiple CSIRO regolith projects.

Important steps in advancing the understanding of Fosterville included recognising the role of the regolith and that the epithermal interpretation was inappropriate, assessing Fosterville as the stand-out opportunity in Victoria, using analogies with giant goldfields globally, and recognising similarities to the Carlin goldfields in USA.

21.4.1   Regolith Geoscience, and Questioning the Epithermal Interpretation

Fosterville was being advanced as an epithermal gold deposit in the 1980s and 1990s with implications for exploration (e.g. Ramsay et al.,

1998). There were also suggestions from the epithermal tag that Fosterville might be the first of many new deposits in the region. Some of the epithermal-like features around Fosterville including clay mineral assemblages and textures were being used globally to interpret other deposits as epithermal. However, these features were not diagnostic meaning that they might indicate an epithermal deposit, but they might instead have other equally valid interpretations. It was also recognised that the term epithermal was being used for gold deposits to encapsulate different sets of features.

The Vicgold Project then took advances in regolith geoscience to understand clay assemblages and textures. In Western Australia, the understanding of the ground surface, soils and weathered rocks advanced rapidly during the 1980s through the regolith geoscience research program of CSIRO designed to assist the local mineral exploration. Despite the project focus being in the Yilgarn Craton of Western Australia, many ideas were transferable to the Victorian Gold Province including regolith geoscience being introduced into the University of Ballarat teaching curriculum by Martin Hughes. The regolith science led to a re-interpretation in 1991-2 that the oxidised (bleached, clay-rich, near-surface, ferric Fe minerals, complete destruction of pyrite, absence of carbonates, bleaching of black shale) interval around the Fosterville goldfield had resulted from weathering and was not indicative of an epithermal deposit. Analogy with well-studies weathering profiles in the Yilgarn goldfields suggested that the bleached material at Fosterville was only the upper part of the regolith profile and that below it lay dark, unoxidised material that was still partially weathered (Fig. 21.3; Fig. 21.4). The lower part of the regolith profile had unoxidised ferrous minerals, local carbonaceous material in shale and partial preservation of pyrite. The boundary from oxidised (bleached) to unoxidised (dark) was routinely being mapped as the base of complete oxidation (BOCO) in Western Australia, with the base of weathering being considerably deeper than BOCO. Postulating that the bleached interval at Fosterville was the product of weathering could account for the lack of carbonate minerals.

Fig. 21.3 Fosterville pit wall looking north along strike showing part of the regolith profile. The metasedimentary stratigraphy is striking north, dipping 80° west (left) and highlighted by colour changes such as the black shale unit at the 'BOCO' arrowhead. Above the base of complete oxidation (BOCO) any Fe is oxidised as ferric minerals conveying cream, yellow, red and orange colours; and sulfides minerals have broken down liberating Au and making it amenable for heap leaching. Below BOCO the dark green colour reflects Fe as ferrous minerals particularly chlorite, as well as the black carbonaceous shale. The regolith extends well below BOCO with an important expression being the partial oxidation of sulfide minerals making the material mostly unsuitable for heap leach processing. BOCO is not horizontal but influenced by rock units particularly to black shale and it is deeper around the ore zone.

Application of regolith geoscience ideas led to questioning that Fosterville goldfield was an *epithermal* gold system.

Vertical changes in quartz vein systems in Yilgarn gold deposits also had analogy with Fosterville and some surrounding deposits in Central Victoria. Major auriferous quartz veins at depth in the primary zone at Yilgarn deposits including Bronzewing and those around Southern Cross were fragmented or even absent in the near-surface weathered zone. This breakdown of the major quartz veins in the regolith was attributed to dissolution of silica along with many other chemical changes to the primary minerals during weathering. Consequently, large gold systems at depth were not always accompanied by major quartz veins at the surface, though clearly some were marked by very prominent outcropping quartz veins.

An outcome of applying regolith geoscience within the Vicgold Project and Fosterville gold-field was a different model on which to base understanding of vertical changes and to at least consider these in terms of weathering and a rego-lith profile. For further exploration, the target no longer needed to be a clay-rich zone with limited depth extent. If there were major quartz veins and carbonate alteration as the expression of deeper mineralisation in a primary zone, they may not necessarily be present near the surface.

21.4.2 Elevating Fosterville Above 7000 Other Small Pits and Old Workings in Victoria

After a century of gold mining in the State, the Geological Survey of Victoria estimated that

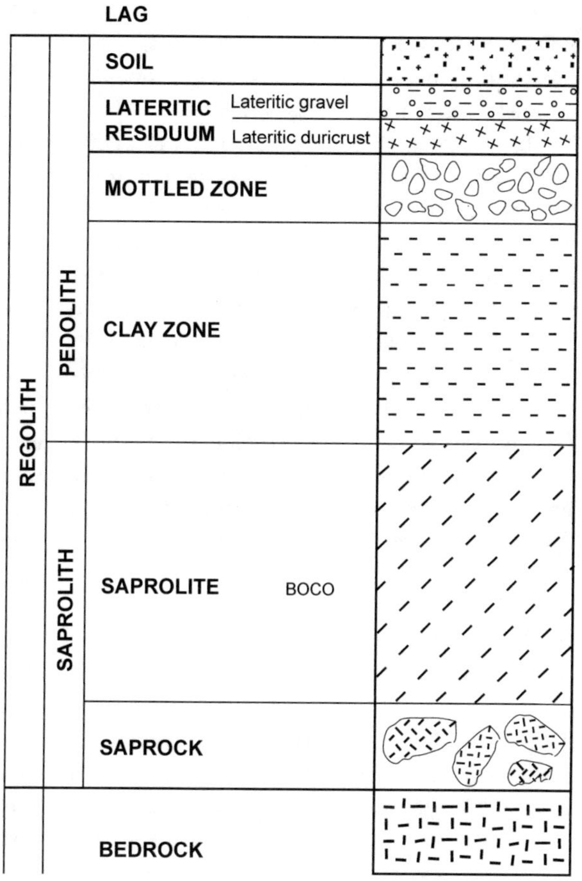

Fig. 21.4 Schematic profile of the regolith based on the Yilgarn Craton of Western Australia where it was developed. Many components of the figure are transferable to other landscape settings and aided in the recognition of the fuller picture of weathering at Fosterville especially beneath BOCO (which is part way down the saprolite).

		LAG	
REGOLITH	PEDOLITH	**SOIL**	
		LATERITIC RESIDUUM Lateritic gravel / Lateritic duricrust	
		MOTTLED ZONE	
		CLAY ZONE	
	SAPROLITH	**SAPROLITE** BOCO	
		SAPROCK	
		BEDROCK	

there were at least 7000 mines, shafts, old workings, and pits in the Victorian Gold Province. By 1990 the challenge for gold exploration was not the finding of gold mineralisation under which to target some drill holes. Rather, the challenge was to prioritise the 7000 and to identify a small number on which to focus some drilling. The key to recognising the potential at Fosterville was the accessing of research globally and using several seemingly minor observations that did not quite fit accepted geological ideas (including on Fosterville). The observations were combined with knowledge of global gold deposits and on-going research on how gold deposits formed, and the continuous documentation by the mine geology team at Fosterville. Dr Martin Hughes of Ballarat was an ideal collaborator as he was familiar with many deposits and old mines in Victoria and likely knew and understood them better than any other geologist. Some early influential

observations included those at the Nagambie deposit 70 km east of Fosterville where there were outcrop samples with 9 g/t Au without major quartz veins; and Fosterville itself where there was fine gold with stibnite and a cross fault at Daley's Hill pit. We were impressed by the extensive open pit area of Fosterville and noted that it had some important similarity of the regolith profile to Camel Creek gold deposit in North Queensland and the Binduli deposit 20 km west of Kalgoorlie. We also came into our collaboration with considerable scepticism regarding the current ideas and interpretations for the major Carlin gold deposits of northern part of Nevada USA based on the literature and field visits. We were both familiar with deeply weathered rocks from our experiences in Western Australia that included learning from the CSIRO regolith geoscience leaders. This Vicgold research focused on the totality of the Fosterville goldfield rather than its many small open pits and was strongly influenced by the complex and repetitive structural setting, black shale and the large footprint covered by the pits and mineralisation.

The Vicgold Project involved collaborative field work including many visits to old mine sites, combined with the extensive research of old records by Hughes including many observations about Fosterville. From this work Fosterville was soon the stand-out for its potential amongst the 7000 old mines and workings in Victoria. By 1995 there was the conviction based on the geology to speak locally and internationally and publish the Fosterville ideas. Despite Fosterville's miniscule production history, short life, low grade and poor existing economics, the ideas appeared to be well received in North America.

21.4.3 Factors in the Formation of Giant Goldfields

The Giant Goldfields Project of University of Western Australia and Great Central Mines Ltd was designed to provide a better understanding of the factors that were common to many of the world's largest goldfields. Although its initial focus was identifying potential amongst modest Archean goldfields in the Yilgarn Craton including Jundee, the findings were very applicable to Fosterville and Central Victoria. The overall tenet was that the giant goldfields do not owe their size to any unique factor absent in small deposits. The large deposit size was attributed to more of the same favourable factors found sporadically at smaller deposits. The project confirmed the importance of structural repetition (structural complexity) and re-affirmed the potential for mineralisation to persist for 1 to 3 km vertically.

21.4.4 Comparisons Between Carlin Gold Province USA and Fosterville

During the investigations of Fosterville in the 1990s, some important similarities were noted with the Carlin Gold Province of USA. These deposits were already major gold producers of global significance, and like Fosterville, were being actively mined in open pits to extract oxidised material; mining typically paused upon reaching the base of the oxidised material (BOCO in regolith science terminology). Unlike Fosterville, several Carlin open pits had been extended well below the oxidised material to extract the dark, unoxidised ores that were processed by pressure oxidation and roasting methods. During a study tour to Carlin in October 1997 with Donald Thomson there was the opportunity to combine regolith geoscience from Western Australia with the considerable local knowledge built up in the Nevada community regarding the Carlin deposits as to what these deposits looked like below the oxidised material. We re-interpreted the profiles of several of the Carlin mines as being within the regolith extending to many 100s metres depth (Fig. 21.5; Fig. 21.6). The Screamer section of the Goldstrike-Betze-Post mine on the Carlin Trend proved a useful analogue with its carbonate minerals and quartz veins that at that time were atypical of the Carlin Gold Province. The mineral assemblages and quartz veins at Screamer were being portrayed in Nevada as being anomalous but using knowledge from Western Australian regolith science allowed the Screamer section to be re-interpreted as potentially the fresh primary ore beneath Carlin deposits and translate such ideas back to Fosterville. It was then a small step to interpret the drill holes of the 1994 Program at Fosterville as likely to be primary Fosterville gold ore.

21.4.5 Transfer and Uptake of Research Ideas

The findings and ideas being generated by the Vicgold Project beginning in 1990 were never confidential and were released regularly through lectures, conference presentations, published papers and workshops in Australia and particularly Victoria but there was limited enthusiasm locally for following up the ideas with exploration.

Fig. 21.5 Idealised
regolith profile developed
in 1997 for deposits of the
Carlin Gold Province
(Phillips et al., 1999).

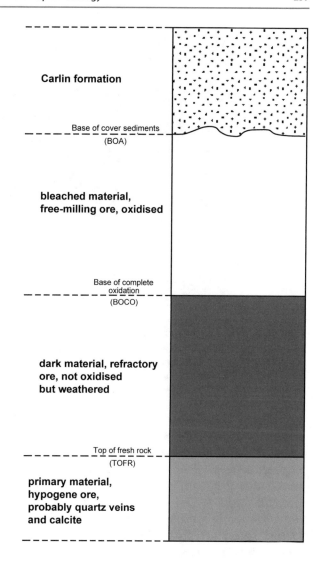

Carlin formation

Base of cover sediments
(BOA)

**bleached material,
free-milling ore, oxidised**

Base of complete
oxidation
(BOCO)

**dark material, refractory
ore, not oxidised
but weathered**

Top of fresh rock
(TOFR)

**primary material,
hypogene ore,
probably quartz veins
and calcite**

An invitation to the Geological Society of America annual meeting in 1994 in Seattle was the start of a two-decade dissemination of the scientific results on Victorian gold and Fosterville in North America including international publications and presentations in Vancouver, Toronto, Sudbury, Montreal, Timmins, Denver and Reno. The unprecedented Canadian interest in Victorian gold and Fosterville included how the science was being used to prioritise targets, though understandably the regolith was not their prime interest.

21.5   Building Confidence to Explore Further Based Upon Geology

The nadir in confidence at the Fosterville operation came in the year 2000 as the company discussed closing the mine operation and shifting its focus and workforce to their Timbarra property 1000 km to the north. Remaining Resources in the open pits were depleted, and the opportunities for additional ores for heap leaching were tested without great success. For the mine site, the 1994

Fig. 21.6 Upper part of the regolith profile in the Gold Quarry open pit of the Carlin Gold Province. Above the base of complete oxidation (BOCO) ferric minerals convey yellow, red and orange colours; and sulfides minerals have been oxidised liberating Au and making it amenable for heap leaching. Below BOCO the dark colour reflects ferrous minerals, arsenian pyrite and black carbonaceous material. The regolith extends well below BOCO but this material is unsuitable for heap leach processing. BOCO is influenced by rock units and particularly by faults. Photo by Juan La Riva Sánchez, with permission.

Program of drilling had not been an exploration success in their aim to extend ores and life of the mine. Although individuals on the mine may have thought that there was potential remaining at Fosterville, there was no firm basis for enthusiasm and there was minimal hard evidence on which to build a case for further exploration after 1994. Within three years the opportunities for exploration had decreased with a global downturn in exploration funding in 1997.

The turnaround in the mine's future can be traced to some discussions in mid-2000 and a major meeting in September 2000 at which the Vicgold Project findings were shared with the Fosterville leadership. The re-interpretation of the drilling from the 1994 Program was provided and then used to make a case that deep Fosterville ore had already been intersected. Further research findings were made available including the global analogies that demonstrated the potential of the Fosterville goldfield. A geological case drawing on global ideas was provided at another meeting in October 2000 supporting the high prospectivity of the Fosterville goldfield at depth.

In January 2001, Fosterville planned to raise its own funds releasing a statement in February 2001 stating that they had formed the opinion that the earlier sulfide initiative may have seriously under-estimated or not optimised the potential at Fosterville. The release went on to use this recent opinion to justify a new Board, and a fund raising to support a deep drilling program at Fosterville. The basis of the Board optimism was the new perception of the Fosterville Deeps opportunities (Fig. 21.7).

The new exploration program from 2001 led to the Phoenix Shoot being discovered in 2003 which became the mainstay of underground sulfide ore mining for several years. The discovery of the Phoenix Shoot cannot be under-estimated because it added over 1 Moz of ore, and more importantly, it changed perceptions of deeper opportunities in the Fosterville goldfield and Victoria for the next decade and more. The process of underground mining of the Phoenix Shoot over several years provided the opportunities to learn more about the deeper ores and effectively test deeper targets. By September 2003, the

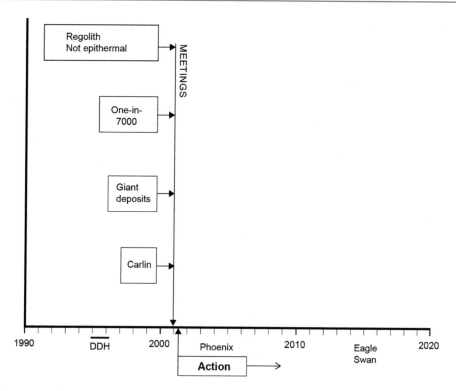

Fig. 21.7 Research ideas pertinent to Fosterville were developed and assembled over many years but once absorbed and accepted at the end of 2000 were acted upon very quickly leading to the discovery of the Phoenix Shoot and an appreciation of the Fosterville Deeps opportunity. Core from diamond drill holes (DDH) during the 1994-96 Drilling Program were revisited as the research ideas led to their re-interpretation as being the deeper expression of ore.

Fosterville Mineral Resource included 2.1 Moz of sulfide ore and 0.2 Moz of oxide ore.

From 2005, deeper mining from open pits and then underground produced sulfide ores that required processing by biological oxidation that has produced 55 t gold to the end of 2018. The all-time production milestone of 1 Moz of gold was reached at Fosterville in December 2015. At the start of 2019, Fosterville was producing gold above 30 g/t with a recovery of 97% at an operating cost below US$200 per oz. In 2019, production exceeded 0.6 Moz pa (20 tpa Au) based on Reserves of 2.7 Moz at 31 g/t but to sustain these tonnes and grades will require ongoing exploration successes. The endowment of the Fosterville goldfield is just over 6 Moz.

Fosterville is an example that our exploration industry often needs to drill deeper. However, this message might be refined to include the need for improved interpretation of the upper 200 m of deeper regolith and into near-fresh rock incorporating data integration and 3D viewing. For Carlin and Fosterville, there were many shallower indicators prior to successful deeper drilling.

21.6   Understanding Fosterville Geology Following the Phoenix, Eagle and Swan Discoveries

The discoveries of Phoenix (Roberts et al. 2003) and Eagle Shoots (Hitchman et al., 2017) are documented, but for the discovery of Swan Shoot in 2017-2018 there is still much knowledge to emerge.

A major mine-based geological project was commenced in 2001 led by Dr Rod Boucher and Simon Hitchman with the aim of subdividing the geologically difficult Ordovician turbidite sequence. This work was concentrated on Bendigo, Ballarat, Lockington, and Fosterville goldfields of Central Victoria and documented the distribution of shale-topped sand units, rarer channel sands and the thicker shale units. The study has become the foundation of core logging practices to resolve the geometry at Fosterville. Importantly the project gave considerable confidence to exploration, aided drill planning and underpinned brownfield exploration success. It also marked a more sophisticated documentation of the geology at the Fosterville mine site (Fig. 21.8).

In the primary zone beneath the effects of weathering the mineralisation at Fosterville occurs as shoots plunging around 30° S with ore zones generally 5 m thick and continuous for 400 m along strike. Ore varies from free-milling material near the surface in which sulfides were destroyed by weathering, a transition zone of partially oxidised sulfide minerals and incomplete liberation of gold grains, and finally black carbonaceous ores at greater depth with sulfide minerals including pyrite, pyrrhotite, arsenopyrite and stibnite. The sulfide minerals, especially pyrite, have trapped some of the gold grains and require biological oxidation to free the gold for processing and recovery. Quartz veins are relatively uncommon in the upper levels of the orebodies but more common at depth in shale and adjacent sandstone. The gold is around 1 – 10 microns in grain size and there is associated arsenopyrite, subordinate pyrite, and sporadic stibnite. Distal feldspar and chlorite minerals in the country rock shale and sandstone have been replaced by white mica, ankerite and siderite proximal to quartz veins. The geochemical signature involves limited elements, mainly CO_2, S,

Fig. 21.8 Stratigraphic cross-section through the Fosterville goldfield, such as in the open pit view in Fig. 21.3. The subdivision of an otherwise monotonous Early Ordovician turbidite succession into these mappable units has played an important role in understanding the structure, and planning brownfield exploration (from Boucher et al. 2015, with permission).

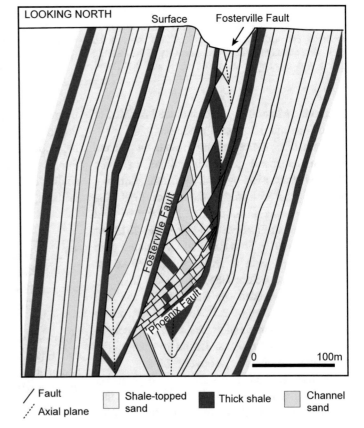

Sb, As, Ag and Au, and with generally low base metals.

21.6.1 Swan Zone Discovery: 2017–2018

Corporate presentations documenting the 2017-18 Swan discovery describe this as an evolving process over many months rather than any epiphany. Eagle zone had been found through drilling a new structure, and Swan zone through drilling another structure near Eagle; all the time the exploration team was gaining a better feel for how to target higher grade areas.

During this evolution visible gold became increasingly common in drilling along with some high-grade intersections that were out of character with much of the drilling history at Fosterville. The host is reported as a sandstone unit with a brittle fault at a high angle to bedding with visible gold, quartz veins and low modal sulfides. On paper there are similarities between Swan zone at depth beneath Fosterville and Screamer zone beneath Goldstrike deposit in Nevada with both marking a strong change to visible gold with carbonate alteration and quartz veins. On the global scale, the geological character of Eagle and Swan zones with quartz veins, free gold and carbonate alteration was predicted well before their discoveries using knowledge of the Carlin deposits and regolith science. As mining continues, knowledge of the orebody geology will inevitably be refined.

21.6.2 Deeper Fosterville Potential and some Interesting Options Requiring Gold Geology

Earlier reference was made to the global examples within the Giant Goldfields Project and how this provided guidance to the potential of ore at depth at Fosterville (Section 21.4.3). Two decades later that research project again has relevance to important decisions at Fosterville including how to best access deep ores in future. The default option, as for many deeper mines today, might be to continue the decline until the mine becomes uneconomic. The results of the Giant Goldfields Project suggested that simply deepening the open pits may leave behind some ore. This may include leaving high grade ores for one kilometre or more below the 2 km depth mark. A vertical shaft is an obvious way to access this deeper ore but is a major investment decision.

Prior to 1980, a new gold deposit in most countries would be developed by a shaft from the surface and this would be extended to whatever depth remained economic. For discoveries after 1980, mining might commence with an open pit. When the pit became too deep the logical short-term progression was to add a decline from the bottom of that pit. This is essentially an underground road allowing large trucks to bring ore from depth by driving up a 1-in-8 gradient. The trade-off inherent in this optimal gradient is that each 100 m of depth requires almost 1 km of driving, a return trip can take hours, and economics deteriorate sharply with depth.

As a major gold mine approaches 1.5 km depth and with ore continuing, a vertical shaft looks attractive in hindsight. However, sinking a shaft to 1.5 or 2 km is a major investment and hardly comparable to sinking an exploratory shaft some tens of metres in 1900 and testing the orebody while sinking progresses. A shaft sunk to 2 km depth today requires considerable confidence that the ore will continue both in grade and depth extent. Starting a shaft when the mine is already accessed by a decline to 2 km depth still faces risk and geological uncertainties, and yet without a shaft there could be millions of ounces of high-grade ore that will never be mined.

The decision whether to opt for a shaft usually involves the mining engineer and financial staff from the mine site and company, but in Australia at least, these professionals are likely to be familiar with pits and declines but probably never supervised the construction of a modern shaft system. One can imagine that planning discussions might centre around types of shaft, cost, cash flow, return on investments and contingencies. However, the greatest risk is that the orebody does not continue as anticipated, and so for this reason the initial input into shaft planning needs to be geological. Simply, no orebody then no shaft.

Returning to the theme of how gold deposits form including their geometry and extent, determining the tonnage and grade of ores below the deepest levels of mining becomes one type of exploration. Drilling is important but drilling holes to 3 km depth is neither cheap nor easy to control and determine the exact position. Learning from deep mines globally also has a role, and one of the most important inputs is geological understanding of the upper 1.5 km of the existing orebody including structural geology and alteration.

Fosterville with its high gold grades, modest present depth and orebodies open at depth has emerged since 2015 as a candidate for a vertical shaft. It is encouraging that the grade of the orebody increases around 800 m depth.

Unlike Australian companies with limited experience in large shaft systems for gold mines, the industries in both South Africa and Canada have a long history of shafts, and at Fosterville the Canadian connection brings important expertise and decision-making skills.

21.7 Learnings from Research and Discoveries at Fosterville: 1990–2020

The importance of discovering the Phoenix Shoot at depth cannot be over-estimated; it provided both confidence and direct guidance to the fuller extent of Fosterville as a major primary goldfield that had substantial depth extension. It represented the beginning of a new era regarding perceptions of the Victorian Gold Province that has persisted with the interest in the government's tenement releases around Fosterville in early 2020.

Some will say that Fosterville was simply lucky around 2015—that the explorers were looking for something like Phoenix with minor visible gold and found the Eagle and Swan zones with abundant quartz veins and visible gold. The other interpretation is that the explorers ensured that they focused within the right district and did so with

conviction from 2001, invested heavily in understanding their host rocks and structure, carried out their exploration diligently, and overall gave themselves every chance to be successful. The considerable focus on stratigraphy and logging since 2001 has meant that important details are much more likely to be noticed and acted upon.

It takes confidence to continue exploring amidst ordinary initial drill results; and it is geology that can form the basis of that confidence whether it be gained from having seen many other deposits, making links that others do not make, introducing ideas from other scientific communities, or through the persistent application of hard thinking. The use of different scales of observation is valuable such as combining careful logging of core integrated with petrological microscope study, and regional synthesis with global analogies (Fig. 21.9). The use of both descriptive and genetic approaches can be powerful. Descriptive geology involves matching features with known deposits and genetic thinking helps to select and deselect information and observations based on what is critical in deposit formation.

Fosterville has had eight owners since the early 1980s, and there are two unusual features of this ownership (Table 21.2). The first is that the last five owners are based in Canada which is not typical amongst smaller Australian-based gold operations. Second, most of the ownership transfers caused minimal staff changes suggesting the operations were regarded as functioning well but required additional capital (which proved correct) to capitalise on brownfield opportunities recognised by the researchers and transferees.

For several decades there has been only minor interest in Victorian gold within Australia. The situation was compounded by negative perceptions of gold in sedimentary host sequences based on the dominance of igneous rocks as hosts of gold deposits in Western Australia in the late 20th century. This last situation was not replicated in North America where there have been accessible examples of large gold deposits in sedimentary rocks.

Fosterville illustrates how modern gold discoveries are rarely made by placing a cross

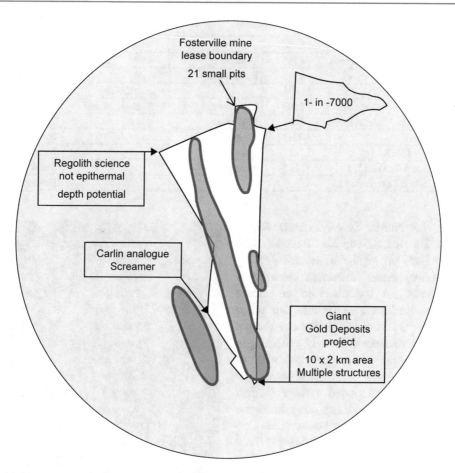

Fig. 21.9 The view from inside the Fosterville lease in 1995-2000 was one of multiple small, low grade open pits. With the benefit of external information including global analogues and regolith geoscience, this goldfield stood out as one of the best exploration opportunities in Victoria.

on a map and then drilling. The critical steps prior to a discovery may only be apparent later and are not always well recorded in the public domain. Fosterville is unusual as a case history of the role of research in critical discovery decisions because the research findings were always in the public domain including peer reviewed publications, government and university funded research projects, conferences, workshops, and public lectures. Far from the research being actively sought by industry, the uptake of some of the important ideas was rather slow. Different versions of a discovery story may be quite valid depending upon who attended various meetings and discussions, and of course any staff turnover will influence who has access to discovery history

and records; this is the value of having contemporary public domain records available. Like many other gold discoveries, Fosterville rewarded a few, but others missed out because of circumstance or because opportunities were not recognised. Without over-emphasis on the number seven, Fosterville adds support to the anecdote that the value often resides with about the seventh owner.

What looks so obvious after the event ('everyone knew there was gold there, all that was needed was to drill beneath the pits'), took several years to evolve into a technically-supported and funded drilling program.

The staff at Fosterville gold mine are thanked for their hospitality, access and sharing of ideas

Table 21.2 Ownership of Fosterville since the 1980s.

Start		End			Corporate base
1988	Bendigo Gold Associates	1989	Take over		Australia
1991	Brunswick Mining NL	1992	Receivership		Australia
1992	Perseverance Corp Ltd	2001		End oxide ores	Australia
2005	Perseverance Corp Ltd	2008	Take over	Sulfide ore mining	Australia
2008	Northgate Minerals Corp	2011	Take over		Vancouver, Canada
2011	AuRico	2012	Fosterville sold		Toronto, Canada
2012	Crocodile Gold Corp	2015	Take over		Toronto, Canada
2015	Newmarket Gold Inc	2016	Merger		Toronto, Canada
2016	Kirkland Lake Gold Ltd				Toronto, Canada

especially Neil Norris, Trevor Jackson, Simon Hitchman, Ian Holland and Rod Boucher. This hospitality included regular mine visits, strong support of the annual Melbourne Geology of GOLD course and invitations to geoscience meetings in Bendigo. Perseverance Ltd board members John Quinn and Chris Roberts showed their faith in the research in 2000-1 by listening to the Vicgold ideas about the potential of Fosterville Deeps and then taking immediate action. This action included refreshing their Board, raising funds and exploring at depth. Access to company reports of Perseverance Corporation and Johnson's Well Mining NL is acknowledged as is the opportunity provided by GeoScience Victoria to document the Fosterville case history in their Special Publication 25. Martin Hughes recognised the outstanding potential of Fosterville by 1990 and supported that opinion with his synthesis of Victorian gold, and his regolith knowledge. From 1995, he shared his ideas and reasoning widely in the community through presentations and scientific publications.

- The Fosterville research was publicly available from 1995 and continued to be disseminated during the periods of major discoveries in 2001–2002 and 2015–2018.
- Corporate changes at Fosterville may have led to much information about the discovery process being lost except that research ideas were routinely and promptly made available publicly.
- These discoveries show the value of combining external ideas on a regional and global scale with mine geology.
- Fosterville is one example where the research was not sought by industry nor immediately appreciated as being of value.
- It is rarely the siting of the discovery drill hole that is the most critical contribution, and the critical moments may happen well before or after drilling.
- Discovery appears simple looking back after the event.

Snapshot

- The discoveries at Fosterville are examples of the importance of three decades of research and its influence on exploration decision making.
- The ideas arising from that research provided the confidence to look beyond initially disappointing results, to raise funds and then to explore after a century of small-scale gold mining.

Bibliography

Arne DC, Lu J, McKnight S, Bierlein FP, Mernagh TP, Jackson T (1998) New developments in understanding the Fosterville gold deposits, Victoria. Aust Inst Geosci Bull 24:87–96

Boucher RK, Rossiter AG, Fraser RM, Turnbull DG (2015) Review of the structural architecture of turbidite-hosted gold deposits, Victoria, Australia. Appl Earth Sci 124:136–146

Hitchman SP, Phillips NJ, Greenberger OJ (2017) Fosterville gold deposit. In: Phillips GN (ed) Australian ore deposits, vol 32. The Australasian Institute of Mining and Metallurgy, Monograph, Melbourne, pp 791–796

Hughes MJ, Phillips GN (2001) The scale of ore fluid circulation in the Victorian Gold Province, southeastern Australia. Glob Tectonics Metallogeny 7:223–225

Hughes MJ, Phillips GN (2015) Mineralogical domains within gold provinces. Appl Earth Sci 124:191–204

Hughes MJ, Phillips GN, Gregory LM (1997) Mineralogical domains in the Victorian Gold Province, Maldon, and Carlin-style potential. Australasian Institute Mining Metallurgy, Annual Conference; pp 215–227

Hughes MJ, Kotsonis A, Carey SP (1998) Cainozoic weathering and its economic significance in Victoria. Aust Inst Geosci Bull 24:135–148

Hughes MJ, Phillips GN, Carey SP (2004) Giant placers of the Victorian Gold Province. SEG Newslett 56:1–18

McConachy GW, Swensson CG (1990) Fosterville gold field. In: Hughes FE (ed) Geology of the mineral deposits of Australia and Papua New Guinea, vol 14. AusIMM Monograph, Carlton, pp 1297–1298

Phillips GN (1991) Gold deposits of Victoria: a major province within a Palaeozoic metasedimentary succession. World gold 91, Aust Inst Min Metal, Melbourne; pp 237–245

Phillips GN (1998) Diversity among gold deposits: examples from the Victorian Gold Province. Aust Inst Geosci Bull 24:1–10

Phillips GN (2010) Victorian Gold Province Australia: a contemporary exploration guide. GeoScience Victoria, Special Publication; p 52

Phillips GN, Hughes MJ (1995) Victorian gold: a sleeping giant. Society Economic Geologists Newsletter 21: 1–13

Phillips GN, Hughes MJ (1996) The geology and gold deposits of the Victorian Gold Province. Ore Geol Rev 11:255–302

Phillips GN, Hughes MJ (1998) Victorian Gold Province. Aust Inst Min Metall Monograph 22:495–504

Phillips GN, Thomson D, Kuehn CA (1999) Deep weathering of deposits in the Yilgarn and Carlin gold provinces. Regolith '98: new approaches to an old continent, CRC-LEME, Perth, pp 1–22

Phillips GN, Hughes MJ, Arne DC, Bierlein FP, Carey SP, Jackson T, Willman CE (2003) Gold, in Geology of Victoria. In: Birch WD, ed. Geological Society Australia, (Victorian Division), Melbourne; pp 377–432

Ramsay WRH, Bierlein FP, Arne DC, VandenBerg AHM (1998) Turbidite-hosted gold deposits of Central Victoria, Australia: their regional setting, mineralising styles, and some genetic constraints. Ore Geol Rev 13:131–151

Roberts C, Jackson T, Allwood K, Shawcross M, Story J, Barbetti L, Tielen R, Boucher R, Norris N (2003) Fosterville - rise of the Phoenix, the emerging goldfield at Fosterville. In New generation gold deposits 2003, Conference Proceedings, Louthean Media, Perth; pp 200–213

Zurkic N (1998) Fosterville gold deposits. In: Berkman DA, Mackenzie DH (eds) Geology of Australian and Papua New Guinean Mineral Deposits, vol 22. AusIMM Monograph, Carlton, pp 507–510

Summary and Conclusions

22

Abstract

Gold deposits form when gold is precipitated from hot waters in the Earth's crust in sufficient quantity and at adequate gold grade to be economic. Hydrothermal fluids, using this term to describe these hot waters, form around 10 km deep in the crust as the surrounding rocks are metamorphosed around 500°C and lose the H_2O that is bonded in mineral structures. These hydrothermal fluids start their life with some dissolved gold, and then rise through the crust to be focused by faults and rocks that can fracture under the high fluid pressures. Ideal situations for fracturing of one rock type will be where rocks of different competence are juxtaposed; an example would be small competent igneous stocks within less-competent metasedimentary sequences. Larger gold deposits coincide with structural complexity and multiple orientations of fault sets each made up of multiple parallel faults. As the auriferous fluid rises to 3-5 km depth and cools to 300 – 400°C it reacts with its surrounding rocks and precipitates gold especially if there is chemical interaction of the fluid with Fe-rich or carbonaceous rocks. It is common for the gold to be deposited with quartz veins and pyrite (FeS_2, also known as fool's gold).

Keywords

Summary of gold deposit formation

Gold is normally insoluble in water but above 300°C there are some special elements it can bond with and then become soluble. In Nature, the two most important elements that can act as ligands and bond with gold in this way are sulfur (S) and chlorine (Cl). Gold has two ionic forms and each works especially well with one of these ligands: Au^{1+} with S, and Au^{3+} with Cl. These pairings lead to the fundamental twofold classification of gold deposits into gold-only and gold-plus deposits. A simple and practical question that separates the two classes is "Does this deposit produce economic base metals?". Gold-only deposits have accounted for 80 % of all-time world gold production, gold-plus deposits account for 20 % and mostly as gold with copper. The gold-only deposits form from hydrothermal fluids with Cl concentrations generally lower than seawater, in which the Au is transported while it is bonded to S in a fluid poor in base metals. Gold-only deposits include types informally described as Witwatersrand, Archean greenstone gold, slate belt or sediment-hosted, Carlin, and low sulfidation epithermal deposits. The gold-plus deposits form from hydrothermal fluids that are much more saline than seawater, transport gold along with abundant Cu bonded to Cl and include types informally described as Cu-Au porphyries, high-sulfidation epithermal- and iron oxide copper gold deposits. The origin of the low salinity fluids is from the metamorphism of basalt and sedimentary sequences; an origin of the high salinity fluids from metamorphism of sequences containing evaporites, i.e. NaCl-bearing, is favoured. Although volcanogenic massive sulfide deposits classify as gold-plus they are only minor producers of gold globally and not a significant topic of this book.

These metamorphic processes confer five fundamental characteristics on gold deposits: a **provinciality** in the distribution of goldfields, extreme **enrichment** of gold above its crustal abundance, a **timing** of formation during metamorphism and tectonism rather than at the sedimentary stage, and specific **ore fluid compositions** (low salinity for gold-only, and very high salinity for gold-plus); and finally, for gold-only but not gold-plus, a marked **segregation** of gold from base metals. From these characteristics arises the commonality between many deposits.

Modifications to deposits after they have formed are responsible for much of the variations and complexity seen today and conveyed by the expression of some frustration "gold is where you find it". These modifications can lead to creating, upgrading, or destroying deposits, and include higher temperature metamorphism, retrogression, weathering, and erosion that in the case of gold-only deposits can sometimes lead to alluvial concentration of gold into placers. Correctly interpreting these changes is of paramount importance in exploration and mining. From these modifications arises diversity between gold deposits.

It has not been possible to identify any essential role for silicate magmas in the formation of gold deposits. However, solidified magmas in the form of igneous intrusions are important for their mechanical and chemical properties that make them significant host rocks, especially when they are susceptible to hydraulic fracturing. Careful descriptive documentation of gold deposits in igneous rocks remains important.

Suggesting how gold deposits form is far from an exact science, and it would be presumptuous to imply that the answer above is the correct one. This book represents a best attempt to address the curiosity and commercial aspects of how gold deposits form, considering the available data, personal observations and the author's individual perspective, choices, biases, and skill sets. *Formation of Gold Deposits* is not meant to be a weighted average of all opinions. That the ideas in this book might differ from others arises for several reasons that are inherent in the research and approach that underpin this book. Two of special importance are **scale** and **classification**.

The **scale** of the process forming gold deposits follows from the extreme several thousand-fold enrichment of gold above its crustal background and necessitates thinking on a scale of cubic kilometres when contemplating the source of the gold. This means that the source of auriferous fluids may be removed by kilometres from the deposit. It also means that the host rocks containing deposits are only a small and final part of the story and studying the host rocks alone is almost certainly the wrong scale with which to understand deposit formation. The importance of scale has necessitated combining global research with detailed mine geology and laboratory analysis that combines the disciplines of mineralogy and petrology with aqueous geochemistry and thermodynamics.

The initial **classification** of all gold deposits has been minimalistic and at first tentative. This book uses gold-only and gold-plus with continued testing of this classification as understanding increases. The author's observational base has included virtually all gold deposit types, and the science base has included a strong emphasis on aqueous fluids in the Earth's crust, combined with magmatic and metamorphic processes. This is a significant difference to many published studies which would start by classifying Carlin and slate-belt deposits, for example, as different, restrict study to one of these only, and never be exposed to important commonalities.

To the perennial question of how readers might judge between many competing genetic models for gold deposits, three ideas are suggested; namely compare **track record**, **unreasonable effectiveness**, and **exploration success**. **Track record** refers to the robustness and resilience of observations and conclusions over time. This might be contrasted with any need to retract observations or make ad hoc modifications to an existing genetic model. A sign of a poor track record might be an overly complex genetic model that is reactive to new information, and here definition broadening would be a form of ad hoc modification. Being **unreasonably effective** refers to simple conclusions that have many more ramifications and greater value than initially expected. Examples include the chemical buffering role of CO_2 in auriferous fluids, forward and backward modelling of metamorphic devolatilisation reactions, and a simple explanation of gold-only—gold-plus classification using undergraduate-level gold chemistry. Discovery and **exploration success** are important outcomes of gold genesis research and examples, such as the discovery of the Fosterville Deeps orebodies in the Victorian Gold Province, are provided of geoscience assisting exploration. However, no one should suggest a one-to-one correspondence exists between discovery, and the efficacy of a genetic model.

There are outstanding issues ahead in the field of gold geology, and some of these will presage revisions and advances of our current knowledge. A small group that stands out include:

- Weathering of gold deposits especially Carlin, epithermal and porphyry types.
- Consideration of igneous rocks as favourable hosts because of their mechanical and chemical properties, particularly porphyry deposits.
- Integration of gold geology globally once a like-on-like comparison of production and endowment can be made with detailed mine geology. The stand-out opportunity here will be comparable data for deposits in China.
- Thermodynamic modelling of evaporites during metamorphism.
- Genesis of Witwatersrand gold, learning from other gold types.
- High grade metamorphism of various types of gold deposits.

Apart from improving our scientific understanding of how gold deposits form, pursuing these and further issues is sure to contribute to further exploration successes.

Appendix A: Production, Endowment, Reserves and Resources

There can be an expectation that measures of the amount of gold, such as production and endowment, are accurate and expressed to the nearest tonne or better. This is not so, and some of the difficulties are discussed to explain why many numbers need to be viewed with caution. Even for Reserves and Resources there are no static, perfect numbers because each requires judgement and includes assumptions such as the availability of new technology and future gold price.

The communications to the public around ore have changed significantly with the development of national reporting codes under the Committee for Mineral Reserves International Reporting Standards (CRIRSCO) framework. In Australia, the Joint Ore Reserves Committee (JORC) was established in 1971, published as the JORC Code in 1989 and has played a crucial role in initiating the development of standard definitions for these codes and guidelines. NI43-101 is used in North America; and equivalent reporting standards are found in almost all other mining countries except China. As well as addressing exploration results, these codes identify Mineral Resources when there is a reasonable chance of economic extraction of material. The component of the Mineral Resource that can be mined economically taking account of several modifying factors is reported as the Ore Reserve. Relevant factors may include mining costs, expected recovery effectiveness and commodity prices. Important components of the reporting process are that the Resources and Reserves are estimated by competent people with relevant experience, the process is transparent so that the identity of the competent person and the methods and assumptions can be gauged, and mine geology is available to support the whole estimate process. As an example of the importance of assumptions, if gold Reserves were calculated using $1000/oz, and then the Reserve recalculated for $2000/oz, the Reserve at almost all mines would increase significantly without further exploration. Similarly, new technology to treat difficult ore might mean some formerly uneconomic ore could become a Reserve. In many jurisdictions it would be illegal to report Resource and Reserve figures without the above rigor or to do so whilst knowing they are not correct; these are checks put in place to protect investors and other parties. Before the 1980s, a mine manager might have provided a verbal gold resource figure to a visitor and that figure could be conveyed to investors and included within scientific publications. Unsubstantiated figures become difficult to revise or eliminate once in the scientific literature.

Economic Demonstrated Resources (EDR) is a term used by Geoscience Australia (www.ga.gov.au) and the US Geological Survey to provide a guide to national and global Resources. EDR, according to Geoscience Australia, combines the JORC categories of Ore Reserves and most of the Measured and Indicated Resources.

Gold production figures for recent years are available from government agencies at national, state and province levels, and from company compilations. Today, 1000 active mines produce most of the world's 100 Moz of gold per year with the largest ten mines producing 10 % of this total. The world's leading verified gold producers

© The Author(s), under exclusive license to Springer Nature Singapore Pte Ltd. 2022
N. Phillips, *Formation of Gold Deposits*, Modern Approaches in Solid Earth Sciences 21,
https://doi.org/10.1007/978-981-16-3081-1

through the 2010s have been Australia, Russia, the USA, Canada, and Peru. For almost all countries it is possible to take a recent national annual gold production figure and identify the contribution that each province has made to that figure, and then the contribution of individual mines to the figure for each province.

All-time production is simply a summation of all annual production figures for a mine or a country. Globally there are 100,000 or more old mines and small workings summing to an all-time production of 6,000 Moz of Au. For some mines and goldfields, there are complications where there are ownership and boundary changes.

A source of uncertainty is a trend in recent years to have larger central mills for processing ore. Such ore may have been transported by truck from many different deposits within 100 km. Gold production as measured as bars leaving the mill is likely to be quite accurate but for geological purposes the amount of gold from each of those deposits may be more informative but less certain.

Endowment reflects the best attempt at defining the pre-mining gold content of a goldfield. It is a particularly useful term in geological studies of deposits and goldfields but comes with the uncertainties inherent in the calculations of Resources and all-time production. For national figures, there are cases where gold may be imported, refined, and then counted when exported, and this may lead to some double counting if the same gold has already been included by another country. The latter is avoided by considering national outputs and the individual mines from which that gold has come. Endowment can increase over time with brownfield exploration success. All-time production and endowment are not JORC terms, nor are they officially sanctified. The approach here has been to access multiple formal and informal sources to provide estimates of all-time production and likely endowment rather than rely too much on a single published figure.

China Gold Figures

Gold figures from China cannot be confirmed independently and are not calculated by JORC or equivalent standards; this means that the figures would not be quoted by other countries. Without published mine geology and documented methodology for calculations on which to base resource estimates, independent verification is not possible.

The national production figure for China has risen significantly during the periods of closure of many small mines for safety reasons. At the same time, there do not appear to be any major mines approaching 1 Moz pa Au. An annual production figure for China from the government applies to the whole country (455 t Au in 2016), and despite extensive searches and enquiries it has not been possible to verify this figure with more detailed figures from each province and producing mine. Where figures for individual mines are cited in the literature (resources or production) they may relate to verbal data from mine managers without any rigorous resource model. Some of these anecdotal resource and production figures for gold in China have entered the scientific literature in articles written by authors who themselves may not qualify as 'competent persons' for JORC or NI43-101 reporting purposes.

On the basis that the China gold production figures of the last decade are unverified, a projection forward from 2010 might suggest an annual production closer to 150 tpa Au whilst noting that this figure has a high degree of uncertainty. Exports of gold from China provide no basis for cross-checking its national production figures as a large and unknown proportion of its gold is used internally (unlike for Australia, Canada, or South Africa where the bulk is exported and can be internationally verified). This situation of uncertainty is not unique to gold nor China as figures from national oil companies have been questioned for some years (Pohl 2020).

Appendix B: Regolith Science

The regolith is described as "the entire unconsolidated or secondarily recemented cover that overlies more coherent bedrock, that has been formed by weathering, erosion, transport and/or deposition of the older material." This is the terminology based on *The Regolith Glossary* produced by the Co-operative Research Centre for Landscape Evolution and Mineral Exploration (Eggleston 2001).

The regolith thus includes fractured and weathered basement rocks, saprolites, and ground water. Weathering refers to any process which, through the influence of gravity, the atmosphere, hydrosphere and/or biosphere at ambient temperature and atmospheric pressure, modifies rocks, either physically, or chemically. Regolith is sometimes defined as everything from fresh air to fresh rock.

Weathering, as used in this book, refers to the breakdown of mineral assemblages and rock texture aided by invasion by meteoric waters from the surface. It is not restricted to the Recent, and in the Yilgarn Craton, for example, weathering profiles are millions of years old. Although the meteoric waters start at surface temperatures, there is ample evidence they are heated after burial as evidenced in Nevada where waters are currently emerging at the surface as hot springs and depositing silica. As these hot ground waters are continuous with near-surface weathering, the term is still used once ground water has been heated above surface temperature; this broader use of 'weathering' is adopted to allow for the meteoric waters to become heated with their descent.

Three terms are useful in describing the regolith profile and trying to constrain it in three dimensions (Anand and Paine 2002). The base of alluvium (BOA) describes the transition from overlying younger cover or alluvium, to weathered bedrock. The base of complete oxidation (BOCO) marks the transition from an upper bedrock sequence where oxidisable minerals such as ferrous minerals and carbon have been oxidised, to the underlying sequence where not all oxidisable minerals are oxidised. The top of fresh rock (TOFR) marks the depth extent of weathering by meteoric water invasion. In the weathered zone beneath BOCO rock textures are relatively well-preserved, the rock is fragmented into clasts with less-weathered cores, primary sulfides may persist, porosity may have developed from dissolution of minerals such as carbonates, and Fe is in its reduced form. In some gold provinces, 3-dimensional models of BOA, BOCO and TOFR are available for greenstone belts covering 100s km^2 and provide important data sources to guide exploration.

Supergene oxidation is not equivalent to weathering. Supergene sulfide enrichment is a special case of weathering where pyrite breakdown generates oxidising, acidic waters above the water table that can mobilise ore metals. Hence, supergene oxidation is a small component of the much broader term of weathering.

© The Author(s), under exclusive license to Springer Nature Singapore Pte Ltd. 2022
N. Phillips, *Formation of Gold Deposits*, Modern Approaches in Solid Earth Sciences 21,
https://doi.org/10.1007/978-981-16-3081-1

Anand RR, Paine M (2002) Regolith geology of the Yilgarn Craton. Australian Journal of Earth Sciences 49:3-162

Eggleton (ed.) RA (2001) The regolith glossary: Co-operative Research Centre for Landscape Evolution and Mineral Exploration; Perth, Australia, 144pp

Appendix C: Disagreements in Gold Geoscience

If one reads the literature or attends a gold geology conference, it is reasonable to conclude that gold geologists are always debating how deposits form. To some extent this is true, but the issues keep changing as we solve some problems and then move on to others as part of a process of continual improvement, correction of errors, and advancement. Some issues are solved easily, whereas others take many years until the ideal observations or new science and investigative techniques become available. Science does not claim absolute truth, so disagreement is not necessarily a failure. Instead, complete agreement, such as in a tight research group, might reflect complacency.

Geoscientists rarely disagree because one or more of them are stupid. It is usually because they come into a discussion having made different observations, and with different training, skills, prejudices, strengths, and weaknesses. Even using similar approaches and sharing access to the same science, differences will remain as each gold geologist will have visited, researched, and been employed at different gold deposits during their career. We all have our scientific areas of strength and weakness; we see things differently and hence we bring some diversity to each question.

In gold geology, as for science generally, there are lumpers and splitters. As no two gold deposits are identical, the splitter will always be able to identify multiple differences. The approach in this book is unashamedly that of a lumper but hopefully supplemented by enough detailed practical work and theory to be aware of the shortcomings of the generalisations.

This book does not reflect a consensus of what gold geoscientists believe nor any weighted mean of the numbers of publications and citations. It is a personal overview, strongly influenced by a group of outstanding colleagues. It addresses some of the data on gold and makes selections as to what information is of greater importance.

© The Author(s), under exclusive license to Springer Nature Singapore Pte Ltd. 2022
N. Phillips, *Formation of Gold Deposits*, Modern Approaches in Solid Earth Sciences 21,
https://doi.org/10.1007/978-981-16-3081-1

Appendix D: Approaches to the Research of Gold Deposits

Descriptive and Genetic Synthesis of Deposits

Descriptive features of gold deposits such as observable features have been noted and used since gold mining first occurred. Genetic aspects, or those related to how gold deposits formed, have become increasingly important with the demands of 20th century exploration. Whether it is for exploration, improving overall understanding of gold, or prioritising what observational data to collect, today the combination of descriptive and genetic focus is beneficial and typically iterative.

To help solve the problem of deposit formation, there are tens of thousands of gold deposits that act as in-situ laboratories from which we can collect information. The early prospectors noted some patterns amongst gold deposits, but they also were aware of striking differences, and this dichotomy of commonality and diversity is discussed in Chapter 9.

The process of observing and documenting the geology of these deposits needs some balance between detailed work at one deposit versus a broad coverage of many; ultimately both approaches are required. An assumption is that the information available on gold and its deposits is infinite, and therefore, decisions are continually being made of the more important and less important data that might help to constrain gold genesis. Such decisions are based on experience, iterative methods, and genetic concepts as to what types of observations will be more useful. It is self-evident that the choices as to what information to focus on will influence opinions as to how gold deposits might form.

Access to Many Gold Deposits

At the start of the 1980s global gold boom, I was fortunate to be able to visit virtually any gold deposit in Western Australia and was given access to numerous collections from closed mines. During the regional work with David Groves, we recognised some critical geological features that were visible and accessible for recording at convenient scales. At the Hunt nickel mine at Kambalda individual stopes exposed footwall metabasalt, nickel sulfide ore and komatiitic ultramafic rocks which were all cut by auriferous quartz veins. At the small Water Tank Hill mine, a single stope exhibited convincing evidence that auriferous pyrrhotite-bearing layers were formed by replacement related to cross-cutting quartz veins. The Mt Charlotte mine in the Kalgoorlie goldfield was on a scale that was easier to understand than the larger adjacent Golden Mile deposit; and Megan Clark used Mt Charlotte to effectively explain the importance of auriferous fluids and favourable host rocks. At the nearby Kanowna goldfield there were multiple deposits in quite different host rocks and Su Ho was able to demonstrate a consistent fluid type. None of these examples provided the final answer to gold deposit formation but all provided vital clues as to the next research steps.

Also, during the 1980s, access to many gold mines was possible in South Africa, Zimbabwe,

N. Phillips, *Formation of Gold Deposits*, Modern Approaches in Solid Earth Sciences 21, https://doi.org/10.1007/978-981-16-3081-1

Canada, Brazil, the Victorian Gold Province, and many other regions globally. In the 1980s, a visiting geologist might arrive at a gold mine unannounced, introduce themselves and be underground within an hour exchanging ideas with the mine geologist. The process could be repeated at another mine each half day in well-mineralised districts. Each visit was recognised as a positive externality in that both parties benefited from the sharing of knowledge.

By 2000, mine visits had become much more difficult to organise, and today there are costs to a company of inductions and staff time with less obvious positive externalities. The integrated regional gold metallogeny studies of the 1980s in Western Australia, Ontario and Quebec in Canada, and Zimbabwe would be difficult to repeat today. Because of the difficulty with underground access, especially for extended underground work to understand multiple deposits, many academic studies are based on state-of-the-art equipment and unconstrained samples from one major gold mine. Many journals will accept such work because the equipment is new and the deposit well known, but the inclusion of geological context always makes the results more applicable.

Classification Can Be Useful or a Hinderance

From the moment the first field map or core logging sheet is developed at a discovery, rocks near a deposit are being classified. The map legend of rock types and structures is one way to communicate the classification scheme. It is not reasonable to expect that the classification scheme will be complete and perfect on its first day; therefore, it should be anticipated that on-going work would expose weaknesses and fill some gaps. In the short term this revising may be inconvenient, but it is a reality that regular, but not too regular, revision of classification schemes and rock legends should be considered a part of learning and improving.

Two criteria of an effective classification scheme are *being practical to apply* and having

a *sound theoretical basis*. The first criterion is not met if years of extensive research is required to classify a sample or deposit. If the second criterion requiring a theoretical basis is not met, then it is likely that new work may require revision or eventual replacement of the scheme. Judgement becomes important in celebrating on-going improvements that build upon a basic legend and classification or recognising when much new data do not fit comfortably and necessitates rather fundamental re-adjustments on an all too frequent basis.

An example of a successful classification system derives from the Golden Mile in Kalgoorlie which was systematically mapped from soon after its discovery in 1893. For many years some vastly different coarse-, medium- and fine-grained rocks from near-ultramafic to near-felsic in composition were each given separate names. In the 1960s Guy Travis recognised that much of this rock variation was from within a 600 m thick differentiated dolerite sill (i.e. a fundamental change to the local understanding) which led to a new classification and legend that has had global influence for half a century. Because of the systematic and careful approach to descriptive mapping in the Kalgoorlie goldfield before Travis, it has been possible to integrate the previous 60 years of data from 1900 into the new Travis legend.

Classification schemes that rely on knowing the genesis of a gold deposit usually fail the criterion of being practical to apply. For example, determining the temperature and depth of formation of an ancient deformed and metamorphosed gold deposit is hardly trivial (i.e. the criterion of *being practical to apply* is not being met). The term 'mesothermal' (medium temperature) is being used less for this reason.

Well-thought-out classification systems can assist in seeing linkages and patterns. A less thoughtful classification may be linking unrelated deposits, failing to appreciate commonality across classes, retarding scientific progress and fostering unproductive debate because of poorly defined classes.

Historical Field-based Terms

Many historic field terms that are commonly used when describing gold deposits have been used in this book sparsely if at all. Examples of such terms include argillic, phyllic, propylitic, skarn, potassic and sericitic. Many of these terms were developed on mines prior to there being regular on-site mineral identification methods and they served their operations well. Today there are scanning electron microscopy, X-ray diffraction and hyperspectral analysis to determine actual mineral species. Each of these analytical methods is costly and can yield very useful information, so it makes no sense to condense this information back to historical terms. The meanings of historic terms vary between mining regions and they convey a mix of mineralogy and texture without being specific about co-existing mineral assemblages.

Where possible minerals and co-existing assemblages are used to take advantage of modern petrology, thermodynamics and geochemistry. As one example of maximising retained information it is useful to compare the use of *co-existing andradite – wollastonite – magnetite – calcite* to the less informative term *skarn*. A further example might be the less informative *chloritic alteration* compared to *coexisting chlorite-albite-muscovite-quartz-magnetite-hematite-pyrite-siderite-ankerite*; the full assemblage here communicates much more information. The pioneering work at the Coronation mine at Flin Flon in Canada by Froese (1969) illustrates the great value that can be obtained from complex gangue mineral assemblages in ore deposits.

Froese E 1969 Metamorphic rocks from the Coronation Mine and surrounding areas. Geological Survey of Canada Paper 68-5 57-77.

Robert F, Brown AC (1986) Archean gold-bearing quartz veins at the Sigma mine, Abitibi Greenstone Belt, Quebec, part II: Vein paragenesis and hydrothermal alteration. Economic Geology 81: 593–616.

Appendix E: Some Useful Reading

Chalmers AF (1980) What is this thing called science? 2nd edn, 179, Buckingham, Open University Press

de Waal S (1988) Of barons and barriers. South African Journal of Geology 91:305–315

Garvin DA (2002) General management: processes and action. 265-286, Boston, McGraw Hill. [Discusses groupthink]

Haynes D (2006) The Olympic Dam ore discovery—a personal view. SEG Newsletter 66:1 and 8–15

Kuhn TS (1970) The structure of scientific revolutions, Chicago, IL, University of Chicago Press

Phillips GN (2011) Gold exploration success: 33rd Sir Julius Wernher Memorial Lecture, Applied Earth Science (Transactions Institute Mining, Mineral and Metallurgy B) 120:7-20 [Discusses groupthink]

Snow GG, MacKenzie BW (1981) The environment of exploration: economic, organizational, and social constraints. Economic Geology 75th Anniversary Volume: 871–896

Suri G, Bal HS (2007) A certain ambiguity: a mathematical novel, Princeton, NJ, Princeton University Press.

Vann J, Stewart M (2011) Philosophy of science: a practical tool for applied geologists in the minerals industry. Applied Earth Science 120: 21–30

White AH (1997) Management of mineral exploration, Moggill, A.White and Associates, Australian Mineral Foundation

Woodall R (1994) Empiricism and concept in successful mineral exploration. Australian Journal of Earth Sciences 41:1–10.

Location Index

Subject Index

A

All-time production (Appendix A), xi, 11, 14, 24–26, 28, 30–32, 34, 36, 37, 40, 42, 131, 252, 261, 274

Alteration, 12–13, 17, 57, 65, 85, 95, 97–99, 104, 109, 116, 125–131, 134–142, 149, 151, 152, 160, 166, 168–175, 177, 181, 186, 188–189, 202, 206, 207, 211–213, 216, 223, 228, 232, 233, 241–244, 248, 256, 263, 264, 281

Alteration halo, 12–13, 17, 19, 85, 116, 125–130, 134–138, 159, 174, 186, 216, 242

Amphibolite facies, 32, 85, 109, 118–122, 166, 168, 169, 171, 173, 175, 177, 232, 233

Andalusite, 169, 172, 173, 217

Antler, 200, 201, 211–213

Aqueous fluid, 56, 92, 102, 103, 108–121, 145, 149, 152, 153, 158, 160, 176, 177, 182, 215, 218, 219, 230–234

Arsenian pyrite, 202, 204, 207, 260

Arsenopyrite, 11, 31, 36, 38, 129, 165, 203, 216, 262

B

Background, 51–58, 87, 93, 97–99, 102, 160, 161, 176, 189, 231, 270

Base metal, ix, 4, 5, 14, 18, 30, 36–39, 42, 45, 48, 52, 59–63, 68, 71, 82–84, 87, 92, 94–96, 99, 103, 108, 112, 131, 141, 152, 154, 155, 158, 160, 164, 197, 198, 207, 212, 217, 221, 223, 229, 230, 253, 263, 269, 270

Basin and Range, 41, 200, 205

BIF banded iron formation, xi

Biotite, 56, 73, 85, 112, 113, 129, 164–175, 177, 182, 183, 209, 231

Black shale, 34, 85, 137, 150, 191, 198, 252, 255, 256, 258

Black smoker, 68

Brine, 68, 110

Brownfield, xii, 8, 32, 33, 36, 79, 131, 240, 241, 262, 264, 274

Buffer, 115, 121, 129–130, 141, 149, 150, 155, 158, 172, 174, 226, 271

By-product, 11, 14, 18, 28, 38, 40, 44, 45, 47, 67, 78, 102, 104, 226

C

Carbonaceous, 54, 79, 81, 85, 132–135, 137, 139, 150, 151, 158, 160, 191, 204, 255, 256, 260, 269

Carbon-in-pulp, 9, 238

Chlorite, 85, 110, 113, 114, 116, 118, 120, 121, 126, 129, 136–140, 142, 151, 160, 164, 168, 170, 182, 183, 188, 200, 231, 253, 256, 262, 281

Class A, 146, 155

Class B, 146, 155

Committee for Mineral Reserves International Reporting Standards (CRIRSCO), 14, 25, 273

Commonality, 4, 19, 83–88, 131, 270, 279, 280

Compatible, 17, 87, 93, 103, 135, 141, 145, 173, 208, 210, 218, 231, 232

Competent, xii, 117, 246, 269, 273, 274

Complexity, 19, 84, 115, 122, 142, 159, 173, 183, 184, 186, 205, 219, 221, 258, 270

Conglomerate, 28, 31, 32, 54, 79–81, 85, 86, 216, 217, 219, 220

Contact metamorphism, 73, 112–114, 118, 119, 163–165, 170, 209–211

Continuum, 174, 176–178, 198, 214, 215, 220

Co-ordination, 146, 147

Co-product, 11, 14, 18, 28, 30, 37, 38, 43, 45, 47, 67, 84

Cordierite, 112–113, 129, 164, 165, 167–170, 172, 178, 183

Covalent bond, 146, 151, 152, 154, 155

Curiosity, ix, 3, 235, 270

Cyanide, 8, 9, 146, 170

D

Dating, ix, 8, 35, 36, 46, 65, 71, 73, 128, 134, 182, 202, 211, 251

Daughter mineral, 78, 79, 147

Decarbonation, 202, 203

Decussate, 168–170, 173, 175

Deep lead, 80, 190

Definition broadening, xii, 18, 110, 155, 215, 230, 271

Dehydration, 110, 114

Desulfidation, 114, 160

© The Author(s), under exclusive license to Springer Nature Singapore Pte Ltd. 2022
N. Phillips, *Formation of Gold Deposits*, Modern Approaches in Solid Earth Sciences 21,
https://doi.org/10.1007/978-981-16-3081-1

Printed in the United States
by Baker & Taylor Publisher Services